Disruptive Technologies for Business Development and Strategic Advantage

Anatoly V. Zhuplev
Loyola Marymount University, USA

A volume in the Advances in
Human Resources Management
and Organizational Development
(AHRMOD) Book Series

IGI Global
DISSEMINATOR OF KNOWLEDGE

Published in the United States of America by
IGI Global
Business Science Reference (an imprint of IGI Global)
701 E. Chocolate Avenue
Hershey PA, USA 17033
Tel: 717-533-8845
Fax: 717-533-8661
E-mail: cust@igi-global.com
Web site: http://www.igi-global.com

Library of Congress Cataloging-in-Publication Data

Names: Zhuplev, Anatoly, 1948- editor.
Title: Disruptive technologies for business development and strategic advantage / Anatoly V. Zhuplev, editor.
Description: Hershey : Business Science Reference, [2018]
Identifiers: LCCN 2017028917| ISBN 9781522541486 (hardcover) | ISBN 9781522541493 (ebook)
Subjects: LCSH: Technological innovations--Economic aspects.
Classification: LCC HC79.T4 D567 2018 | DDC 338/.064--dc23 LC record available at https://lccn.loc.gov/2017028917

This book is published in the IGI Global book series Advances in Human Resources Management and Organizational Development (AHRMOD) (ISSN: 2327-3372; eISSN: 2327-3380)

British Cataloguing in Publication Data
A Cataloguing in Publication record for this book is available from the British Library.

All work contributed to this book is new, previously-unpublished material.
The views expressed in this book are those of the authors, but not necessarily of the publisher.

For electronic access to this publication, please contact: eresources@igi-global.com.

Advances in Human Resources Management and Organizational Development (AHRMOD) Book Series

ISSN:2327-3372
EISSN:2327-3380

Editor-in-Chief: Patricia Ordóñez de Pablos, Universidad de Oviedo, Spain

MISSION

A solid foundation is essential to the development and success of any organization and can be accomplished through the effective and careful management of an organization's human capital. Research in human resources management and organizational development is necessary in providing business leaders with the tools and methodologies which will assist in the development and maintenance of their organizational structure.

The **Advances in Human Resources Management and Organizational Development (AHRMOD) Book Series** aims to publish the latest research on all aspects of human resources as well as the latest methodologies, tools, and theories regarding organizational development and sustainability. The **AHRMOD Book Series** intends to provide business professionals, managers, researchers, and students with the necessary resources to effectively develop and implement organizational strategies.

COVERAGE

- Job Enrichment
- Upward Feedback
- Employment and Labor Laws
- Collaborative Method
- Outsourcing HR
- Strategic HRM
- Coaching and Mentoring
- Change Management
- Compliance
- Employee Evaluation

IGI Global is currently accepting manuscripts for publication within this series. To submit a proposal for a volume in this series, please contact our Acquisition Editors at Acquisitions@igi-global.com or visit: http://www.igi-global.com/publish/.

The Advances in Human Resources Management and Organizational Development (AHRMOD) Book Series (ISSN 2327-3372) is published by IGI Global, 701 E. Chocolate Avenue, Hershey, PA 17033-1240, USA, www.igi-global.com. This series is composed of titles available for purchase individually; each title is edited to be contextually exclusive from any other title within the series. For pricing and ordering information please visit http://www.igi-global.com/book-series/advances-human-resources-management-organizational/73670. Postmaster: Send all address changes to above address.

Titles in this Series

For a list of additional titles in this series, please visit:
https://www.igi-global.com/book-series/advances-human-resources-management-organizational/73670

Statistical Tools and Analysis in Human Resources anagement
Dipak Kumar Bhattacharyya (Xavier University, India)
Business Science Reference • ©2018 • 295pp • H/C (ISBN: 9781522549475) • US $175.00

Management Strategies and Technology Fluidity in the Asian Business Sector
Patricia Ordóñez de Pablos (University of Oviedo, Spain)
Business Science Reference • ©2018 • 338pp • H/C (ISBN: 9781522540564) • US $225.00

Arab Women and Their Evolving Roles in the Global Business Landscape
Ebtihaj Al-A'ali (University of Bahrain, Bahrain) Minwir M. Al-Shammari (University of Bahrain, Bahrain) and Hatem Masri (University of Bahrain, Bahrain)
Business Science Reference • ©2018 • 321pp • H/C (ISBN: 9781522537106) • US $200.00

Creativity in Workforce Development and Innovation Emerging Research and Opportunities
Sally Blake (Flagler College, USA) and Candice M. Burkett (University of Illinois at Chicago, USA)
Business Science Reference • ©2018 • 152pp • H/C (ISBN: 9781522549529) • US $145.00

Evaluating Media Richness in Organizational Learning
Albert Gyamfi (Aalborg University Copenhagen, Denmark) and Idongesit Williams (Aalborg University Copenhagen, Denmark)
Business Science Reference • ©2018 • 328pp • H/C (ISBN: 9781522529569) • US $185.00

Teaching Human Resources and Organizational Behavior at the College Level
John Mendy (University of Lincoln, UK)
Business Science Reference • ©2018 • 305pp • H/C (ISBN: 9781522528203) • US $195.00

Handbook of Research on Organizational Culture and Diversity in the Modern Workforce
Bryan Christiansen (PryMarke LLC, USA) and Harish C. Chandan (Argosy University, USA)
Business Science Reference • ©2017 • 506pp • H/C (ISBN: 9781522522508) • US $255.00

For an entire list of titles in this series, please visit:
https://www.igi-global.com/book-series/advances-human-resources-management-organizational/73670

IGI Global
DISSEMINATOR OF KNOWLEDGE

701 East Chocolate Avenue, Hershey, PA 17033, USA
Tel: 717-533-8845 x100 • Fax: 717-533-8661
E-Mail: cust@igi-global.com • www.igi-global.com

Table of Contents

Detailed Table of Contents

the ultimate solution to challenges the society faces. This scholarly analysis builds on the theory of disruptive innovation to categorize and analyze the impact of disruptive innovations on key sectors of an economy in terms of their impact on the strategies and commercialization from a sectorial perspective. This analysis offers insightful information in terms of disruptive innovations which can be categorized into market-driven, product-driven, and competency-driven disruptive innovations.

The aftermath of the 2007-2009 financial crisis and European debt crisis proved to be a transformative event to financial markets, primarily in the areas of fixed-income asset yields and asset price appreciation due to wide-spread credit contraction at the retail level. After worldwide markets collapsed for equities and many corporate debt securities, frightened investors liquidated holdings at absurdly low prices. However, the worldwide recovery, while uneven, remains underway. Yet despite the recoveries in the broader indices, individual investors at the retail level have not widely shared in market gains after exiting from their investments during the crises. As a result of uncomfortably high asset prices and the lack of income potential to investment-grade fixed income investments, new alternatives are sought which may present better yield-earning potential amidst the current credit market environment made available by certain Fintech companies.

Years after the 2007-2009 financial crisis and European debt crisis, the European Union's (EU) banking system sustained persistent strain due to those two shocks, austerity and economic contraction, political events, poor banking operations, enhanced regulation, and litigation. The European Central Bank's response was significant: short-term interest rates collapsed and markets were flooded with money via quantitative easing programs. Consequently, investors fled risky assets for the safety of government debt. Yet as banks recovered, savers sacrificed asymmetrically: yields on bank deposits and bonds were decimated. As macroeconomic challenges subsided, Fintech increasingly threatened legacy financial institutions' business models to the benefit of the EU savings public. This study assesses the impact by Fintech companies on legacy banks in the EU with respect to savings, lending, and wealth management. This study also assesses and makes recommendations on a strategy by Fintech to benefit savers, and the measures legacy institutions must take to survive amidst this new competitive landscape.

This chapter discusses the important socioeconomic role of financial technologies in the emerging market which is Russia today. While the issues of financial inclusion are of recognized importance for the developing markets, until recently they were seen largely as areas of affirmative regulatory action, not of competitive play by private market actors. However, the advent of fintech companies changes the paradigm. Many fintech companies in Russia view the gaps in financial inclusion as attractive market niches and formulate relevant consumer offers. This chapter reviews their strategic approaches based on the study of five business cases, and introduces an analytical matrix mapping the approaches to existing inclusivity gaps. The model strengthens the existing policy aimed at developing financial inclusion as it allows a targeted cost-benefit analysis of market players' actions. As Russia demonstrates many of the financial inclusivity challenges seen in other countries, the findings of this chapter have certain applicability in the context of both emerging and advanced economies.

Shifts in the disruptive innovations and the growing impact of consumer cleantech are best understood by looking at megatrends in water management. Trends are global long-term forces of development that impact society, economy, environment, culture, and personal lives. Trends are continuously changing the dynamic landscape in which companies operate. This chapter examined commercial opportunities for the internet of things in the market of cleantech with respect to water treatment. The chapter theoretically looked at the four clusters of water management as relates to the application of internet of things: metering, piping, reclamation, and agricultural clusters. An analysis of market strategies and competitive strategies is conducted leading to a brief development of innovation frontiers in water management.

Preface

INTRODUCTION

In its recent history, the world has experienced numerous paradigm shifts propelled by four major stages of revolutionary developments. *The Industrial Revolution* (from approximately 1760 to about 1820-1840) facilitated the transition to new manufacturing processes. This transition signified a shift from manual production methods to machines, new chemical manufacturing and iron production processes, the increasing use of steam power, the development of machine tools and the rise of the factory system. *The Second Industrial Revolution*, a.k.a. the Technological Revolution (from approximately 1870 and 1914 up to the beginning of WWI). At its core, the Second Industrial Revolution is significantly rooted in earlier innovations in manufacturing, such as the emergence of a machine tool industry, methods for manufacturing interchangeable parts, the invention of the Bessemer steel production process, and others, based on widespread adoption of preexisting technological systems such as telegraph and railroad networks, or gas and water supply. In the early 20th century, Henry Ford mastered the moving assembly line and ushered in the age of mass production. *A third revolution* is under way: under it, manufacturing and the whole economy are going digital. A number of remarkable technologies are converging: smart software, novel materials, more dexterous robots, new manufacturing processes (notably 3D printing) and a whole range of web-based services, such as Airbnb (a major company in the peer-to-peer property lodging/rental sector) and Uber (a passenger hauling service), to name a few. The factory of the past was based on mass manufacturing of identical products. Ford once infamously said that car-buyers could have any color they liked, as long as it was black. However, the cost of producing much smaller batches of a wider variety, with each product tailored precisely to each customer's preferences, is falling in the digital world. The factory of the future will focus on mass customization—and may look more like the weavers' cottages of centuries past than Ford's assembly line. *The Fourth Industrial Revolution* signifies a wide range of new technologies that are fusing the physical, digital and biological worlds, and affecting all disciplines, economies, and industries. It involves emerging technological advances in the areas of artificial intelligence, robotics, the Internet of Things, autonomous vehicles, 3D printing, quantum computing and nanotechnology, among others.

In 1942, Joseph Schumpeter, an Austrian-American economist, coined the concept of "creative destruction." It refers to the continuous product and process innovation mechanism by which new production units and platforms replace outdated ones. In 1995, Clayton Christensen of Harvard University and his associates introduced the concept of disruptive innovation. In the field of business administration, this refers to an innovation that creates a new market and value network and eventually disrupts an existing market and value network, displacing established market-leading firms, products, and alliances. Some observers go as far as calling the concept of disruptive innovation the most influential business idea of the early twenty-first century.

A sustaining type of innovation that can be either evolutionary or revolutionary does not significantly affect existing markets. For example, replacing steel with aluminum in automotive manufacturing (evolutionary innovation) was a significant advancement, but was not disruptive in nature. The first automobiles in the late 1900s were initially expensive luxury items, and as such enjoyed limited market appeal. The emergence of the lower-priced Ford Model T, made even more affordable through the proliferation of consumer credit, eventually displaced the horse-drawn carriage. The introduction of a more affordable model had a disruptive effect on the transportation market.

Disruptive innovation in its technological and socio-economical entirety has exerted profound, wide-ranging impacts on various aspects of life—human activities, natural and social environments. To mention a few examples, the advent of Wikipedia has had a disruptive impact on print encyclopedias and crippled their customer appeal; users embrace web-based encyclopedias because they are easily accessible and cost efficient (in many cases, there are no costs involved). The proliferation of smartphones, with their amazing computing capability, imaging power, and universal Internet connectivity, disrupted the watch industry, micro-calculator industry, physical area maps, and GPS-based navigation systems in cars. Digital photography obliterated Polaroid technology, film-based photography and many related industries in the supply chain. Computer word processing exterminated the typewriter. Web-based video streaming technology and the rise of Amazon, Netflix, and Hulu is eroding the power of Hollywood studios, movie theaters, and cable companies. The list goes on.

As the churn continues, innovative products and technologies of the future continue emerging: robotics, cultured meat and seafood in agri/aquaculture; civilian and military drones in aviation; driverless vehicles; new technologies and materials in material sciences; new types of electronic devices, imaging and displays; innovative materials and methods in medicine; 'clean'/renewable energy sources; virtual reality;

artificial intelligence and deep learning; automatic multi-language translation, voice and image recognition, to name just a few.

As noted earlier, these innovative products and technologies exert massive impacts on life and work, natural and social environments. Perhaps one of their most profound disruptive impacts is on jobs.

According to a recent report by McKinsey's Global Institute (2017),[1] automation technologies including artificial intelligence and robotics will generate significant benefits for users, businesses, and economies, lifting productivity and economic growth. The extent to which these technologies displace workers will depend on the pace of their development and adoption, economic growth, and growth in demand for work. Stunningly, even as it causes declines in some occupations, automation will change many more: 60% of occupations have at least 30% of constituent work activities that could be automated. It also will create new occupations that do not exist today, much as technologies of the past have done.

While about half of all work activities globally have the technical potential to be automated by adapting currently demonstrated technologies, the proportion of work actually displaced by 2030 will likely be lower because of technical, economic, and social factors that affect adoption. McKinsey's scenarios across 46 countries suggest that between almost zero and one-third of work activities could be displaced by 2030, with a midpoint of 15% (McKinsey Global Institute, 2017). The proportion varies widely across countries, with advanced economies more affected by automation than developing ones, reflecting higher wage rates and thus economic incentives to automate.

Meanwhile, even with automation, the demand for work and workers could increase as economies grow, partly fueled by productivity growth enabled by technological progress. Rising incomes and consumption especially in developing countries, increasing health care for aging societies, investment in infrastructure and energy, and other trends will create demand for work that could help offset the displacement of workers. Additional investments in infrastructure and construction for instance, while beneficial in their own right, could be needed to reduce the risk of job shortages in some advanced economies.

Even if there is enough work to ensure full employment by 2030, major transitions lie ahead that could match or even exceed the scale of historical shifts out of agriculture and manufacturing. McKinsey's scenarios suggest that by 2030, 75 million to 375 million workers (3% to 14% of the global workforce) will need to switch occupational categories (McKinsey Global Institute, 2017). Moreover, all workers will need to adapt as their occupations evolve alongside increasingly capable machines. Some of that adaptation will require higher educational attainment, or focusing on activities that require social and emotional skills, creativity, high-level cognitive capabilities and other skills that are relatively difficult to automate.

Income polarization could continue in the U.S. and other advanced economies where demand for high-wage occupations may grow the most while middle-wage occupations decline, assuming current wage structures persist. Increased investment and productivity growth from automation could spur enough growth to ensure full employment, but only if most displaced workers find new work within one year. If reemployment is slow, frictional unemployment will likely rise in the short term and wages could face downward pressure. Meanwhile, in China and other emerging economies, middle-wage occupations such as service and construction jobs will likely see the most net job growth, boosting the emerging middle class.

To achieve good outcomes, policy makers and business leaders will need to embrace the benefits of automation and, at the same time, address the worker transitions brought about by these technologies. Ensuring robust demand growth and economic dynamism is a priority; history shows that economies that are not expanding do not generate job growth. Mid-career job training will be essential, as will enhancing labor market dynamism and enabling worker redeployment. These changes will challenge current educational and workforce training models, as well as business approaches to skill-building. Another priority is rethinking and strengthening transition and income support for workers caught in the crosscurrents of automation (McKinsey Global Institute, 2017).

ORGANIZATION OF THE BOOK

Authorship of this book spans geographically from Russia, Spain and Turkey in Eurasia to Canada and the United States in North America. Thematically, the book begins with a chapter exploring business strategies and disruptive technologies in the context of disruptive innovation theory. It is followed by one chapter examining the concept of innovation and identification and categorization of disruptive innovations under the strategic scope of the firm. The next three chapters look at innovative approaches and applications in finance and banking. The topics include peer-to-peer lending as an alternative investment asset class in the EU, fintech impact on EU retail savings and investment, and the role of financial technologies in filling financial inclusion gaps in Russia. Technological aspects of disruptive technologies are represented in the chapters examining cleantech and water treatment and the diffusion of 3D printing technology.

A brief description of the chapters follows:

Chapter 1: Business Strategies and Disruptive Technologies – An Overview of Disruptive Innovation Theory

The chapter presents key findings from a review of existing literature on disruptive technologies and discusses their importance for successful business strategies. The chapter provides a historical overview of the topic from predecessors to the most recent developments as well as an analysis of its predictive side. It also examines whether the theory allows for the development of business models. The last section is devoted to the identification of strategies that are applicable in technological disruptive innovations and discusses relevant issues to develop such strategies. Disruptive innovations are widely recognized as key engines for competitive advantage, so it is important for companies to recognize the opportunities that emerge from developing disruptive technological innovations.

Chapter 2: Identification and Categorization of Disruptive Innovations According to the Strategic Scope of the Firm

This chapter seeks to categorize and analyze the impact of disruptive innovations on established companies from key sectors of an economy in terms of their impact on strategies and commercialization from a sectorial perspective. The author makes an effort to examine and categorize the impact of disruptive innovations according to three drivers of the strategic scope of the business. The chapter seeks to provide insightful information in terms of disruptive innovations, which can be categorized into market-driven, product-driven and competency-driven disruptive innovations.

Chapter 3: Peer-to-Peer Lending as an Alternative Investment Asset Class in the EU

The aftermath of the 2007-2009 financial crisis and European debt crisis proved to be a transformative event for financial markets, primarily in the areas of fixed-income asset yields and asset price appreciation due to widespread credit contraction at the retail level. After worldwide markets collapsed for equities and many corporate debt securities, frightened investors liquidated holdings at absurdly low prices. However, the worldwide recovery, while uneven, remains under way. Yet despite recoveries in the broader indices, individual investors at the retail level have not widely shared in market gains after exiting from their investments during the crises. Because of the uncomfortably high asset prices and the lack of income potential to investment-grade fixed income investments, new alternatives are sought which may present better yield-earning potential amidst the current credit market environment made available by certain fintech companies.

Chapter 4: Fintech Impact on EU Retail Savings and Investment

Years after the 2007-2009 financial crisis and European debt crisis, the European Union's (EU) banking system sustained persistent strain due to those two shocks as well as austerity and economic contraction, political events, poor banking operations, enhanced regulation, and litigation. The European Central Bank's response was significant: short-term interest rates collapsed and markets were flooded with money via quantitative easing programs. Additionally, investors fled risky assets for the safety of government debt. Yet as banks recovered, savers sacrificed asymmetrically: yields on bank deposits and bonds were decimated. As macroeconomic challenges subsided, fintech increasingly threatened legacy financial institutions' business models and benefit the EU savings public. This study assesses the effects of fintech companies on legacy banks in the EU with respect to savings, lending, and wealth management. The study also makes recommendations on a strategy by fintech to benefit savers, and the measures legacy institutions must take to survive amidst this new competitive landscape.

Chapter 5: Inclusive Disruption – The Role of Financial Technologies in Filling Financial Inclusion Gaps in Russia

This chapter discusses the important socioeconomic role of financial technologies in the emerging market, which is Russia today. While the issues of financial inclusion are deemed important for the developing markets, until recently, they were seen largely as areas of affirmative regulatory action, not of competitive play by private market actors. The advent of fintech companies, however, changes the paradigm. Many fintech companies in Russia view the gaps in financial inclusion as attractive market niches and formulate relevant consumer offers. This chapter explores their strategic approaches based on five business case studies and introduces an analytical matrix mapping the approaches to existing inclusivity gaps. The model strengthens the existing policy aimed at developing financial inclusion as it allows a targeted cost-benefit analysis of market players' actions. Because Russia demonstrates many of the financial inclusivity challenges seen in other countries, the findings of this study have certain applicability in the context of both emerging and advanced economies.

Chapter 6: Cleantech and Water Treatment as a Case of Disruptive Innovation

Shifts in disruptive innovations and the growing impact of consumer cleantech can be best understood by looking at megatrends in water management. This chapter examines commercial opportunities for the Internet of things in the market of cleantech with respect to water treatment. From a theoretical perspective, the study looks at the four clusters of water management as related to the application of the Internet of Things: metering, piping, reclamation and agricultural clusters. An analysis of market strategies and competitive strategies is conducted leading to a brief development of innovation frontiers in water management.

Chapter 7: 3D Printing Technology Diffusion – A Revolution or an Illusion?

3D printing technology is one of the most potentially groundbreaking technologies of the future, as customer expectations, market requirements and the competition grow in a global scale. In order to understand the potential effect of 3D printing technology and whether it is a disruptive innovation that will change the traditional manufacturing paradigm, it is essential to examine the diffusion of knowledge in this area. In this study, 3D printing technology has been reviewed and patent analysis regarding 3D printing technology has been conducted in order to understand the diffusion of 3D printing technology. The results of the patent analysis indicate that the diffusion of 3D printing technology, which is represented by the patents of four key methods is not expected to fit with Bass diffusion model. The findings of the study lead to the conclusion that a state of maturity has yet to be reached with respect to 3D printing technologies, yet, growth continues.

Chapter 8: Impacts of the Peer-to-Peer Lodging Platform on the Traditional Lodging Industry – California vs. Southern Europe

The study presents a comparative analysis of trends in peer-to-peer lodging and their impacts on the established lodging industry in the metropolitan regions of Greater Los Angeles, California, USA, and Southern Europe (Athens, Greece; Barcelona, Spain; and Vienna, Austria). The main research question is whether the emerging peer-to-peer lodging platforms, part of the exploding sharing economy, are competitive or complementary to the traditional lodging industry. Data was obtained from secondary research and collected in 24 field interviews. The study found variations in the complementarity/competitiveness equilibrium, depending on the market segment in the hospitality industry and regulatory environment. The paper draws recommendations for competitiveness in the traditional hotel sector.

Anatoly Zhuplev
Loyola Marymount University, USA

REFERENCES

McKinsey Global Institute. (2017). *Jobs Lost, Jobs Gained: Workforce Transitions in A Time of Automation.* Retrieved from https://www.mckinsey.com/~/media/McKinsey/Global%20Themes/Future%20of%20Organizations/What%20the%20future%20of%20work%20will%20mean%20for%20jobs%20skills%20and%20wages/MGI-Jobs-Lost-Jobs-Gained-Report-December-6-2017.ashx

ENDNOTE

1 This report covers 46 countries comprising almost 90 percent of global GDP. The report focuses on six countries that span income levels (China, Germany, India, Japan, Mexico, and the United States). For each of these countries, the researchers modeled the potential net employment changes for more than 800 occupations, based on different scenarios for the pace of automation adoption and for future labor demand.

Chapter 1
Business Strategies and Disruptive Technologies:
An Overview Within the Disruptive Innovation Theory

Sucet Jimena Martínez-Vergara
Universitat de Barcelona, Spain

Jaume Valls-Pasola
Universitat de Barcelona, Spain

ABSTRACT

This chapter presents key issues taken from the review of existing literature on disruptive technologies and their importance for successful business strategies. The review is done within the framework of the disruptive innovation theory. The chapter covers an historical overview of the topic from the precursors to the most recent developments as well as an analysis of its predictive side. It also finds out whether the theory allows for the development of business models. Last section is devoted to the identification of strategies that are applicable in technological disruptive innovations and discusses relevant issues to develop such strategies. Disruptive innovations are widely recognized as key engines for competitive advantage, so it is important for companies to recognize the opportunities that emerge from developing disruptive technological innovations.

DOI: 10.4018/978-1-5225-4148-6.ch001

INTRODUCTION

In the mid-1990s, the winds of change blew with great force and intense competition, even threatening some of the strongest companies, according to Clayton M. Christensen, a professor at Harvard Business School who is now considered one of the world's main experts in innovation and growth. In his 1997 book, *The Innovator's Dilemma*, Christensen provided an explanation for the failure of respected and well-managed companies. Good managers are faced with a dilemma, he argued, because by doing the same things they need to do to be successful (listening to customers, investing in the business, and creating distinctive capabilities), they run the risk of ignoring rivals with "disruptive innovations."

Some innovations have the potential to disrupt the competition's products and services market, while others maintain the competitive position of established companies, making it possible for new participants to identify business opportunities and opportunities for creation via disruptive innovation (Hang et al., 2015). Disruptive innovations (DIs) are addressed in numerous research contributions, product contests, and advertisements. They carry a high risk of failure and they often require a lot of investment in research and development (Watty., 2013). Another important aspect to consider is that in the context of the globalization of innovation, many companies have begun to travel this disruptive innovation path, which involves a series of possible underlying mechanisms such as new business models, incentive systems, and cost structures (Christensen & Bower, 1996).

On the other hand, it is important to point out that disruptive innovation (DI) allows businesspeople and entrepreneurs with limited resources to use this type of innovation that is not commonly applied by the companies already established in the market, promoting business initiatives to fight against poverty in an effective way, producing and offering simpler, smaller, and cheaper products at a payable price, as in the case of the study carried out in the Chinese city of Yiwu (Si et al., 2014), where entrepreneurs with limited resources successfully applied DI.

In this context, and with the goal of offering a reference framework for the literature relating to technological innovations from the perspective of disruptive innovation theory, this chapter aims to show the process this theory has followed, as well as its usefulness for identifying strategic aspects and opportunities related to R&D in the field of disruption; with the understanding that disruptive technological innovations have a broad scope of action not only in the academic world, but also in business.

This chapter is organised as follows: the first section presents the historical perspective of the theory, showing its evolution, an overview of disruptive technologies, some key characteristics of disruptive innovation, some recent examples, contributions of other researchers and as well as three main actors of disruptive technology under disruptive innovation. In the second section, the predictive value of the theory is analysed and the third section shows if disruptive innovation allows for the development of business models. Section fourth deals with the strategies that can be used in the DI world and some issues to develop successful strategies. Lastly, in the fifth section, some final remarks on the elements presented in this chapter are put forward.

HISTORICAL PERSPECTIVE

The Journey to Disruptive Innovation

The DI theory presented by Christensen (1997, 2006; Christensen & Bower, 1996; Christensen & Raynor, 2003) was developed on the basis of a series of prior studies on technological innovation. Its precursors include the work of economist Joseph Schumpeter (1942): the upheaval of business competition has often been compared to a stormy sea ('gales of creative destruction'), which periodically sweeps away industries, sinking weak and old-fashioned companies (Yu & Hang, 2010). Figure 1 summarizes publications of disruptive innovation theory from 1942 to 2003.

When Christensen successfully studied businesses whose fortune had declined, he discovered a category of companies that collapsed or suffered a decrease in their success when they were overtaken by others that had initially offered less attractive products to customers from the mainstream market, perhaps because they could not allow themselves the products of successful companies or because they were the least profitable customers for those companies (Vasquez et al., 2016).

Figure 1. Background of disruptive innovation theory from 1942 to 2003
Source: Adapted from Yu & Hang, 2010

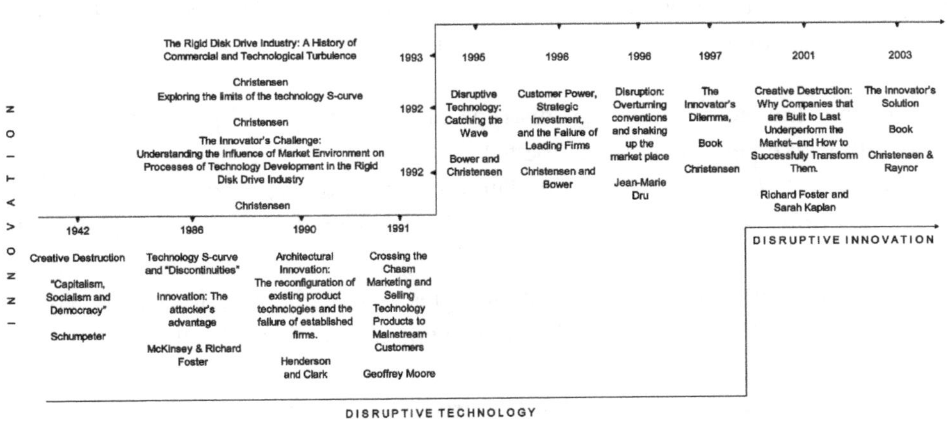

Christensen's initial research, which formed the nucleus of DI theory, was based mainly on the hard disk drive industry in the 1970s and 1980s, publishing sole-authored articles about this industry (Christensen, 1993) or jointly with his colleagues (Christensen & Bower, 1996), where also explain the difference between disruptive and sustaining innovation, which has implications for theories on the detection of business and creative opportunities.

Based on the prior studies, in 1997, he published his influential book titled *The Innovator's Dilemma*, which made him well-known in the study of technological innovation in commercial enterprises, and where he puts forward the basic theory of disruptive technology in an exhaustive and detailed way (Yu & Hang, 2010).

In these early studies Christensen focus on disruptive technology as "innovations that result in worse product performance in relation to mainstream markets," highlighting their characteristics as "typically cheaper, simpler, smaller, and, frequently, more convenient to use than the existing product" (Corsi & Di Minin, 2014).

Christensen specifies that in the initial development phase of DIs, each product based on a certain disruptive technology can only serve segments of a smaller market niche and that after further development they can increase the technology to a sufficient level to satisfy the mainstream market. He documented these technological and market dynamics in numerous contexts, such as hard disk drive units, earthmoving equipment and motor controls.

Christensen also describes sustaining and disruptive technology. A sustaining technology improves product or service performance through sustained steps and incremental changes. On the other hand, the introduction and marketing of a disruptive technology requires a radical change. Technologies of this type alter the relationship of a company with its customers and suppliers and disrupt the financial arrangements that companies have achieved to support their current market and technology. It is therefore natural to resist such changes. He also indicates that disruptive technologies have various disadvantages:

- In the short term, adopting this type of technology will result in a product with worse performance, not better.
- They attract a much smaller market, until they are widely accepted.
- They offer less profit since they are cheaper, smaller, simpler, and easier to use.
- One obstacle for disruptive technologies is that their performance will be uncertain, as with anything that is new and not yet fully-developed.

In order to remain in the market, Christensen recommends that companies establish an independent business that exploits the new technology and develops new customers around it. Given that disruptive technologies generally open the door to new markets and developments, the companies that enter soon have a clear advantage. Some principles that Christensen claims must be recognized and harnessed include:

1. Listen to what consumers want to determine how to use resources.
2. Small markets are not important.
3. The ultimate uses and applications of disruptive technologies are unknowable in advance.
4. There may not be enough demand for the new technology.

Christensen also advises large companies that want to use disruptive technologies to follow these four essential steps:

1. Assign the development and marketing work for the new technologies to a company that is interested in doing it; for example, a small company that is starting out in the new field.
2. Place the project in an organization that is "small enough to get excited about small opportunities and small wins". In fact, a company could set up an internal department or a small separate company to explore the options.
3. Recognize the possibility of failure. Be prepared to "fail early and inexpensively" throughout the repeated trial-and-error tests, while seeking a market for the new technologies.
4. Look for ways to identify or develop new markets that value products based on these disruptive technologies.

From Technological Innovation to Disruptive Innovation

With the aim of resolving the innovator's dilemma around how well-established and well-managed companies can avoid being toppled by developing the cutting-edge technologies of their economically competitive contexts, Christensen and Raynor (2003) published another book called *The Innovator's Solution*, where they replace the term disruptive technology with disruptive innovation, broadening the theory to include not only technological products, but also services and business models, such as large discount/low-price department stores, point-to-point airlines, and online education. They fine-tuned the theory and highlighted that DIs can be classified in low-end markets (the ones that focus on less profitable consumers and those overlooked by the mainstream market) or new markets (where there were no consumers even in the mainstream market) (Yu & Hang, 2010).

Christensen et al. (2015) provide a summary of the idea "disruptive," describing it as a process through which a smaller company with less resources is capable of successfully challenging established companies that focus on improving their products and services for their most discerning (and in general most profitable) customers, exceeding the needs of some segments and ignoring the needs of the rest.

Companies that start out with DI begin by successfully advancing upon those overlooked segments, gaining a foothold in functionality and often at a lower price. Established companies pursuing greater profitability in the most discerning segments tend not to respond forcefully. Therefore, the disruptive products later achieve a greater market share, offering the performance that mainstream consumers require and preserving, at the same time, the advantages that led them to early success. When mainstream consumers begin to buy in masse, the disruption has been produced. Table 1 shows important characteristics between low-end markets and new markets.

Overview of Disruptive Technologies

Some technological developments, most notably the Internet, provoke innovations within a growing number of industries that are considered radical by some (Abernathy & Clark, 1985) or disruptive by others (Gilbert, 2003) to the industry, and often involve reshaping existing markets, or the creation of entirely new markets

Table 1. Characteristics between low-end markets and new-markets

<table>
<tr><th colspan="3">Disruptive innovations are possible because they start out in two types of markets that incumbents overlook</th></tr>
<tr><td rowspan="2">DIs originate in:</td><td>Low-end markets</td><td>New-markets</td></tr>
<tr><td>Here, traditional operators normally offer products/services to their most profitable and discerning customers.
They pay less attention to less-discerning customers.
Traditional operators' offers often exceed the performance requirements of low-end customers.
This opens the door to a disruptor focused (at the beginning) on low-end customers with a "good enough" product, to later move to the mainstream market.</td><td>Disruptors create a market where there was none.
They find a way to turn non-consumers into consumers.</td></tr>
<tr><td>DI does not impose itself on mainstream customers until the quality is up to date with their standards.</td><td colspan="2">A product or service's DI is considered initially inferior by most of the customers of a traditional operator.
The customers are not willing to change to the new offer simply because it is less costly. Once it fulfils their quality expectations, they adopt the new product and happily accept its lower price. (That is how disruptions lead to lower prices in a market).</td></tr>
<tr><td>DI theory is different to "sustaining innovations"</td><td colspan="2">With sustaining innovations, products that are better in the eyes of an incumbent's existing clients are developed; e.g. clearer TV image.
These improvements can be incremental advances or significant advances, but all of them allow companies to sell more products to their most profitable customers.</td></tr>
</table>

Source: Own elaboration based on Christensen et al., 2015

(Callaway & Hamilton, 2008). While there is some disagreement in the literature about the definitions of these types of innovations, Crockett, et al, 2013, refer to them as disruptive innovations. In that sense, then, disruptive innovations refer to technological innovations, new products, or new services that require a "disruptive" strategy reaction that often serves to overtake the prevailing dominant technologies or status quo products in a market (Christensen, 2006).

Such innovations typically follow the introduction of a new technology, product, or service that exhibits performance characteristics initially viewed by the mainstream customer base as inferior to existing products. This new technology, product, or service may actually perform better on other dimensions, but over time, the disruptive innovation improves on the primary performance dimension to the extent that it eventually appeals to those mainstream customers who initially viewed the innovation as inferior (Crockett et al., 2013).

The core of disruptive technology is to change the performance metrics along which firms compete (Danneels, 2004). Since disruptive technologies introduce different product features that were of no or little value before, they therefore change the basis for competition. Disruption occurs when the performance trajectories of disruptive technology intersect with that of the performance in the mainstream market (Christensen & Raynor 2003, Alpkan & Gemici, 2016).

To be able to adapt successfully to the rapidly changing and uncertain external environment, foreseeing and adapting to the changers are very critical but not enough; because it just makes us followers but not prospectors or proactive entrepreneurs. It is also possible to change the rules of the game to one´s own advantage instead of just trying to adapt the established norms. This is not only innovation but also disruption. Especially those small or newly established organizations might try to disrupt the already established way of doing business dominated by the more powerful, larger, and older competitors. Since they are already integrated and adapted to the status quo it would be difficult for them to admit the importance or radical change; but for the smaller and potentially more dynamic and adaptable newcomers, disruption is not only easier but also a must for survival (Alpkan & Gemici, 2016).

Representative cases of technological DI and their new value prepositions were chose across a wide range of industries, the study specific technology, function and performance of products in order to understand the new value propositions (NVPs) and the R&D strategies applied to offer these NVPs: 11 categories can be considered as principal components, which are (i) small (ii) light weight; (iii) less power consumption ; (iv) portability; (v) customization; (vi)ease of usage/ design/ production: (vii) time-saving; (viii) cost reduction: (ix) augment disruptive features; (x) explore applicability and (xi) other unique values for specific products (Yu & Hang, 2011). Figure 2 presents examples of disruptive products.

Figure 2. Examples of disruptive products
Source: Adapted from Yu & Hang, 2011

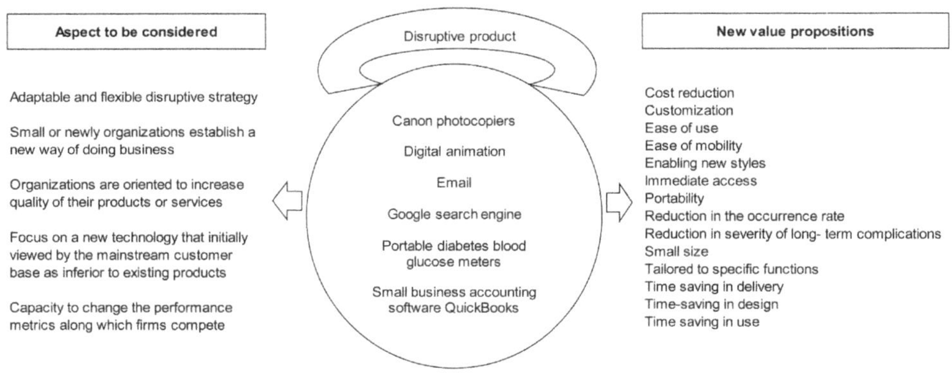

Some Key Characteristics of Disruptive Innovation

What constitutes a real disruptive innovation deserves examination through different lenses, and it is crucial to discuss the definition in any further research on disruptive innovation, as well as clarification of some potential misunderstanding. Three of these are highlighted here (Yu & Hang, 2010):

- First, disruption is a relative phenomenon.
- Second, disruptive innovation does not always imply that entrants or emerging business will replace the incumbents or traditional business; it does not imply that disruptors are necessarily start-ups. In fact, an incumbent business with existing high- end technologies can still survive by concentrating on how to satisfy its most demanding but least price- sensitive customers.
- Third, disruptive innovation is not equal to destructive innovation. A technological innovation that has superior performance in key dimensions with a relatively low-cost structure would directly invade the mainstream market and cause more serious destructive innovation that focuses on low cost but initially lower performance.

Innovations not only influence the technology but have wider implications; new customer segments with different preferences (Christensen & Bower, 1996; Tripsas, 2008); and the need for new complementary assets (Tripsas, 1997); business models (Chesbrough & Rosenbloom, 2002; Baden-Fuller & Morgan, 2010); industry architectures (Baldwin, 2008; Jacobides et al., 2006); and alliance networks (Ansari & Munir, 2008; Rosenkopf & Nerkar, 2001; Rothaermel & Boeker, 2008). As a result, incumbent firms often either fail to adapt to technological change and successfully enter new markets, or enter too late to be able to effectively compete (King & Tucci, 2000; Ansari & Krop, 2012).

Although scholars generally agree that entrepreneurial new businesses are appropriate response to disruptive innovations and have a positive impact on overall corporate performance (e.g. Christensen & Raynor, 2003; Gilbert, 2005; Thornhill & Amit, 2000; see Markides, 2006, or Danneels, 2004, for a countervailing view) there is some ambiguity as to which contingences (e.g. environments, strategies, managerial characteristics, and corporate relationships) are most likely to lead to successful new business units (Crocket et al., 2013).

Disruptive innovations have an impact on important sectors of the economy. Identifying potencially disruptive innovation is important to be aware about threats by enabling incumbents to assign resources to be competitive in the market and grow increasingly their oportunities to survive. Improvement of cognitive capabilities is the basis of strategic activities in every organization (Rad, 2017). Figure 3 identifies some recent examples of disruptive innovation.

Figure 3. Examples of disruptive innovation
Source: self-elaboration

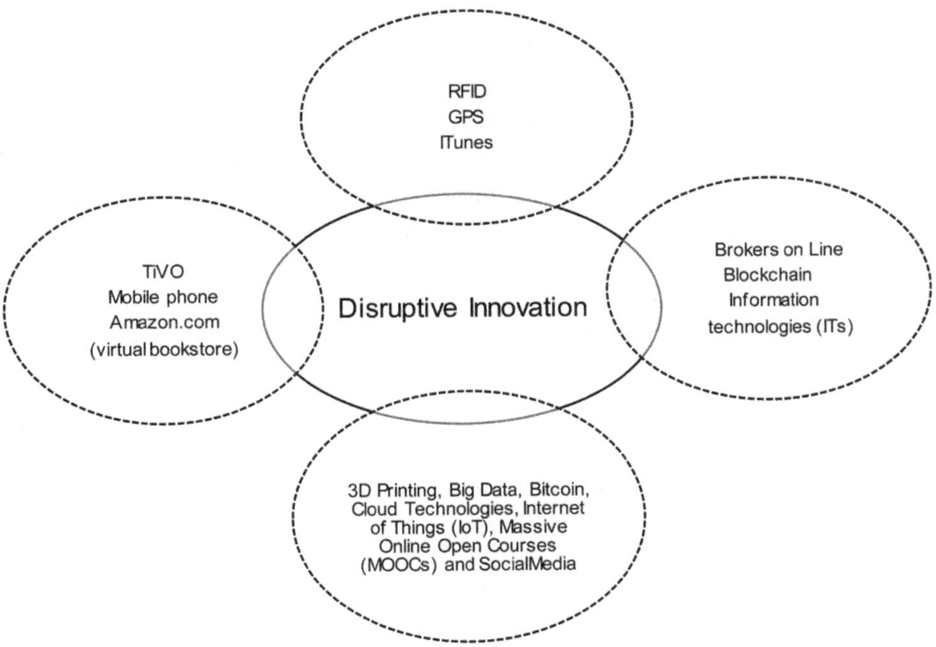

Internet. As Latzer (2009) said, Internet is conceived as the "mother of all disruption" and the future of threat is forever (Levina, 2016). The Internet has been disruptive to traditional newspapers, book retailers, and travel agencies. Catalog retailers, on the other hand, have benefitted greatly (Crocket et al., 2013).

- **TiVo:** A company that pioneered the Digital Video Recorder- TiVo navigated coopetitive tensions by continually adjusting its strategy, its technology platform, and its relational positioning within the evolving U.S. television industry ecosystem (Ansari et al., 2015).

Blockchain may be conceptualized as a DI, however, its practical application is, as yet, rather limited and the actual impact of the approach remains to be seen (White, 2017). Blockchain has the potential to become a significant source of disruptive innovations in business and management. A blockchain is a form of digital ledger consisting of "blocks" of information. Each "block" contains a record of the transactions that occur within a network. In the case of cryptocurrencies, these transactions are typically movements of currency in exchange for goods or services. Once a predetermined number of transactions are recorded, that "block" of information is added to the ledger, thus forming a "chain of blocks," hence the term blockchain.

Information technologies (ITs) have long been recognized as seeds of DI. The case of radio- frequency identification (RFID) has been widely cited (Zhou & Piramuthu, 2013). The impact of IT as a DI has proliferated to the extent that IT-specific models of DI have emerged (Carlo et al., 2014). Lyytinen & Rose's (2003) seminal model classifies three different ways in which IT innovations may be disruptive: through developments in the base technology, in the systems development processes, and in the services that may subsequently be provided (White, 2017).

From their side, Dotsika and Watkins (2017) selected seven technologies identified as disruptive, current and influential by business research on top trends (Gartner, 2015; Forbes, 2015; Frost et al., 2015; McKinsey, 2015; KPMG, 2015) and government policy publications (EU Commission, 2015a; EU Commission, 2015b; NESTA, 2015; OECD, 2015). The technologies considered were: 1. 3D Printing, 2. Big Data (the potential of their analysis as opposed to the data itself) 3. Bitcoin 4. Cloud Technologies 5. Internet of Things (IoT) 6. Massive Online Open Courses (MOOCs) 7. Social Media.

Contributions Made by Other Researchers

One noteworthy contribution is Govindarajan and Kopalle's (2006a, 2006b), which introduces low-end and high-end DI, providing a more general view of DIs and exploring beyond the case of low-cost/low-performance. DI (initiated in an inferior market with the traditional attributes) with a high price, which they call high-end DI, is in fact a blank space where Christensen's theory (lower performance and low unit cost) has not ventured. They use the mobile phone that was first accepted by corporate executives that appreciated its convenience and portability, despite its relatively high price, as an example of a DI with a higher initial price. When the mobile phone was introduced to the mainstream market, consumers continued to prefer landlines due to their reliability, cost, and coverage. However, over time, the new advances in mobile technology allowed for reliable coverage at a price that satisfied the needs of consumers in general, resulting in the disruption. Table 2 shows technological innovations based on cost, market and performance

Specifically, Govindarajan and Kopalle (2006a, 2006b) make a clear distinction between low-end disruptions and high-end disruptions based on the level of radicalness of the DIs (high-end disruptions are technologically more radical, low-end disruptions are technologically less radical) (Corsi & Di Minin, 2014).

Another contribution to DI theory was made by Hang et al. (2015), who indicate that DI intended for a low-end market segment is based on the existence of unsatisfied needs among a certain group of customers, regardless of the actions of businesspeople. Although, only those who have experience, whether technical or industrial, in the field are capable of detecting unsatisfied needs of this type and aligning DI more with the context in order to "discover opportunities." On the other hand, new markets show characteristics that are closer to those sought for the creation of opportunities to produce things intended for low-end markets. In this case, DI involves opportunities that depend on making potential customers aware that they have needs of which they may not have been aware. Table 3 shows types of DI with some important aspects to consider in relation to this.

Table 2. Technological innovations based on cost, market and performance

Product performance	Market	Cost	Example	Effect
Low	Low-end of the market	High	Cellular phone	High-end disruption
		Low	Hard disk drives	Low-end disruption

Source: Adapted from Yu & Hang, 2010

Table 3. Types of disruptive innovation associated with opportunity discovery and creation

Type of disruptive innovation	Context	Opportunity
Low-end DI	Discovery	Opportunities exist in low end market where entrepreneurs with technical expertise or industry experience can develop a disruptive product
New-market DI	Creation	Opportunities exist in the latent marker where entrepreneurs experiment and are prepared to change strategies

Source: Adapted from Hang et al., 2015

On the other hand, Markides (2012), in order to distinguish DI from other types of innovation, proposed two criteria: first, it must start out as inferior in terms of the performance that existing customers expect, but superior in price; second, it should become "good enough" in performance while at the same time remaining superior in price. He indicates that these criteria reflect the idea that DI challenges the value propositions and business models of traditional operators.

Moreover, another important point is that DIs do not necessarily have to be involved in advances in the product's technology or functionality in order to be disruptive. In fact, many innovations that otherwise could be described as "imitations" (Huang et al., 2010) can be disruptive because they challenge the existing value propositions and business models in the market.

Main Actors of Disruptive Technology

In many ways, the potential of disruptive technology under disruptive innovation theory appears and grows to make changes to the way that businesses are conducted. The real impact of the approach remains to be seen and research activities are underway. Figure 4 provides a summary about the characteristics of this theory organized in three main actors: customers, companies and market. Each actor identifies

Figure 4. Analysis of characteristics of disruptive technology innovations
Source: Self-elaboration

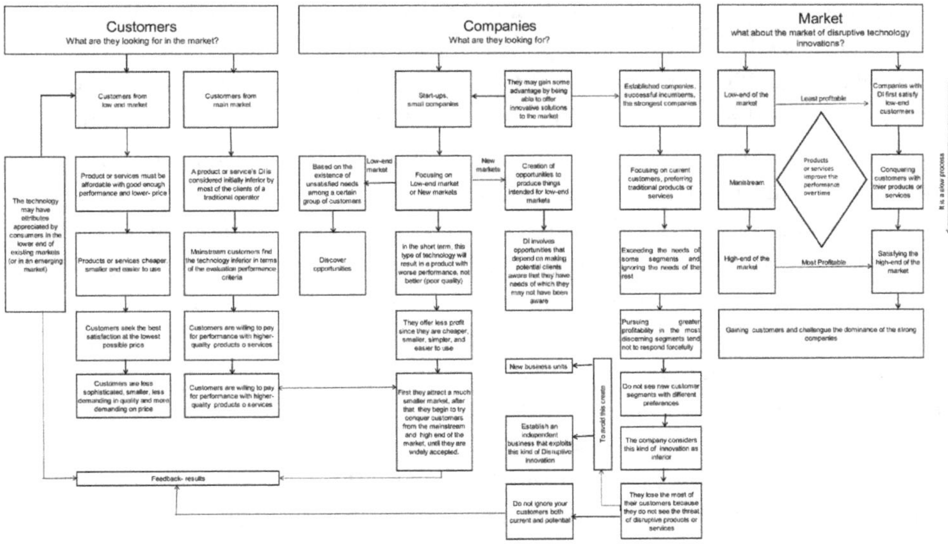

some important features of disruptive technology, specifying basic elements in a comprehensive way and usefulness for an easy understanding of this theory. As well as offering an overview of disruptive innovation.

THE PREDICTIVE VALUE OF DISRUPTIVE INNOVATION

The predictive power of the theory has been challenged and there are debates around whether DI can be used to predict things; that is, whether it can be used to predict if an innovation at an early stage could later become disruptive (Yu & Hang, 2010; Klenner et al., 2013).

Managers need to be alert to the fact that DI can stem from a much broader set of sources than breakthrough technologies alone. In particular, some of the innovations that are potentially most disruptive to incumbents with legacy products, processes and assets have their antecedents in new processes for R&D, product design and production as well as in market conditions that encourage and facilitate the development of innovative business models (Wan et al., 2015).

Many research scholars have challenged the predictive use of disruptive innovation theory. Barney (1997) argued that "it may simply be the case that some firms are lucky in their technology choices," those firms with lucky choices were subsequently scrutinized, and a retrospective rationale for their success was formed. Ex ante predictions involve predicting what performance the market will demand along various dimensions and what performance levels technologies will be able to supply (Yu & Hang 2010). It is not entirely clear what methods exist for such predictions (Danneels 2004). If one must wait until the disruption has occurred, what predictive value is there in the concept (Tellis 2006, Yu & Hang 2010), and Paap and Katz (2004) pointed out general guidance to predict future disruption, such as "do not ignore your customers both current and potential."

Yu and Hang (2010) regarded different perspectives: (a) the internal perspective, the business models and organizational challenges of incumbent firms; (b) the external perspective, the context and environment; (c) the marketing perspective, customer orientation under disruptive change; and (d) the technology perspective, technological strategies for disruptive innovation. These four perspectives try to answer two valuable questions, namely why a well-managed incumbent firm could fail in technology innovations, and how a seemingly inferior technology could grab the market shares and even replace the dominant technology.

While a number of DIs have been analyzed ex-post by Christensen and others, the problem of identifying DIs in advance has not been sufficiently addressed in the academic area of the literature (Klenner et al., 2013).

BUSINESS MODELS AND DISRUPTIVE INNOVATION

Within the literature on disruptive innovation, important contributions have been made concerning the initial emergence of disruptive technologies (Adner, 2002; Govindarajan & Kopalle, 2006; Sandström, 2011) and how the concept relates to business models (Christensen, 2006; Markides, 2006). Previous work on disruptive innovation has resulted in a quite simple framework that provides a persuasive argument for how and why incumbents fail under conditions of discontinuous change. However, this simplicity has been achieved partly at the expense of generality and accuracy (Sandström et al., 2014).

In this context previous work on disruptive innovation could only address a limited part of the business model, namely whether segments of a firm's market demanded a new offering or not. When focusing on individual firms and sub-groups within firms instead of market segments as the unit of analysis, it becomes possible to discuss business models in a much more nuanced way and to point out how incumbent firms can actually change and manage their network to achieve alignment (Sandström et al., 2014).

Companies that emerge with DIs have a marked inclination towards interpreting and addressing the needs expressed by a market niche or a new market segment (Corsi & Di Minin, 2014). This is why research on DI has paid more attention to the role of business models (Habtay, 2012); that is, the ways in which companies appropriately create, deliver, and set value. Markides, 2006, has even claimed that disruptive innovation is a "business model problem, not a technology problem" (Christensen, 2006, Sandström et al., 2014).

A successful response probably involves both a shift in mindset and an extension and re-engineering of innovation capabilities (Wan et al., 2015):

- First, incumbents may need to reassess their exposure to DI and broaden the focus of their innovation efforts to give greater weight to opportunities for cost, application and business model innovation relative to pure technological advancement.
- Second, to improve the likelihood of delivering these kinds of innovation, R&D and innovation processes may need to be re-engineered to create new processes that can run in parallel with established approaches. This re-engineering is likely to involve more industrialization of some R&D processes, greater use of parallel processing in more development stages with greater timing overlaps, more modularization of the development processes for products aimed at mainstream and value segments and the adoptions of less process-bound and more pragmatic and rapid decision making in respect of R&D projects.

Established firms can respond to disruptive innovations in a number of ways (Danneels, 2004; Markides, 2006; Markides & Charitou, 2004; Siggelkow & Levinthal, 2003; Tushman & O'Reilly, 1996), but a solution frequently offered is the creation of a separate business unit (Gilbert & Bower, 2002). This is particularly true when the exploitation of the threatening innovation requires a different cost structure, or when the opportunity afforded by the innovation fails to satisfy the overall growth goals of the mainstream business (Christensen, 2000). Moreover, such an organizational structure not only prevents the parent company´s existing processes and culture from suffocating the new business by breaking the rigidities of "doing it the old way" but also provides the opportunity for exploiting possible synergies between the parent company and its autonomous business units (Crockett et al., 2013).

Christensen and his co-authors (2015) highlight four important points that are overlooked and misunderstood when taking strategic decisions:

1. **Disruption Is a Process:** Disruptors tend to focus on getting the business model, rather than merely the product, just right; in fact, that disruption can take time helps to explain why incumbents frequently overlook disruptors.
2. Disruptors often build business models that are very different to those of incumbent firms.
3. Some DIs are successful, and others are not: failures are not proof of the deficiencies of disruption theory; they are simply indicators of limits to the application of the theory.
4. The mantra "disrupt or be disrupted" is misleading. Established companies need to respond to disruption if it is happening, but they should not react in an exaggerated way by dismantling a still profitable company. They should continue to strengthen relationships with mainstream customers through investment in maintaining innovations and create a new department focused exclusively on the growth opportunities resulting from the disruption.

The main concern must be to be able to concentrate at the same time on different dimensions of a product or service feature which in turn, attract different customer needs, then it is crucial to determine the conflicts between two businesses and how they strategically are similar to one another, once this is done, separation can be an option. However, it is also found that separate division does not necessarily secure or guarantee success especially when mechanistic and rigid structures and routines of the traditional way of doing business are replicated. The key to success is to be able to launch the new business model in a creative manner (Alpkan & Gemici, 2016).

On the other hand, the demand for low prices drives managers to find better ways of taking advantage of the precursors of DI, having to redesign their existing research and development activities to get closer to creating new disruptive products (Reinhardt & Gurtner, 2015); DI provides the stimulus that encourages and promotes a business model (Schulman & Thomas, 2014).

STRATEGIC ADVANTAGE AND DISRUPTIVE INNOVATIONS

When talking about business strategy and disruptive innovations is a key reference on the topic. The four strategies proposed by Yu and Hang, 2011 are miniaturization, simplification, augmentation and exploitation.

- **Miniaturization:** Strategy to design or to construct products on a smaller scale to increase the benefits of portability and convenience and that are two routes to archive: one is by heavy investment of time and resources to archive a much superior product with significant reduction in size but without degradation in key performance-this type of breakthrough R&D is adopted in radical innovation; the other one is modest investment of time and resources to achieve a smaller but good enough product to facilitate DI, two classical cases that applied the miniaturization strategy are the transistor radio and hand disk. Finally, miniaturization of established products involves miniaturization of most if not all modules, which can demand substantial efforts from not only the firms that attempt to create disruptive technology, but also the component suppliers of these firms. Miniaturization is the second most frequently used strategy in the "Consumer Electronics" category and the "Healthcare and Medical Equipment" category. This implies that with continuing technology advances in areas such as microelectronics, nanotechnologies and optimization of architecture designs, there exists much opportunities for firms to miniaturize their products into compact portable disruptive.

- **Simplification:** Strategy to create potential disruptive products based on the current products of overly high complexity and excess functionality, so that the population that historically lacked the skills to use current products can become new customers. Simplification is not only to reduce redundant features or the degree of performance in certain feature(s) but also to focus on the new value proposition that is most delightful to its targeted customers. Examples- two cases: a simplified tank gauging machine in the Chinese market and an automated external defibrillator (AED). This strategy was the most frequently used. It is anticipated that companies will find this strategy to be in line with the trend of open innovation (Chesbrough. 2003) as the high-end technology (before simplification) could be sourced externally from universities or other organizations.
- **Augmentation:** Suggest that firms can augment the existing sustaining product with a disruptive feature. This means that the traditional primary performances that mainstream customers historically valued could still be maintained at as satisfactory level of performance, while the R&D could focus on adding new disruptive features, examples two cases, Nintendo Will Game Console and Intel Centrino chipsets. Maintaining the key performances in mainstream market to satisfy current customers while augmenting a sustaining product with new disruptive features to attract potential new customers is one effective way to develop both orientations in parallel, and to create technology candidates for potential DI.
- **Exploitation for Another Application:** When attacked by disruptive entrants, incumbents could also think of new applications to reposition their existing products in addition to the option to fight back. The strategy of exploitation for another application covers two perspectives: repositioning

existing technology and borrowing existing technologies for targeted applications. The two illustrative cases are war game products in military training industry and MEMS enabled controller in Wii Game console. With significant performance improvements in gaming consoles such as network playability, local scenario storage and graphic quality, games of Spearhead, America's Army gradually emerged in military training applications. As discussed by Roger (2007), several potential disruptions have started to take place in the low-end of the military training industry by exploiting products from the games industry MEMS in Wii. The successful DI of Nintendo Wii used Strategy Augmentation, Nintendo Wii also used Strategy Exploitation for another application. The disruptive module of Wii is its controller. There is a possibility of leveraging on sustaining progress of certain established products which has already created affordable, high-tech components. The use of the light weight, good quality and yet low-cost head phones in the Walkman (Nayak & Ketteringham, 1994) was one classical example.

Consciously, adopting these four R&D strategies would help more firms to become disruptors. However, the R&D strategies are not intended to be prescriptive and no "one" or "Best strategy" exists. Rather, managers may use them as input for their own strategy crafting efforts or as tools to benchmark their own efforts. Likewise, the R&D strategies are specific for DI. In order not to confuse with R&D for radical innovation, it is important for the R&D managers to keep three points in mind when applying them:

1. The DI must be affordable with good enough performance.
2. It should create new features valued by the current non-consumers.
3. The time and resources required are not exhaustive-unlike the much more demanding radical R&D.

Especially noteworthy, R&D strategies for purposefully creating technologies for DI will be increasingly valuable tools in two high-growth mass-markets in future (Yu & Hang, 2011).

Another strategy could be collective efficacy may be particularly important in the face of the unrest and uncertainty resulting from the threats that disruptive technologies pose for established mainstream businesses. Teams with higher collective efficacy have been found to persist longer in the face of adversity (Cannon-Bowers, Tannenbaum, Salas, & Volpe, 1995) and perceive more opportunities in an uncertain environment (Krueger & Dickson, 1994). Because increased initiative and persistence are a product of high efficacy (Shepherd & Krueger, 2002), collective efficacy appears to be a critical competence for the management team of a new business unit struggling within an existing business (Crockett et al., 2013).

Some Issues to Develop Successful Strategies

Hang et al. (2015) presented a summary of criteria to help identify a DI according to those identified by Christensen (1997, 2003) that could serve the companies in the development of their strategy:

1. Successful incumbents often fail to recognize new opportunities presented by an emerging disruptive innovation because of their focus on existing customers.
2. They do not see the opportunity, or choose not to pursue it, because their mainstream customers find the technology inferior in terms of the performance criteria they value.
3. The technology may have attributes appreciated by consumers in the lower end of existing markets (or in an emerging market) because they are cheaper, smaller, easier-to-use, etc.
4. To secure their market foothold, innovators must improve the technology continuously in the face of strong competition from other potential disruptors or imitators.
5. Once the performance is improved sufficiently to overcome competition at the lower end of the market the technology may also be able to meet the requirements of customers in the mainstream market.

Other DI theory researchers indicate that it provides a general useful warning about the short-sightedness of management; many managers overlook or misunderstand the importance of an emerging threat (King & Baatartogtokh, 2015).

When DI appears, there are three main factors that play a significant role in leading companies to the failure. This failure stems from the methodological difference in pursing sustaining and disruptive technologies, the rate of improvement that customers can utilize or absorb in the market and the revenue and the cost structure of established companies (Christensen & Raynor, 2003), and as explained by Alpkan & Gemici, 2016:

- All of a sudden, low end or high-end value propositions may become much more attractive for the customers instead of continuing to buy mainstream products. Such radical shifts in the market norms stem first of all from disruptive innovations. Without the penetration of such disruptive value propositions, just because of market saturation, radical technology push or other environmental developments market norms would not change at the expense of sustaining innovations. Therefore, disruptive competitive moves of newcomers initiate the encroachment effect.

- Only after this attack, established firms and newcomers would move towards extremes and turn marginal solutions into the new normal. In this process of evolution of the market conditions, every single player should explore a suitable strategic position to exploit the future. Will remain the same, some will move towards the low end and imitate the disruption, and some will prefer to try to concentrate on the high-end segments which are not attacked for a while by the first disruptors.
- Established firms facing the unexpected or marginal innovations coming from some successful disruptors need to develop counter attacks to preserve their market shares. In such a marketplace, the old normal standardized value propositions begin to disappear, and new innovative avenues emerge in the margins causing a duality for selection: either low quality and cheaper novelties or luxurious ones. These response strategies may force firms to choose one of these once marginal segments and try to enlarge these segments to new markets.

Another aspect is the importance of a shared organizational vision between corporate entities and ventures and how important integration mechanisms are to the success of a venture (e.g., Burgers, Jansen, Van den Bosch, & Volberda, 2009). A shared vision contributes to a consistent culture and identity that permits employees to more readily communicate and coordinate activities across perceived organizational boundaries (Camerer & Vepsalainen, 1988). Specially, a shared vision brings about a common language and mutual understanding, which facilities the legitimacy and acceptance of the new venture by the mainstream business (Ashforth & Mael, 1989; Subramaniam & Youndt, 2005; Crockett et al., 2013).

CONCLUSION

This chapter summarizes key issues of the review of existing literature on disruptive technologies and their importance for successful business strategies. The review has been done within the framework of the disruptive innovation theory. The work has mainly focused on academic articles, as well as the two main books published on the topic by Clayton Christensen, the author of the disruptive innovation theory.

The chapter has tried to identify main precursors of the theory, the characteristics that define it and its evolution over time. In addition, an overview of disruptive technologies and the key characteristics of disruptive innovation have been developed. We have also find out whether the theory allows for the development of business models. Last section has been devoted to the identification of strategies that are applicable in technological disruptive innovations and the relevant issues to develop such strategies. Disruptive innovations are widely recognized as key engines for competitive advantage, so it is important for companies to recognize the opportunities that emerge from developing disruptive technological innovations. Main conclusions arising from the review done are summarized in the following paragraphs.

Regarding the "birth" of the theory and its evolution, we indicate that it has precursors and elements for its identification described by its authors, who at first defined it as technological innovation (Christensen, 1997) and later expanded it to disruptive innovation (2003), with the aim of not only considering technology, but also other business areas. Along its trajectory, this theory has been complemented by other researchers exploring the issue in their studies. However, the existence of a first and second definition and the term "disruptive" have led some disruptive innovation researchers to use only the first definition or have led to confusion with the second one. Over time, the same theory has been used to explain all kinds of disruptive innovations, resulting in mistakes because different kinds of innovations have different competitive effects and produce different kinds of markets (Markides, 2006). In addition, the importance of understanding the differences between low-end markets and new-markets has been raised up.

In what concerns its predictive value, it is important that managers remain alert to the fact that disruptive innovations can stem from a much broader set of elements than just technological advances. They should pay attention to the different products, processes, and assets they have and to the development of new product through R&D activities. Design, and production processes are also key as well as the observation of the market conditions that boost and facilitate the development of innovative business models (Wan et al., 2015). There is no agreement between researchers on the possibility of predicting a future disruption, but they do agree that current and potential customers should not be ignored.

Moreover, in terms of whether this theory allows us to develop business models, we can point to the fact that implementing disruptive technological innovation promotes the development of a business model because it considers the forms in which companies appropriately create, deliver and price the products they wish to sell (Markides, 2006). Companies with a business model focused on disruption can understand, address, and compete to tackle the needs of their current or future customers.

Concerning the strategies that are applicable to technological disruptive innovations, our work has presented some that can be considered very useful for companies interested in the R&D of strategies that are geared towards the creation of disruptive technologies, and which could be valuable tools within their organizations in their efforts to earn, survive, stay competitive, or be market leaders. Furthermore, in order to develop effective strategies, it is important for companies to recognize the opportunities that emerge in market niches that at first are not very profitable, or to try to unlearn what they have learned and become risk-takers in order to possibly develop disruptive technological innovations.

REFERENCES

Abernathy, W. J., & Clark, K. B. (1985). Mapping the winds of creative destruction. *Research Policy*, *14*(1), 3–22. doi:10.1016/0048-7333(85)90021-6

Adner, R. (2002). When are technologies disruptive? A demand-based view of the emergence of competition. *Strategic Management Journal*, *23*(8), 667–688. doi:10.1002mj.246

Alpkan, L., & Gemici, E. (2016). Disruption and Ambidexterity: How Innovation Strategies Evolve? *Procedia: Social and Behavioral Sciences*, *235*(October), 782–787. doi:10.1016/j.sbspro.2016.11.080

Ansari, S., Garud, R., & Kumaraswamy, A. (2015). The Disruptor ' s Dilemma : TiVO and the U. S. Television Ecosystem. *Strategic Management Journal*. doi:10.1002mj

Ansari, S., & Krop, P. (2012). Incumbent performance in the face of a radical innovation: Towards a framework for incumbent challenger dynamics. *Research Policy*, *41*(8), 1357–1374. doi:10.1016/j.respol.2012.03.024

Ansari, S., & Munir, K. (2008). How valuable is a piece of the spectrum? Determination of value in external resource acquisition. *Industrial and Corporate Change*, *17*(2), 301–333. doi:10.1093/icc/dtn004

Ashforth, B. E., & Mael, F. (1989). Social identity theory and the organization. *Academy of Management Review*, *14*, 20–39.

Baden-Fuller, C., & Morgan, M. (2010). Business models as models. *Long Range Planning*, *43*(2-3), 156–171. doi:10.1016/j.lrp.2010.02.005

Baldwin, C. Y. (2008). Where do transactions come from? Modularity transactions and the boundaries of firms. *Industrial and Corporate Change*, *17*(1), 155–195. doi:10.1093/icc/dtm036

Barney, J. B. (1997). *On flipping coins* and *making technology choices: luck as an explanation of technological foresight and oversight*. In R. Garud, P. Nayyar, & Z. Shapira (Eds.), *Technological Innovation: Oversight and Foresights* (pp. 13–19). New York: Cambridge University Press. doi:10.1017/CBO9780511896613.004

Burgers, J. H., Jansen, J. J. P., Van den Bosch, A. J. F., & Volberda, H. W. (2009). Structural differentiation and corporate venturing: The moderating role of formal and informal integration mechanisms. *Journal of Business Venturing*, *24*(3), 206–220. doi:10.1016/j.jbusvent.2009.01.006

Callaway, S. K., & Hamilton, R. D. III. (2008). Managing disruptive technology: Internet banking ventures for traditional banks. *International Journal of Innovation and Technology Management*, *5*(1), 5–80. doi:10.1142/S0219877008001242

Camerer, C., & Vepsalainen, A. (1988). The economic efficiency of corporate culture. *Strategic Management Journal*, *9*(S1), 115–126. doi:10.1002mj.4250090712

Cannon-Bowers, J., Tannenbaum, S., Salas, E., & Volpe, C. (1995). Defining competencies and establishing team training requirements. In R. Guzzo & E. Salas (Eds.), *Team effectiveness and decision making in organizations* (pp. 333–380). San Francisco, CA: Jossey-Bass.

Carlo, J. L., Gaskin, J., Lyytinnen, K., & Rose, G. M. (2014). Early vs late adoption of radical information technology innovations across software development organizations: An extension of the disruptive information technology innovation model. *Information Systems Journal*, *24*(6), 537–569. doi:10.1111/isj.12039

Chesbrough, H. (2003). *Open Innovation: The New Imperative for Creating and Profiting from Technology*. Boston: Harvard Business School Press.

Chesbrough, H., & Rosenbloom, R. S. (2002). The role of the business model in capturing value from innovation: Evidence from Xerox Corporation's technology spin-off companies. *Industrial and Corporate Change*, *11*(3), 529–555. doi:10.1093/icc/11.3.529

Christensen, C. M. (1993). The Rigid Disk-Drive Industry: A History of Commercial and Technological Turbulence. *Business History Review*, *67*(4), 531–588. doi:10.2307/3116804

Christensen, C. M. (1997). *The Innovator's Dilemma: When New Technologies Cause Great Firms to Fail*. Boston: Harvard Business School Press.

Christensen, C. M. (2000). *The innovator's dilemma: The revolutionary book that will change the way you do business*. New York: Haper Business.

Christensen, C. M. (2006). The Ongoing Process of Building a Theory of Disruption. *Journal of Product Innovation Management*, 39–55. 10.1111/j.1540-5885.2005.00180.x

Christensen, C. M., & Bower, J. L. (1996). Customer power, strategic investment, and the failure of leading firms. *Strategic Management Journal*, *17*(3), 197–218. doi:10.1002/(SICI)1097-0266(199603)17:3<197::AID-SMJ8043.0.CO;2-U

Christensen, C. M., & Raynor, M. (2003). *The Innovator's Solution: Creating and Sustaining Successful Growth*. Boston: Harvard Business School Press.

Christensen, C. M., Raynor, M. E., & McDonald, R. (2015). What Is Disruptive Innovation? *Harvard Business Review*, 44–53. PMID:17183796

Commission, E. U. (2015a). *Digital Single Market*. Retrieved on February 16, 2016 from http://ec.europa.eu/prio-rities/digital-single-market/docs/dsmcommunication_en.pdf

Commission, E. U. (2015b). *Bitcoin Regulation*. Retrieved on February 16, 2016 from http://coinbrief.net/bitcoinr-egulation-european-commission/

Corsi, S., & Di Minin, A. (2014). Disruptive innovation in reverse: Adding a geographical dimension to disruptive innovation theory. *Creativity and Innovation Management*, *23*(1), 76–90. doi:10.1111/caim.12043

Crockett, D. R., McGee, J. E., & Payne, G. T. (2013). Employing new business divisions to exploit disruptive innovations: The interplay between characteristics of the corporation and those of the venture management team. *Journal of Product Innovation Management*, *30*(5), 856–879. doi:10.1111/jpim.12034

Danneels, E. (2004). Disruptive technology reconsidered: A critique and research agenda. *Journal of Product Innovation Management*, *21*(4), 246–258. doi:10.1111/j.0737-6782.2004.00076.x

Dotsika, F., & Watkins, A. (2017). Identifying potentially disruptive trends by means of keyword network analysis. *Technological Forecasting and Social Change*, *119*, 114–127. doi:10.1016/j.techfore.2017.03.020

Forbes. (2015). *20 Digital Trends For 2015*. Retrieved on February 16, 2016 from http://www.forbes.com/sites/georgedeeb/2015/01/16/the-top-20-digital-trends-of-2015/

Frost & Sullivan by Marc Einstein. (2015). *2015 Global and Japan Information and Communication Technology (ICT) Outlook*. Retrieved from http://www.frost.com/sublib/displayreport.do?id=9ABE-00-1A-00-00

Gartner. (2015). *Top 10 Strategic Technology Trends for 2015*. Retrieved from http://www.gartner.com/newsroom/id/2867917

Gilbert, C. (2003). The disruption opportunity. *Sloan Management Review*, *44*(4), 27–33.

Gilbert, C. (2005). Unbundling the structure of inertia: Resource versus routine rigidity. *Academy of Management Journal*, *48*(5), 741–763. doi:10.5465/AMJ.2005.18803920

Gilbert, C., & Bower, J. (2002). Disruptive change: When trying harder is part of the problem. *Harvard Business Review*, *80*(5), 94–101. PMID:12024762

Govindarajan, V., & Kopalle, P. K. (2006a). Disruptiveness of Innovations: Measurement and an Assessment of Reliability and Validity. *Strategic Management Journal*, *27*(2), 189–199. doi:10.1002mj.511

Govindarajan, V., & Kopalle, P. K. (2006b). The Usefulness of Measuring Disruptiveness of Innovations. Ex-post in Making ex-ante Predictions. *Journal of Product Innovation Management*, *23*(1), 12–18. doi:10.1111/j.1540-5885.2005.00176.x

Habtay, S. R. (2012). A Firm-Level Analysis on the Relative Difference between Technology-Driven and Market-Driven Disruptive Business Model Innovations. *Creativity and Innovation Management*, *21*(3), 290–303. doi:10.1111/j.1467-8691.2012.00628.x

Hang, C. C., Garnsey, E., & Ruan, Y. (2015). Opportunities for disruption. *Technovation*, *39–40*, 83–93. doi:10.1016/j.technovation.2014.11.005

Huang, J., Chou, T., & Lee, G. (2010). Imitative innovation strategies. *Management Decision*, *48*(6), 952–975. doi:10.1108/00251741011053488

Jacobides, M., Knudsen, T., & Auger, M. (2006). Benefiting from innovation: Value creation, value appropriation and the role of industry architectures. *Research Policy*, *35*(8), 1200–1221. doi:10.1016/j.respol.2006.09.005

King, A., & Tucci, C. L. (2000). Responding to technology-created market niches: Do industry-specialized capabilities facilitate or hamper entry? *Academy of Management, TIM:* D1–D6.

King, A. A., & Baatartogtokh, B. (2015). How useful is the theory of disruptive innovation? *MIT Sloan Management Review*, *57*(1), 77–90. doi:10.1017/CBO9781107415324.004

Klenner, P., Husig, S., & Dowling, M. (2013). Ex-ante evaluation of disruptive susceptibility in established value networks - When are markets ready for disruptive innovations? *Research Policy*, *42*(4), 914–927. doi:10.1016/j.respol.2012.12.006

KPMG. (2015). *KPMG Survey Reveals Top Disruptive Technologies*. Retrieved from https://home.kpmg.com/im/en/home/media/press-releases/2015/12/survey-reveals-disruptivetechnologies. html

Krueger, N., & Dickson, P. (1994). How believing in ourselves increases risk taking: Perceived self-efficacy and opportunity recognition. *Decision Sciences*, *25*(3), 385–401. doi:10.1111/j.1540-5915.1994.tb01849.x

Latzer, M. (2009). Information and Communication Technology Innovations: Radical and Disruptive? *New Media & Society*, *11*(4), 599–619. doi:10.1177/1461444809102964

Levina, M. (2016). Disrupt or Die: Mobile Health and Disruptive Innovation as Body Politics. *Television & New Media*. doi:10.1177/1527476416680451

Lyytinnen, K., & Rose, G. M. (2003). The disruptive nature of information technology innovations: The case of Internet computing in systems development organizations. *Management Information Systems Quarterly*, *27*(4), 557–595. doi:10.2307/30036549

Markides, C. (2006). Disruptive innovation: In need of better theory. *Journal of Product Innovation Management*, *23*(1), 19–25. doi:10.1111/j.1540-5885.2005.00177.x

Markides, C., & Charitou, C. D. (2004). Competing with Dual Business Models – A Contingency Approach. *The Academy of Management Perspectives*, *18*, 22–36.

Markides, C. C. (2012). How Disruptive Will Innovations from Emerging Markets Be? *MIT Sloan Management Review*, *54*, 23–25.

McKinsey. (2015). Disruptive Technologies: Advances that will Transform Life. *Business, and the Global Economy*. Retrieved from http://www.mckinsey.com/insights/business_technology/disruptive_technologies

Nayak, R., & Ketteringham, J. M. (1994). *Breakthroughs*. San Diego, CA: Pfeiffer and Co.

NESTA. (2015). *Growing a Digital Social Innovation Ecosystem for Europe*. Retrieved from http:// www.nesta.org.uk/sites/default/files/dsireport.pdf

OECD. (2015). *Digital Economy Outlook OECD.* Retrieved from http://ec.europa.eu/euro-stat/documents/42577/3222224/Digital+economy+outlook+2015/dbdec3c6-ca38-432c-82f2-1e330d9d6a24

Paap, J., & Katz, R. (2004). Anticipating disruptive innovation. *Research Technology Management, 47*(5), 13–22. doi:10.1080/08956308.2004.11671647

Rad, M. G. (2017). Disruptive innovation in media industry ecosystem and need for improving managerial cognitive capabilities in polymediation era. *Cogent Business & Management, 23*(1), 1–24. doi:10.1080/23311975.2017.1352183

Reinhardt, R., & Gurtner, S. (2015). Differences between early adopters of disruptive and sustaining innovations. *Journal of Business Research, 68*(1), 137–145. doi:10.1016/j.jbusres.2014.04.007

Roger, S. (2007). The disruptive potential of game technologies. *Research Technology Management, 50*(2), 57–64. doi:10.1080/08956308.2007.11657431

Rosenkopf, L., & Nerkar, A. (2001). Beyond local search: Boundary spanning, exploration, and impact in the optical disk industry. *Strategic Management Journal, 22*(4), 287–306. doi:10.1002mj.160

Rothaermel, F. T., & Boeker, W. (2008). Old technology meets new technology: Comple- mentarities, similarities, and alliance formation. *Strategic Management Journal, 29*(1), 47–77. doi:10.1002mj.634

Sandström, C. (2011). High-End Disruptive Technologies with an Inferior Performance. *International Journal of Technology Management, 56*(2/3/4), 109–122. doi:10.1504/IJTM.2011.042977

Sandström, Ch., Berglund, H., & Magnusson, M. (2014). Symmetric Assumptions in the Theory of Disruptive Innovation: Theoretical and Managerial Implications. *Creativity and Innovation Management, 23*(4), 472–483. doi:10.1111/caim.12092

Schulman, S., & Thomas, L. (2014). New Frontiers of the Innovation Economy and Education. *Entrepreneurship Research Journal, 4*(1), 137–146.

Schumpeter, J. A. (1942). Capitalism, Socialism, and Democracy. Harper Perennial.

Shepherd, D. A., & Krueger, N. (2002). An intentions-based model of entrepreneurial teams' social cognitions. *Entrepreneurship Theory and Practice, 27*(2), 167–185. doi:10.1111/1540-8520.00005

Si, S., Yu, X., Wu, A., Chen, S., & Su, Y. (2014). Entrepreneurship and poverty reduction: A case study of Yiwu, China. *Asia Pacific Journal of Management*, *32*(1), 119–143. doi:10.100710490-014-9395-7

Siggelkow, N., & Levinthal, D. A. (2003). Temporarily divide to conquer: Centralized, decentralized, and reintegrated organizational approaches to exploitation and adaptation. *Organization Science*, *14*(6), 650–669. doi:10.1287/orsc.14.6.650.24840

Subramaniam, M., & Youndt, M. A. (2005). The influence of intellectual capital on the types of innovative capabilities. *Academy of Management Journal*, *48*(3), 450–463. doi:10.5465/AMJ.2005.17407911

Tellis, G. J. (2006). Disruptive technology or visionary leadership? *Journal of Product Innovation Management*, *23*(1), 34–38. doi:10.1111/j.1540-5885.2005.00179.x

Thornhill, S., & Amit, R. (2000). A dynamic perspective of internal fit in corporate venturing. *Journal of Business Venturing*, *16*(1), 25–50. doi:10.1016/S0883-9026(99)00040-3

Tripsas, M. (1997). Unraveling the process of creative destruction: Complementary assets and incumbent survival in the typesetter industry. *Strategic Management Journal*, *18*(S1), 119–142. doi:10.1002/(SICI)1097-0266(199707)18:1+<119::AID-SMJ9213.3.CO;2-S

Tripsas, M. (2008). Customer preference discontinuities: A trigger for radical technological change. *Managerial and Decision Economics*, *2*(9), 79–97. doi:10.1002/mde.1389

Tushman, M. L., & O'Reilly, C. A. III. (1996). Ambidextrous organizations: Managing evolutionary and revolutionary change. *California Management Review*, *38*(4), 8–30. doi:10.2307/41165852

Vázquez, J. P., Bienenstock, M. J., & Zuckerman, E. W. (2016). Debating Disruptive Innovation. *MIT Sloan Management Review*, *57*(3), 25–31.

Wan, F., Williamson, P., & Yin, E. (2015). Antecedents and implications of disruptive innovation: Evidence from China. *Technovation*, *39-40*, 94–104. doi:10.1016/j.technovation.2014.05.012

White, G. R. T. (2017). *Future applications of blockchain in business and management: A Delphi study*. Wiley. 10.1002/jsc.2144

Yu, D., & Hang, C. C. (2010). A Reflective Review of Disruptive Innovation Theory. *International Journal of Management Reviews*, *12*(4), 435–452. doi:10.1111/j.1468-2370.2009.00272.x

Yu, D., & Hang, C. C. (2011). Creating technology candidates for disruptive innovation: Generally applicable R&D strategies. *Technovation*, *31*(8), 401–410. doi:10.1016/j.technovation.2011.02.006

Zhou, W., & Piramuthu, S. (2013). Technology regulation policy for business ethics: An example of RFID in supply chain management. *Journal of Business Ethics*, *116*(2), 327–340. doi:10.100710551-012-1474-4

KEY TERMS AND DEFINITIONS

Augmentation Strategy: Type of strategy related to the improvement of the existing products where R&D is done in order to develop new disruptive characteristics.

Disruptor: It refers to the key issue that provokes a disruption. From a technological point of view the disruption occurs when the new technology crosses the performance of the dominant technology and leads to significant increases of market share. But business disruptors can be of a very diverse nature.

Exploitation for Another Application (Strategy of): It refers to the development of new applications in order to reposition existing products in addition to the fight back option.

High-End Market: Markets with the most lucrative customers, willing to spend money for good products or services. Incumbents or traditional companies have most of the market and ignore low end markets.

Low-End Market: It refers to markets of less-discerning customers called low end customers where affordable product or services are frequently at a lower price. Start-ups or small companies often tend to focus on these types of markets that are far from the mainstream markets.

Miniaturization Strategy: It refers to the strategy based on good product that are created from disruptive technology without high capital investments.

New Value Proposition: Can be defined as a business statement that a company uses to summarize why a consumer should buy a product or use a service. The aim is to convince a potential consumer that one particular product or service will add more value or better solve a problem than other similar offerings.

Predictive Value of Disruptive Innovation: The concept is linked to the analysis of the the mechanisms and effects of disruptive technologies on firms and markets. There is an academic debate on the capability of the disruptive innovation theory to predict trends and future developments in order to help companies to be competitive in the future.

Simplification Strategy: The focus of this strategy is strongly oriented to the definition of a new value proposition that focus on current products and tries to lower some characteristics or the degree of presentation in certain features.

Chapter 2
Identification and Categorization of Disruptive Innovations According to the Strategic Scope of the Firm

Vincent Sabourin
University of Quebec, Canada

ABSTRACT

The concept of innovation has for a long time been considered as the drive behind the ever-changing positions and functioning of the global community. Since the beginning of twenty-first century, creative individuals and entrepreneurs have without influence declined to accept the existing products and services as being the ultimate solution to challenges the society faces. This scholarly analysis builds on the theory of disruptive innovation to categorize and analyze the impact of disruptive innovations on key sectors of an economy in terms of their impact on the strategies and commercialization from a sectorial perspective. This analysis offers insightful information in terms of disruptive innovations which can be categorized into market-driven, product-driven, and competency-driven disruptive innovations.

DOI: 10.4018/978-1-5225-4148-6.ch002

INTRODUCTION

Among the concepts connected with the revolutions provided by the field of sciences of technology, one of the most important for research in strategic management is "disruptive innovations". In his 1997 book, The Innovator's Dilemma, Christensen offered an elaboration for the collapse of well-managed and respected and companies under the influence of disruptive innovations (Christensen, 2003). These disruptive innovations are entering the market on a marginal basis but create significant disturbances in the functioning of the market as soon as the process of conquest of the early buyers to attack the early majority of buyers will have been completed (Moore, 2014). In 2006, Tellis showed that among all the concepts dealing with innovation in management, it is the expression "disruptive innovations" which had received most entries with the Google search engine. The concept had even more entries than the concept of "radical innovation (Tellis, 2006).

In this regard, Bower (2012) has explained that the process of how disruptive technology, through its requisite support net, dramatically transforms a certain industry. According to Bower (2012) "When a technology that has the potential to revolutionize an industry emerges, established companies typically see it as unattractive: it's not something their mainstream customers want, and its projected profit margins aren't sufficient to cover big-company cost structure. As a result, a new technology tends to get ignored in favor of what's currently popular with the best customers. But then another company steps in to bring an innovation to a new market. Once disruptive technology becomes established there, smaller-scope innovation rapidly raises the technology's performance to attribute what mainstream customers' value." According to the proponent of this theory, disruptive innovation describes a phenomenon in which the innovation process revolutionize existing markets and or industries by developing affordability and accessibility in a more simple and convenient manner. Disruptive innovation stems from the fact that there is high cost of products in markets offered by companies that seek to maintain their status quo. Disruptive innovation starts from a niche market which is usually considered less attractive with little consequence to disrupt existing incumbents; however a new idea or product comes in and shifts the industry status quo. The economic press has highlighted how disruptive innovations could substantially transform the competitive context of established players and their position in terms of market share.

Thus the main objective of this chapter is to tentatively categorize disruptive innovations using the conceptual framework of the strategic scope of the firm to categorize disruptive innovations according to the perspective of the incumbents defined by players that occupies a particular position or place in a market with a survey of examples from the economic press. The chapter wishes to study the impact of disruptive innovation of the strategic scope of activities of companies called incumbent in their market to develop a categorization of impact.

BACKGROUND

According to Schumpeter (2003), these disruptive patterns of creative destruction," typically flow through industries, and in the process sink outdated, outmatched and weak sinking companies. In this chapter, the author notes that few academic and scholarly theories on management have not had as much influence in the world of business as Clayton M. Christensen's theory of disruptive innovation. Before proceeding, this chapter therefore poses this question, what role does disruptive innovation play in business?

In his 1997 book, The Innovator's Dilemma, Christensen presented a description for the collapse of well-managed and respected companies (Christensen, 2003). According to Christensen, best performing managers face a dilemma, because by "doing the very things they need to do to succeed — listen to customers, invest in the business, and build distinctive capabilities" — they operate with the risk of ignoring competitors with "disruptive" technological innovations. In this regard, it is important to identify the key elements, which hold the greatest value for the theory of disruptive innovation.

In *The Innovator's Solution*, Christensen and Raynor (2016) argue that the disruptive innovation has four (4) characteristics/roles. The first characteristic is Incumbent companies are improving along a path of sustaining innovation. One of the key factors of disruptive innovation is that "in every market there is a distinctly different path of improvement that innovating companies provide as they introduce new and improved products." Therefore, an incumbent company's improvement path emanates from what is termed as "sustaining innovation" — the periodic improvements that all great companies seek to bring out." Usually, sustaining innovations improve products in a few established value areas. For instance, automotive companies might continue improving their "horsepower or torque" of their engines as a result of sustaining innovations. As Christensen and Raynor (2016) explain, great managers strive "to make better products that they can sell for higher profit margins to not-yet-satisfied customers in more demanding tiers of the market.

The second attribute is that incumbent players are exceeding customer needs. A second element of Christensen & Raynor's theory is that the pace of sustaining innovation along the path of particular value propositions "almost always exceeds the ability of customers in any given category of the market to consume it. Thus, a company whose products are basically positioned on mainstream customers' current needs will probably over exceed what current customers are able to consume in the future."

The third characteristic is that incumbent companies possess the capability to respond to disruptive threats. Christensen & Raynor argue that incumbent companies frequently possess the capabilities needed to succeed in the market; however managers fail to utilize them effectively to deal with potential disruptors in the market. According to Christen and Raynor (2016), "disruption has a paralyzing effect on industry leaders. With resource allocation processes designed and perfected to support sustaining innovations, managers are constitutionally unable to respond to the emerging trends and challenges in the disruptive process". The researcher notes that although these threats are considered to be inferior when measured against value propositions on which sustaining innovation has been focused, these disruptive products have other hidden characteristics such as being less expensive, simpler and more convenient. Thus from the value of disruptive innovation theory, managers should be to adopt this theory to improve the sustaining innovation process of their companies and in the long run sustain their competitive position in the market.

The fourth feature of this theory is that Incumbent companies develop as a result of disruption. Christensen argues that "performance oversupply "opens the door for simpler, less expensive, and more convenient and almost disruptive technologies to enter (Christensen, 2003) Companies with these disruptive technologies, "will always improve performance of their products and in so doing they can take over older markets." Christensen and Raynor (2016) write that "Once the disruptive product gains a foothold in new or low-end markets, the disruptors are on a path that will ultimately crush the incumbents."

The theory of disruptive innovation has some contributions that offer some managerial and practical usefulness. This research has established that new business models and technology can bring massive changes to an industry where disruption is yet to occur, for instance as seen from the case of Airbnb in the hotel industry. Airbnb is a typical case of disruption. It began as a low end business targeting those customers who could not afford hotel costs or those who could not get a hotel in the most crowded market. The theory also helps emerging companies to develop competitive responses by creating those products that appeal to the neglected and the usually considered low end markets.

According to Christensen, the theory of disruptive innovation does not seek to help companies achieve growth in their market or product dimensions but it rather seeks to help incumbent companies to strengthen their core business. In this regard, any company that feels its market niche is threatened by new business models should establish a different distinct company that is different from the core to engage in disruptive businesses. From the perspectives of this theory, incumbent companies should not just focus on the disruptive process but on the expected level of perceived value and the degree of preference for change in the mainstream through price variations.

The theory of disruptive innovation shows that innovation is always going to be the order of the day in every industry and as such incumbent companies should focus on the product improvement rather than focusing on the threats posed by new entrants. For any innovation to occur in the industry, Christensen provides that it must provide a unique value/attribute to customers that are different from the existing product. Therefore through the theory of disruptive innovation, incumbent companies are able to focus their resources on product improvement through value proposition and quality improvement.

MAIN FOCUS OF THE CHAPTER

Literature Review

Since its introduction by Christensen in 1997, the concept of disruptive innovation has been studied in various fields of the management (Daneels, 2004). Danneels (2004) indicated how research on this concept can facilitate a better and deeper understanding of the nature and the impact of radical technological changes. Scholars and practitioners have also indicated that disruptive innovations have had a significant effect on research and the approaches used to study the technological change and the phenomenon of the competition (Adner, 2002).

However, empirical validation did not allow the generalization the initial phenomenon proposed by Christensen (1997) for a majority of innovations of disruptive innovations.

The authors questioned the methodology and the selection of the studied cases. But in a general manner, the main criticism came from the lack of generalization and the external validity of the concept. For instance, Leopore (2014) surveyed examples of companies which were victims at the origin of disruptive innovations in bid to protect their market domination. This was in particular, the case of American companies: Seagate Technology, U.S. Steel, and Bucyrus. It can therefore be argued in this chapter, that the generalization of the concept and the external validity are explained among other things, by the complexity of the phenomenon of disruptive innovations. The economic and technological phenomenon associated with the disruptive innovations has revealed complexities in terms of technological economic, technological and commercial perspectives. The research issue could be examined with three perspectives: understanding the disruptive character of innovation, the impact on the business of the incumbents and the need for a categorization on the impact of the incumbent.

The Disruptive Character of Innovation

This chapter highlights the fact that the disruptive character has proved to have different meanings according to the authors (Yu & Hang, 2011). In a general way, research on the concept revealed certain criticisms in particular as for the definition of the disruptive character of the innovations identified by Christensen (Rao & Cull 2006). According to these authors, one of the critics of this concept results from the fact that the analyses which research has, are essentially ex-ante and that the robustness of the concept of disruptive innovations remains fragile to study and forecast disruption. The fragile character of the predictability of the concept regarding disruption highlights the necessity of studying this phenomenon in a finer way to take into account the variety of disruptive innovations with a more structured categorization.

In lieu of this, the author notes that from 2002, ADNER et al., proposed a methodology to better understand the disruptive process when innovations appear in a market. This is also similar to the case of Rafii and Kampas (2002) who propose a model in six phases allowing the explanation of the various passages to understand he disruptive character of innovation.

Impact on the Business of the Incumbents

What is the impact of disruptive innovation on the business of the incumbents? The literature has built a perspective based on the work by Tushman and Anderson (1986), who wrote about competence-enhancing and competence- destroying innovations to analyze the impact on the incumbents. Several authors have examined the impact on the business and the scope of their activities from the incumbents' perspective. Henderson and Clark (1990) advanced the argumentation by introducing the concept of architectural and component level innovation from the incumbents' perspective. Similar explanations have been provided by Clark (1985), Tripsas (1997) and Cooper and Schendel (1976). In his seminal work, Christensen (1997) proposed an alternative explanation of incumbent failure. Authors have also focused on the external environment to understand the impact on the incumbents, drawing upon resource dependence theory (Pfeffer & Salancik, 1978) and the concept of value networks (Christensen & Rosenbloom, 1995). Sandstrom (2010) explained that consequently, a key determinant of the probability of success for an incumbent firm to innovate is the extent to which the new product addresses the needs of actors in the incumbent firm's current value network that refers to the set of internet and external competencies. For instance, several scholars have pointed out the importance of both looking at performance attributes and how these are translated into the market scope of the business and value for the customer (Oskarsson & Sjöberg, 1991;

Saviotti & Metcalfe, 1984). One of the important questions in the construction of the impact on the incumbents concerned the potential of an innovation on existing business (Keller & Husing on 2009). Sandsrom (2011) explicitly mentioned that more knowledge is needed regarding how incumbent firms can successfully manage disruptive innovations. Keler and Hüsig (2009) analyzed the question whether web applications pose a disruptive threat to incumbents or a disruptive growth opportunity for entrants in the application software industry using a novel method for ex ante.

Categorization of Disruptive Innovations for the Incumbents

Markides (2006) indicated that the categorization of disruptive innovations of Christensen was not fine and granular enough to underrate these technological changes. Following the seminal studies of Christensen (1997), many scholars have tried to refine the initial definition of disruptive innovation, trying to differentiate the concept of "disruption" in different typologies, i.e. technological, radical product and business model (Danneels, 2004; Markides, 2006). According to Markides (2006), the disruptive technological innovations presented by Christensen limit themselves to a single category while two other categories can be identified in that concerning the business model and that concerning products. Markides proposes two other categories of disruptive innovations in complement to Christensen that are business models disruptive innovation and product disruptive innovations. According to Baiyere (2014), too few studies have however focused on examining the lessons that could be learnt from innovations that were considered as potentially disruptive on several measures which eventually end up not being disruptive or where they were disrupted. The lack of adequate categorization led several authors to question the generalization and the external validity of the concept of disruptive innovation (Leopore, 2014; Weeks, 2015). Danneels indicated how research on this concept to categorize the impact can allow a better understanding the nature and the impact of radical technological changes on the incumbents. The lack of adequate categorization led several authors to question the generalization and the external validity of the concept (Leopore 2014; Weeks 2015). To facilitate the categorization, the research suggests a better focus on specific sectors (Roy; 2014) and complementary concepts such as the pre-disruption component experience in firm response (Roy & Cohen, 2015).

Sandstrom (2010) proposed to shift the focus of attention towards the utility that it creates for customers to analyze the impact of the business of the incumbents.

In conclusion from the review of literature, the "disruptive character" of innovations needs to better examine the disruptive character of innovations, a focus on the perspective of the incumbents and more specifically to analyze in more depth the impact on the business and the scope of activities by the incumbents to develop categorizations.

Methodology

This research design framework is based on action-response research design (Chen et al., 1991; Chen et al., 1992). This Chapter's analysis focuses on the strategic disruptive innovations responses adopted by key competitors in the industry to challenge the incumbents. This research relied on a similar research design that other scholars in previous analysis on competitive responses (Chen et al., 1992; Smith et al., 1991; Tang & Shapira 2011) had adopted in their researches.

Conceptual Framework

The author relied on the conceptual framework of Abell's (1980) but also Allaire and Firsirotu (1993) generic categorization scheme is built on the idea that a business can be defined by the scope of its offerings, its degree of competitive differentiation, and the extent to which its offerings are differentiated across different product/market segments (segment differentiation).

Abell matrix is a three-dimensional tool most often is referred to as the three dimensional business definition model. The model is used to analyze the scope of operation for a business. This may include areas such as the technologies and products a business operates in a market or the audience that it targets. A detailed analysis of the business's current activities can help create strategies for the future that will help the business stay tuned to the changes that may occur within the market.

The three dimensions of the business are the customer groups (who will be served by the business), customer needs (what are the customer needs that will be met) and technology or distinctive competencies (how are these needs going to be met). A major point of importance in this matrix is to focus on understanding the customer rather than the industry and its products and services. Through these three dimensions, this tool helps define a business by its competitive scope (narrow or broad) and the extent of competitive differentiation of its products/services.

According to Martin (2015), the major point of importance in this matrix is to focus on understanding the customer rather than the industry and its products and services. Through these three dimensions, this framework helps define a business by its competitive scope (narrow or broad) and the extent of competitive differentiation of its products/services. From this view, it can be noted that a detailed analysis of the business's current activities can help create strategies for the future that will help the business remain focused to the disruptive changes that may occur within the market. This research used this framework to analyze the impact of these three independent variables for a set of 20 disruptive innovations identified in the literature. These variables include: Product driven disruptive innovations; Market driven disruptive innovations and Competency driven disruptive innovations

Data Collection and Sample

To collect information on this analysis' cases the author followed five steps:

Step 1: The author collected information and surveyed 20 innovations for their disruptive character from a technological and economic perspective. The following table illustrates in a preliminary manner of these 20 disruptive innovations. The data collection was completed with a survey of reports and articles on the topic of disruptive innovation with the following lists and sources: Accenture; EAI Catalyst Cleantech Sustainability, (2015); CNBC Disruptors 50; Forbes, Interaction Design Foundation; McKinsey Global Institute, Fortune; Intelligent HQ; Observer, 2005; Mashable.com; PC World (2008); Solis (2015) & MIT Technology Review.

Step 2: All these reports were supported with economic data (generally quantitative) and technological data (generally descriptive) describing their disruptive nature. The McKinsey Global Institute has produced an integrative and very complete report and also a full set of detailed reports surveying and assessing specific disruptive innovation for specific innovations. These reports were completed but other sources to reach a list of 20. Since the objective of the research was not to provide an assessment the technological and economic impact, the author listed these disruptive innovations by alphabetic order.

Step 3: The author also provided a categorization using the dominant drivers according to the description of the technological and economic impact of the literature. Several of the twenty five disruptive innovations combined several aspects and were at the same time product but also market driven. Using the Abel's matrix, the researcher selected the dominant factor among the three factors of the model.

Step 4: The disruptive character of the innovation was assessed from a technological perspective and from the economic perspective. From a technological perspective, the disruptive character was assessed by the radical nature of the substitution provided by the innovation. When the innovation could be accommodated and easily integrated by the incumbent it was not considered as being disruptive. When it has to be assimilated and not easily integrated (measure by the ease of substitution) and therefore required a transformation of the strategic scope of the incumbent of the incumbent, it was considered as being "disruptive".

Step 5: The selection was completed using a technological and an economic perspective. From a technological perspective the author assessed the disruptive character of these innovations in terms of functionalities (products), distribution and competencies based on the sources from engineering, computer science and software and life science. For an economic perspective, this scholarly analysis relied on economic impact assessment. For instance, the McKinsey Global Institute has provided a full set of economic impact for a large number of innovations into its report and the author relied on these economic evaluations to assess the disruptive aspect of the innovations.

Data Selection

The author provided a categorization using the dominant drivers according to the description of the economic and the technological impact provided by the economic press. The selection was completed using both a technological and an economic perspective. From a technological perspective the author assessed the disruptive character of these innovations in terms of functionalities (products), distribution and competencies based on the sources from engineering, computer science and software and life science. For an economic perspective, the author relied on economic impact assessment. For instance, the McKinsey Global Institute has provided us with a full set of economic impact for a large number of innovations into its report and the author relied on these economic evaluations to assess the disruptive aspect of the innovations. For the final selection, the author only kept innovations that combined these three dimensions: Firstly, the author rejected some of the innovations that were not considered for the study had a high technological impact but no significant economic impact, and were not really disruptive from a technological point of view such as OLED/LED technology and advanced photonics.

Secondly, for instance, the author included Carbone captation but several innovations such as the next generation nuclear, flexible displays or advanced water treatment were not considered for final evaluation. The author corroborated selection with this research's information sources.

Thirdly, some innovations had a large economic impact but were not disruptive from technological perspective and were thus not considered. This was the case of new feature of smart phone or the IPad or the smart televisions that has economic impact but was not really disruptive from a technological perspective.

Data Analysis

The researcher analyzed each component of the conceptual framework drivers for each innovation using the economic literature on the subjects. For each innovation

Table 1. Survey of disruptive innovations

1.	3D printing
2.	Artificiel Intelligence
3.	Advanced Materials
4.	Advanced Oil and gas exploration and recovery
5.	Advanced Robotics
6.	Augmented Reality
7.	Automation of knowledge work
8.	Automotive Technology
9.	Autonomous and near-autonomous vehicles
10.	BlockChain
11.	Captation CO2
12.	Cloud technology
13.	Crowdfunding
14.	Energy storage devices or systems
15.	Fintech (Financial Technology)
16.	Immune Engineering
17.	Internet of Things
18.	Mobile Payment
19.	Nanotechnology
20.	Next-generation Genomics

Source: compiled from Accenture; EAI Catalyst Cleantech Sustainability, (2015); CNBC Disruptors 50; Forbes, Interaction Design Foundation; McKinsey Global Institute, Fortune; Intelligent HQ; Observer, 2005; Mashable.com; PC World (2008); Solis (2015); MIT Technology Review.

the author identified a predominant driver among the product scope, the market and the scope of competencies on the incumbents of the dominant market.

1. Product scope was defined as having an impact on the substitution of product or services functionalities for the incumbents.
2. Market scope was defined as an impact on the substitution of channels of distribution or the segmentation for the market (such as the long tail effect) the incumbents.
3. Scope of competencies was identified as having an impact on substitution of the core competencies the incumbents.

Finally, the author developed short background case studies to build on the systematic review of literature.

Categorizing Disruptive Innovations Based on Strategic Scope

This section provides three cases that are describing the impact of disruptive innovations on the strategic scope of the incumbents.

Product Driven Disruptive Innovations

Product driven disruptive innovation could be understood as a new substitute for all functionalities that means replacing existing functionalities by new disruptive innovative functionalities for incumbents. In this category, examples to be found include: Wearables, Captation Co2, Crowdfunding, and Renewable Energy though the chapter focuses more on three cases that are: energy storage, autonomous vehicles such as Google cars and the advanced robotics.

Wearables

The fast emerging field of wearables is poised to disrupt current technologies and processes across industries while creating new market opportunities and business models. The market continues to evolve as various start-ups and established companies take a keen interest in product research and development, particularly in creation of multiple functions for individual devices (Frost & Sullivan, 2016).

1. *Dominant Impact of Wearables on Product Scope*

Wearable Technology involves body worn electronics that can perform various functionalities such as sensing, communicating, and computing. Wearable's is the result of convergence of various technologies in sensors, microelectronics, and ICT. The disruption of wearables comes mainly for the multiplication of new product with innovative functionalities that are emerging as devices but that are progressively integrated to the clothing. With the continued miniaturization of enabling technologies, wearable devices have hit the mass market in a diverse variety of forms, ranging from glasses to even jewellery (PR Newswire, 2016). Key enabling technologies including low cost sensors, wireless connectivity, active materials and energy have converged to make wearable technology mainstream (PR Newswire, 2016).

2. *Secondary Impact on the Market Scope*

Wearables are also having a secondary impact on the market scope with the connectivity features coming from the internet of things that established new marketing and communication channels among separate markets. For instance new electronic gloves and plant uniform that are connected through the Wi-Fi and the internet are in direct relationship with the concept of Factory 4.0. The connected feature of clothing is transforming the traditional boundaries of the clothing market. Wearable technology impacts the lives of millions of people who are adopting this

Figure 1. Range of product application for wearables
Source: Beecham Research Ltd. & Wearables Technologies AG

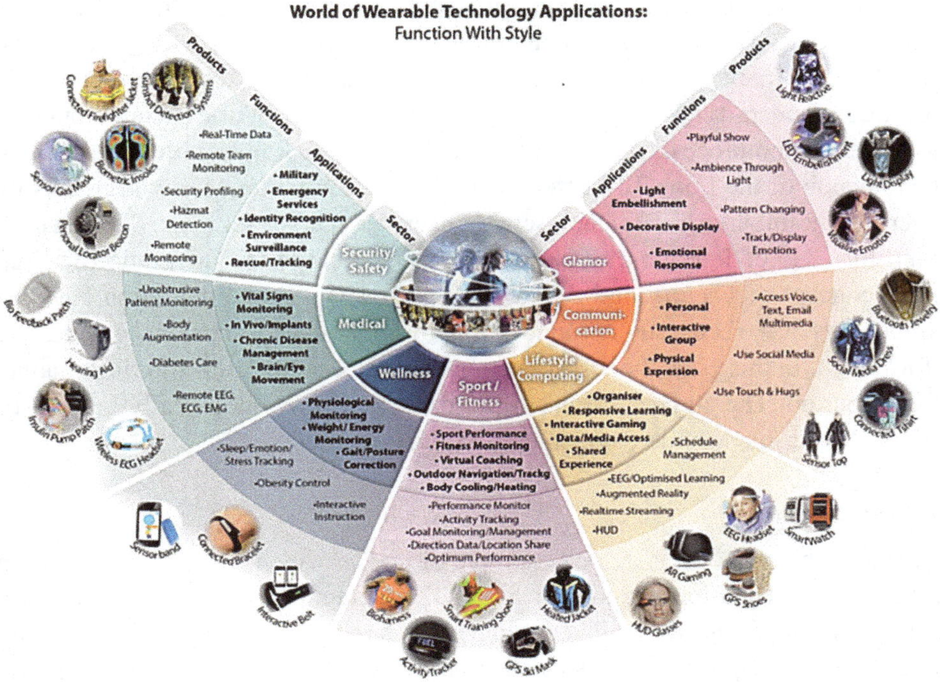

technology primarily by improving their lifestyles through monitoring of health and fitness (Report Linker Review, 2015).

3. *Secondary Impact on the Scope of Competencies*

Wearables also have a secondary impact of transforming the traditional set of competencies and skills that are required to succeed in the clothing market, by introducing key electronics and computing skills to help produce not only smart fabrics and clothing but also manage data coming from inter-connected clothing ecosystem. Wearables are introducing radically new skills in areas such electronic and software leading to technologies such as Artificial intelligence and advanced simulations. This is why suppliers of technological components play a strategic role in the development of new products.

Market Driven Disruptive Innovations

This is the case where the disruptive innovation becomes a substituting channel of distribution. This category of disruptive innovations deals with a disruptive impact on the market scope of incumbents. The case of market driven disruptive innovations concerns not only the market scope of the distribution channels but also the segmentation process. It is in particular the category linked with the economic phenomenon of the Long Tail Effect that allows the creation of new segments by lowering the minimum efficiency scope to serve a market segment. This chapter found the case of Internet of Things interesting because the integration leading to the concept of Factory and Service 4.0 with a real time integration between components of the ecosystem of innovation, by creating new channels to distribute information. In this case, the disruptive innovation becomes a substituting channel of distribution. The author of this chapter examined more specifically Internet of Things, Electronic Mobile Payment and Mobile Internet

Internet of Things

Internet of things (IOT) is a good example of how a market disruptive innovation brings new distribution channel to life. IOT is creating a new communication channel by integrating vertically components of a market. Driven by the capability of a variety of supporting devices such as inbuilt sensors and actuation nodes, the Internet of Things has moved out of its infant state and is now changing and reshaping the current communication models (Lopez et al., 2012). The channels of distribution and decision making are now being enabled by internet of things. This research has established that the adoption of large-scope diverse and independently managed platforms of Internet of Things Infrastructure can help in this process by inculcating data collection, analysis, and actuation and management tools in real time (Carnot, 2011). This new ability is now being used by companies to engage internal and external stakeholders on all matters operations, decision making and implementation facilitating a redefinition of the market and the segmentation process.

1. *Dominant Impact of Internet of Things on the Market Scope*

This chapter has established that the adoption of large-scope in diverse and independently managed platforms of Internet of Things Infrastructure can help in this process by inculcating data collection, analysis, and actuation and management tools in real time (Carnot, 2011). This new ability is now being used by companies to engage internal and external stakeholders on all matters operations, decision making and implementation facilitating a redefinition of the market and the segmentation process. Internet of things (IOT) is therefore a good example of how a market disruptive innovation brings new distribution channel. Driven by the capability of a variety of supporting devices such as inbuilt sensors and actuation nodes, the Internet of Things has moved out of its infant state and is now changing and reshaping the current communication models (Lopez et al., 2012). The channels of distribution and decision making are now being enabled by internet of things.

2. *Secondary Impact of Internet of Things on the Product* Scope

A secondary impact is being experienced at the product scope. IOT introduced new product and service functionalities. The Smart home and building market deals with multiplication of devices competing among themselves. In terms of product and service functionalities, building automation is an example of a distributed control system – the computer networking of electronic devices designed to monitor and control the mechanical, security, fire and flood safety, lighting (especially emergency lighting), HVAC and humidity control and ventilation systems in a building.

3. *Secondary Impact of Internet of Things on the Scope of Competencies*

Another impact of IOT is to transform the scope competencies of the incumbent. While consumer IoT devices have drawn most press attention, it is the enterprise-level adoption of the technology that is having an impact on the core competencies of the incumbents. Indeed, the industry is positioned to implement the technology, using IoT-enabled building management systems (BMS) to make building performance more efficient and also use sensor-generated data to enhance building user experience (Deloitte University Press, 2016). Gartner's (2015) recent smart city forecasts highlight the potential: "Smart commercial buildings will be the highest user of Internet of Things (IoT) until 2017, after which smart homes will take the lead with just over 1 billion connected things in 2018. In terms of scope of competencies, ability to leverage data created by IoT-enabled buildings. According to Deloitte, if a company wishes to use IoT applications to do more than improve efficiency and offer tenants' new products and services, it must overcome a different hurdle: analyzing the vast volumes of data that applications generate and finding insights there. The volume and types of data make this analysis difficult for traditional real state owners in commercial buildings (Deloite University Press, 2016).

Competency Scope Driven Disruptive Innovations

The competency scope driven innovation makes references to the transformation of the core competencies that are essential to design, production or distribution in the internal activity relating to the outsourcing. In this third categorization, the author notes that the disruption of these innovations is located within the existing set of competencies of the established companies. They have no direct applications on the feature of the products or services but require a transformation in background at the level of basic competencies of the firm to keep its position in the market. The author of this chapter finds several cases such as artificial intelligence, artificial intelligence, genomics and nanotechnologies.

These competency scopes of disruptive innovations are substituting the value of the intangible and tangible assets of competencies and accelerating the obsolescence of the competencies of the incumbents. However these competency scope market disruptions is not having per se an immediate and direct impact on the features of products and offered services or on the market of applications. It is moreover the reason why the majority of the discussions in the economic perspective concern the development of products and services connected with these category have a concern for potential application of products and markets, for instance, in the case of Artificial intelligence and Artificial Intelligence, innovations or still in the identification of market application allowing a segmentation according to usages. In sum, this category represents the case of competency driven disruptive innovation and a substitution of actual competencies of established firms by new set of skills. In this category, the author examines in more details, the case of Artificial intelligence and advanced simulation, Nanotechnology and Next generation genomics.

Artificial Intelligence and Advanced Manufacturing

The case of artificial intelligence and advanced simulation is a disruptive innovation that is introducing new competencies in algorithmic and mathematics to develop new disruptive application to serve new markets. According to Hagen et al., (2016) artificial intelligence and simulation models have emerged as new frontier in providing insightful information to facilitate decision making. The development of deep-learning technology will drive the growth prospects for the global AI market for the next four years. The increasing use of AI to combat cyber security threats will drive the growth prospects for the global AI market in the coming years. Some of the issues that can be addressed by such systems include the failure of firewall and signature-based techniques to stop current threats, a large volume of security threat data, and scalability challenges in addressing security threat data with people. For instance, algorithmic trading now analyses massive amounts of market data on a real-time basis, identifying opportunities to capture value almost immediately. In the retail sector, simulation and artificial intelligence executes analysis of in-store purchasing behaviors on real time basis. With such quick insight into demand movements, stores can adjust merchandise, stock levels, and prices to maximize sales. Across industries,

Figure 2. Evolution of data science and Artificial Intelligence

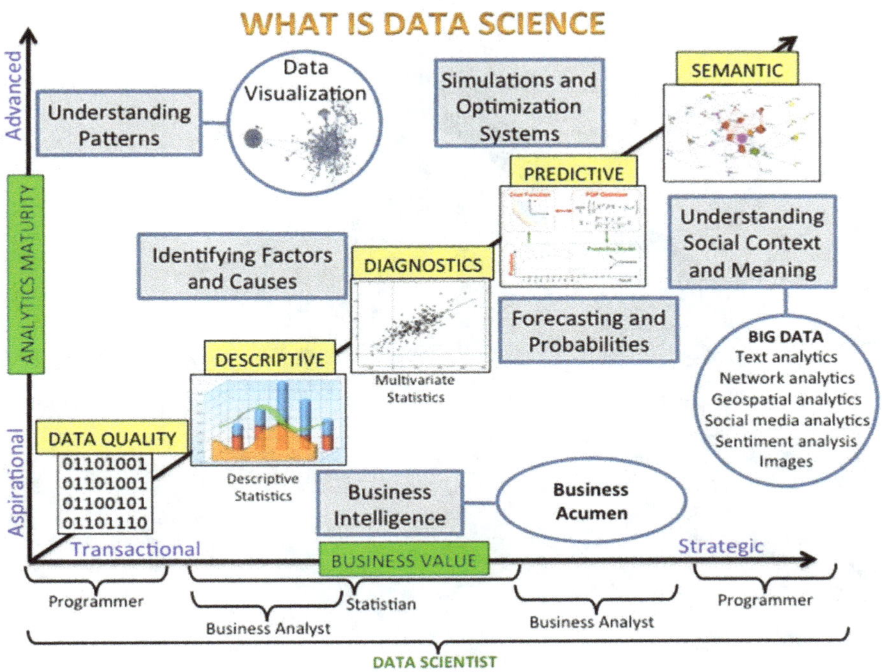

artificial intelligence can produce large data sets coupled with massive processing capabilities to spur growth and reveal cost-reducing opportunities.

1. *Impact of Artificial Intelligence on the Scope of Competencies*

Artificial intelligences along with big data are transforming the aerospace industry by increasing automation of piloting as well as safety by predicting when airplanes need maintenance. The top strategic focus for advanced technology manufacturers is around creating an integrated and connected closed loop design and build process, with 'Smart Factories (IoT)' as their top focus. Figure 3 below illustrates intelligence building framework

2. *Secondary Data of Artificial Intelligence on the Product Scope*

In terms of product applications, Intelligent Systems in Aerospace Engineering Intelligent System (IS) is the new development. This gathers sets of applications that have gained popularity among aerospace professionals in the last decade due to the ease with which several to be implemented. The role of intelligent systems in

Figure 3. Illustrates intelligence building framework
Source: Intelligent Buildings Institute (IBI), blog.econocom.com, Wikipedia

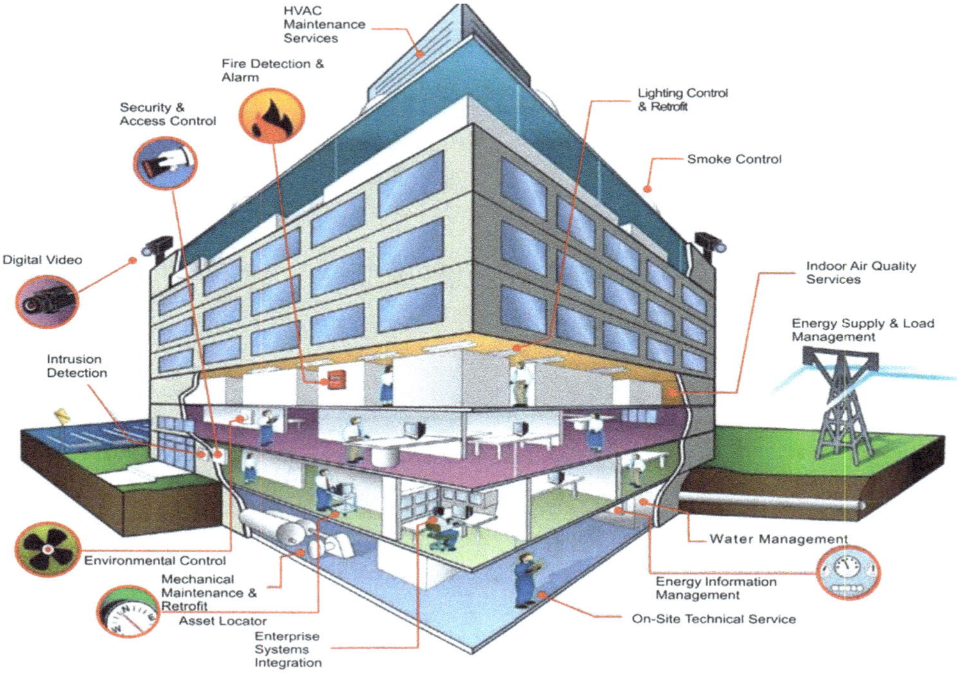

aerospace engineering is two-fold: (1) function as intelligent assistants to augment human expertise; and (2) act as a substitute for human expertise in endeavors that save cost, time, and life. Artificial Intelligence is leading to a range of new product functionalities in aeronautic such as Intelligent Systems for Modeling, Intelligent Systems for Search Application, Search for Optimal Air Combat Tactics, Intelligent Systems for Design of Aerospace systems, Engine Estimator Intelligent Systems for Control, Intelligent Systems for Training Integrated artificial immune system.

DISCUSSION

This chapter's discussion and thoughts provide insights in terms of impact on the strategic scope of the incumbents. It sought to categorize the impact from the perspective of the incumbents using the strategic scope framework of Abell (1980) and Allaire and Firsirotu 1993) and its three components: product driven, market driven and competency driven. In a general manner, this chapter has indicated that product driven disruptive innovation plays a key role on the strategic scope of the incumbents by substituting functionalities indicators of product driven. It means replacing or adding new functionalities to existing product. The author also noted from the case of market driven disruptive innovation that disruptive innovation is having an impact by substituting alternate market channels for the incumbents or ways to distribute the product along the product chain. Examples to be found include: online payments, Internet of things and sharing economy disruptive innovation such as Uber or ArnB

The last case in this chapter represented competency driven disruptive innovations which the impact is at the competency level of the incumbents by introducing a substitution of actual competencies of established firms and corporation by new set of skills. For instance, several disruptive innovations such as the next generation genomics, artificial intelligence and advanced simulation or nanotechnology are essentially disruptive because they are driven by new set of competencies which have been introduced to transform this rigid scope of business.

The author presented the impact of disruptive innovations under each of the strategic scope drivers identified above. The author offers to explain key thoughts on each strategic component and gave three brief descriptive cases for each component. It was established that Abell's model is a conceptual tool that might facilitate the categorization of disruptive innovation and lead us to better study this complex economic phenomena in more and granular perspective using a market/ industry perspective.

POTENTIAL CONFIGURATIONS AND IMPLICATIONS

The theory of disruptive innovation has some implications that offer some managerial and practical usefulness. This research has established that new business models and technology can bring massive changes to an industry where disruption is yet to occur, for instance as seen from the case of Airbnb in the hotel industry. Airbnb is a typical case of disruption. It began as a low end business targeting those customers who could not afford hotel costs or those who could not get a hotel in the most crowded market. The theory also helps emerging companies to develop competitive responses by creating those products that appeal to the neglected and the usually considered low end markets. According to Christensen, the theory of disruptive innovation does not seek to help companies achieve growth in their market or product dimensions but it rather seeks to help incumbent companies to strengthen their core business. In this regard, any company that feels its market niche is threatened by new business models should establish a different distinct company that is different from the core to engage in disruptive businesses.

The theory of disruptive innovation shows that innovation is always going to be the order of the day in every industry and as such incumbent companies should focus on the product improvement rather than focusing on the threats posed by new entrants. For any innovation to occur in the industry, Christensen provides that it must provide a unique value/attribute to customers that are different from the existing product. Therefore through the theory of disruptive innovation, incumbent companies are able to focus their resources on product improvement through value proposition and quality improvement. This research has demonstrated that in the case of wearables, the researcher attended an impact mattering from the point of view of the scope produces follow-up of an impact from the point of view of the scope market and finally on the scope of skills. Additional researches will be necessary to identify cases where an impact on the scope of skills but rather a secondary impact on the transformation of the skills rather than that of the market.

In the case of the Internet of things, this study shows that the main impact concerns the scope of markets by transforming the border of the traditional markets working in silos and by allowing an integration of these various markets. In this second case, the secondary impact was at first from the point of view of the scope produced followed by the scope of skills. A question of potential research could be there are innovations of break having an impact in a dominating way on the scope market but for which the impacts on the scope of skills would be dominating with regard to the scope of the market.

Finally in the 3rd case of the artificial intelligence, this research showed that it is the transformation of skills that is the most important impact from the point of view of the strategic scope of activities of companies. This impact was followed of one secondary incident on the produced scope than on the market scope via in particular the implementation initiatives connected with the Internet of objects. It exists among various disruptive innovations, from which the impact finds itself mainly in a new market scope rather than product scope.

RESEARCH IMPLICATIONS

As a matter of fact, this analysis highlights the complexity of the subject and the research needed to better categorize the transformation of disruptive innovations in various industries of the economy. In a general manner, this scholarly analysis has indicated that the Abell model (1980) facilitates a categorization of disruptive innovations and a better understanding of its disruptive nature of the strategic scope of the incumbents. A contribution of this analysis is to highlight the need to study complex configurations of disruptive innovations. In this chapter's analysis, the author noticed that the disruptive character of these disruptive innovations combines several aspects leading to complex configurations. It could be argued that the low generalization of the concept (Leopore 2014; Weeks, 2015) is partially understandable by the variety of the phenomenon of the innovations of disruptive innovation in a large number of sectors. The economic phenomenon of disruptive innovation represents a challenge of complexity for the categorization of disruptive innovations. In this regard, it could examine the links between innovations between product and process disruptive innovations.

FUTURE RESEARCH DIRECTIONS

This analysis brings insight and opens a perspective for additional work to better categorize disruptive innovation according the impact of their disruptive character. But significant additional work will have to be provided to structure this research perspective.

CONCLUSION

Based on the model of strategic scope in disruptive innovation, this chapter concludes the strategic scope of Abell's model is a conceptual tool that might facilitate the categorization of disruptive innovation and leads to better study this complex economic phenomenon in more and granular perspective using a market/ industry perspective.

REFERENCES

Abell, D. F. (1980). *Defining the Business: The Starting Point of Strategic Planning*. Englefield Cliffs, NJ: Prentice Hall.

Adner, R. (2002). When are technologies disruptive? A demand-based view of emergence of competition. *Strategic Management Journal*, *23*(8), 667–688. doi:10.1002mj.246

Baiyere, A. (2014). *Disrupted Disruptions: Lessons from Potential Disruptive Innovations that barely disrupted The XXV ISPIM Conference – Innovation for Sustainable Economy & Society*. Dublin, Ireland: Academic Press.

Bower, J. (2002, May). Disruptive Change. *Harvard Business Review*, *80*(05), 95–101. PMID:12195923

Center for Responsible Nanotechnology (CRN). (2016). *The meaning of Nanotechnology*. Retrieved from: http://www.crnano.org/whatis.htm

Centers for Disease Control and Prevention. (2006). *Human Genetic Epidemiology Network*. Available at: www.cdc.gov/genomics/hugenet/default.htm

Chen, M. J., Smith, K. G., & Grimm, C. M. (1992). Action Characteristics as Predictors of Competitive responses. *Management Science*, *30*(30), 439–455. doi:10.1287/mnsc.38.3.439

Christensen, C. M. (2003). *The Innovator's Dilemma: When New Technologies Cause Great Firms to Fail*. New York: HarperCollins.

Christensen, C. M., & Raynor, M. E. (2003). *The Innovator's solution: Creating and sustaining successful growth*. Harvard Business School Press.

Cohen, H., & Lefebvre, C. (Eds.). (2005). *Handbook of Categorization in Cognitive Science*. Elsevier.

Cooper, A., & Schendel, D. (1976). Strategic Responses to Technological Threats, Business Horizon. *Corporate Change*, *5*(4), 1143–1180.

Crowd funding's Potential for the Developing World. (2013). *Info Dev, Finance and Private Sector Development Department*. Washington, DC: World Bank.

Danneels, E. (2004). Disruptive Technology Reconsidered: A Critique and Research Agenda. *Journal of Product Innovation Management*, *21*(4), 246–258. doi:10.1111/j.0737-6782.2004.00076.x

Danneels, E. (2004). Disruptive technology reconsidered: A critique and research agenda. *Journal of Product Innovation Management*, *21*(4), 246–258. doi:10.1111/j.0737-6782.2004.00076.x

EAI Catalyst Cleantech Sustainability. (2015). Retrieved from http://www.eai.in/blog/2015/07/big-list-of-disruptive-innovations-excellent-list.html

Frost & Sullivan. (2016). The way of the future: Wearables poised to disrupt existing technologies and processes across industries. *PR Newswire*. Retrieved from https://search.proquest.com/docview/1758135273?accountid=14719

Hagen, C., Evans, H., Ciobo, M., Miller, J., Wall, D., & Yadav, A. (2016). *Big DataArtificial intelligence and the Creative Destruction of Today's Business Models*. Adopted from ATkearney: https://www.atkearney.com/strategic-it/ideas-insights/article/-/asset_publisher/LCcgOeS4t85g/content/big-data-and-the-creative-destruction-of-today-s-business-models/10192

Henderson, R. M., & Clark, K. B. (1990). Architectural innovation: The reconfiguration of existing product technologies and the failures of established firms. *Administrative Science Quarterly*, *35*(1), 9–30. doi:10.2307/2393549

Innovation: Evidence from Xerox Corporation's Technology Spin-off Companies. (2012). *Industrial and Internet of Things*.

Keller, A., & Stefan, H. (2009). Ex ante identification of disruptive innovations in the software industry applied to web applications: The case of Microsoft's vs. Google's office applications. *Technological Forecasting and Social Change*, *76*(8), 1044–1054. doi:10.1016/j.techfore.2009.03.005

LeoporeJ. (2014). Retrieved from http://www.newyorker.com/magazine/2014/06/23/the-disruption-machine

Lopez, T. S., Ranasinghe, D. C., Harrison, M., & McFarlane, D. (2012). Adding sense to the Internet of Things An architecture framework for smart object systems. *Personal and Ubiquitous Computing*, *16*(3), 291–308. doi:10.100700779-011-0399-8

Markides, C. (2006). Disruptive innovation: In need for better theory. *Journal of Product Innovation Management*, *23*(1), 19–25. doi:10.1111/j.1540-5885.2005.00177.x

Martin, M. (2015). *Strategic Framework: Abell's Framework for Strategic Planning*. Retrieved from https://www.cleverism.com/abells-framework-for-strategic-planning/

MIT. (2016). *10 Disruptive innovations through Technologies 2016*. Adopted from: https://www.technologyreview.com/lists/technologies/2016/

Oskarsson, C., & Sjöberg, N. (1994). Technology strategy and competitive analysis: The case of mobile phones. *Technology Analysis and Strategic Management*, *6*(1), 3–20. doi:10.1080/09537329408524149

Pfeffer, J., & Salancik, G. R. (1978). *The External Control of Organizations: A Resource Dependence Perspective*. New York: Harper & Row.

PwC. (2014). The new hire: How a new generation of robots is transforming manufacturing. *PwC*. Retrieved from http://www.pwc.com/us/en/industrial-products/publications/next-manufacturing-robotics.jhtml

Rafi, F., & Kampas, P. (2002). How to identify your enemies before they destroy you? *Harvard Business Review*, *80*(11), 115–123. PMID:12422795

Roy, R., & Cohen, S. K. (2015). Disruption in the US machine tool industry: The role of in house users and pre-disruption component experience in firm response. *Research Policy*, *44*(8), 1555–1565. doi:10.1016/j.respol.2015.01.004

Roy, R. (2014). Exploring the Boundary Conditions of Disruption: Large Firms and New Product Introduction with a Potentially Disruptive Technology in the Industrial Robotics Industry. *Engineering Management*, *61*(1), 90–100.

Saviotti, P. P., & Metcalfe, J. S. (1984). A theoretical approach to the construction of technological output indicators. *Research Policy*, *13*(3), 141–151. doi:10.1016/0048-7333(84)90022-2

Schumpeter, J. A. (2003). *Capitalism, Socialism, and Democracy*. London: Routledge.

Smith, K. G., & Gannon, C. M. (1991). Organizational information processing competitive response and performance in US Domestic Airline industry. *Academy of Management Journal*, *34*(Winter), 60–85. doi:10.2307/256302

Solis, B. (2015). *25 Disruptive Technology Trends For 2015–2016*. Retrieved from http://www.briansolis.com/2015/01/25-disruptive-technology-trends-2015-2016

Tang, L., & Shapira, P. H. (2011). Regional development and interregional collaboration in the growth of nanotechnology research in China. *Scientometrics*, *86*(2), 299–315. doi:10.100711192-010-0274-9

Thomsen, M. (2015, February 19). Microsoft's Deep Learning Project Outperforms Humans In Image Recognition. *Forbes*.

Tripsas, M. (2000). Commercializing emerging technology through complementary assets. In G. S. Day, P. J. H. Schoemakers, & R. E. Gunther (Eds.), *Wharton on managing emerging technologies* (pp. 172–186). John Wiley & Sons.

Tushman, M., & Anderson, P. (1986). Technological discontinuities and organizational environments. *Administrative Science Quarterly*, *31*(3), 439–465. doi:10.2307/2392832

Wearable technology ecosystem industry 2016 to 2030 - analysis of 6 region and 465 companies profiles. (2016, May 03). *PR Newswire*. Retrieved from https://search.proquest.com/docview/1786275049?accountid=14719

Weeks, M. (2015). Is disruption theory wearing new clothes or just naked? Analyzing recent critiques of disruptive innovation theory. *Innovation: Management. Policy & Practice*, *17*(4), 417–428. doi:10.1080/14479338.2015.1061896

KEY TERMS AND DEFINITIONS

Artificial Intelligence (AI): Artificial intelligence is a branch of computer science that seeks to mimic human behavior and intelligence to perform high involving tasks and responsibilities. These innovations simulate intelligent computer behaviors which are closer if not equal to human behavior.

Disruptive Innovation: Disruptive innovation is a term that was developed by Christensen to explain the process through which a product/service goes through right from development to commercialization in the market place, to a period of displacing competitors.

Internet of Things: This is an ever-growing technology involving physical computer structures, machines and digital objects that are assigned unique identifications to help transmit data over an identified network or ecosystem with human interaction.

Strategic Scope: This indicates the number of products and services that a company plans to offer over a given time frame. This term also specifies the type and nature of market to be served by the company.

Chapter 3
Peer-to-Peer Lending as an Alternative Investment Asset Class in the EU

Manuel Gutierrez
Loyola Marymount University, USA

ABSTRACT

The aftermath of the 2007-2009 financial crisis and European debt crisis proved to be a transformative event to financial markets, primarily in the areas of fixed-income asset yields and asset price appreciation due to wide-spread credit contraction at the retail level. After worldwide markets collapsed for equities and many corporate debt securities, frightened investors liquidated holdings at absurdly low prices. However, the worldwide recovery, while uneven, remains underway. Yet despite the recoveries in the broader indices, individual investors at the retail level have not widely shared in market gains after exiting from their investments during the crises. As a result of uncomfortably high asset prices and the lack of income potential to investment-grade fixed income investments, new alternatives are sought which may present better yield-earning potential amidst the current credit market environment made available by certain Fintech companies.

DOI: 10.4018/978-1-5225-4148-6.ch003

PEER-TO-PEER LENDING (P2P): BACKGROUND

Fintech's rise was catalyzed by the aftermath of the 2007-09 Global Financial Crisis and the Euro Debt Crisis' damage to the brand and business of many large legacy financial institutions. After the collapse of large institutions such as Lehman Brothers, Lloyd's and Northern Rock in the UK, Bankia Bank in Spain, Anglo Irish Bank, Allied Irish, Bank of Ireland, and others throughout the continent, people doubted the soundness of their trusted institutions. From that doubt and despair, the search for alternative houses of savings began in earnest. Episodes of suspected malfeasance at institutions like Deutsche Bank, Goldman Sachs, and other institutions came to light:

- Banks played the role of supporting both parties in transactions they helped foster and even participated in
- The role of some banks in originating and promoting flawed asset-backed securities and derivatives
- Trading scandals such as the Jérôme Kerviel trading scandal at Société Générale, and
- The Libor rate manipulation scandal involving banks like UBS, Barclays, and Royal Bank of Scotland – the consequences of which involved the daily credit transactions and major purchases of almost everyone on the continent.

Amidst the damage to the financial industry and its reputation came the collapse of fixed-income asset yields. Investors of all stripes bolted "risky" assets like stocks and non-investment grade bonds, pouring their money in safer assets like sovereign debt of safe havens like US Treasuries, German, Swiss bonds, and the Yen. As the EU economy collapsed from the aftermath of the 2007-09 Global Financial Crisis (and again in 2012 amidst the Euro Debt Crisis), the flight to safety precipitated a drop in sovereign debt yields (see Figure 1) compelled individual retail investors to liquidate and avoid any further pain from persistent losses. Outflows of European mutual funds and ETFs, for instance, peaked in the aftermath of both crises (see Figure 2). Much of the post-crisis gains in mutual fund and ETF investment levels, however, did not come from increased investment flow but through asset appreciation. Indeed, while large institutional investors including hedge funds and private equity funds benefitted from the recovery in asset prices, many household investors missed out on the market rally. Only now with the passing of time and the waning of memories, the hype over gains in more speculative assets in the media, and the dearth of investment opportunities, are household investors slowly (and reluctantly) seeking yields through increased investments.

As the supply of available sovereign debt to the public contracted (in part due to increased buying of bonds via quantitative easing programs by major central

Figure 1. Sovereign long-term bond interest rates for EU-28 nations

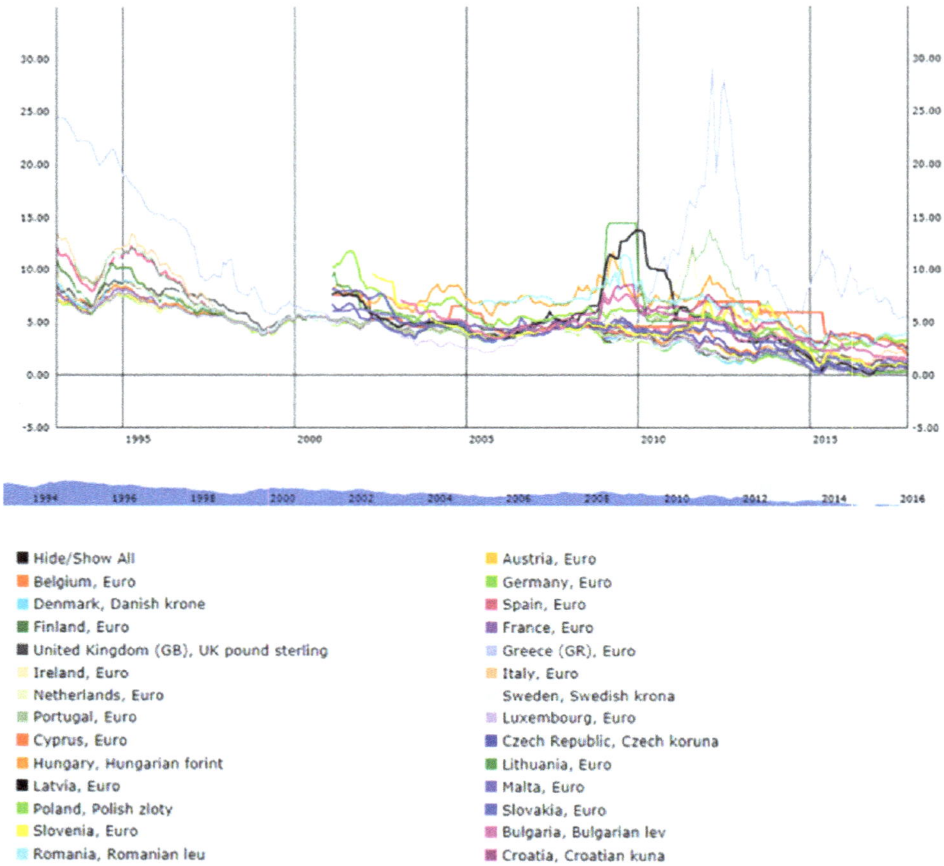

banks), the yields on similarly-rated corporate bonds collapsed as demand bid prices higher. Those investors who participated in the flight from stocks and mutual funds were further hit by low savings interest rates and collapsing Euro exchange rates versus such foreign currencies as the US Dollar and the Japanese Yen (due to a combination of the European recession and other central bank actions). The sheer size of investable funds worldwide, especially in Europe, sought investment in an environment where prudent investment opportunities were scarce. In fact, even speculative investments were bid higher, as illustrated by the collapse of non-investment grade (junk) corporate bond yields. Yields on interest-bearing savings and money market accounts (accounts investing in corporate commercial paper) were anemic – typically on the order of 5 basis points (0.05%) as of October 2017 (see Figure 3).

Figure 2. Mutual fund and ETF holdings by EU retail investors

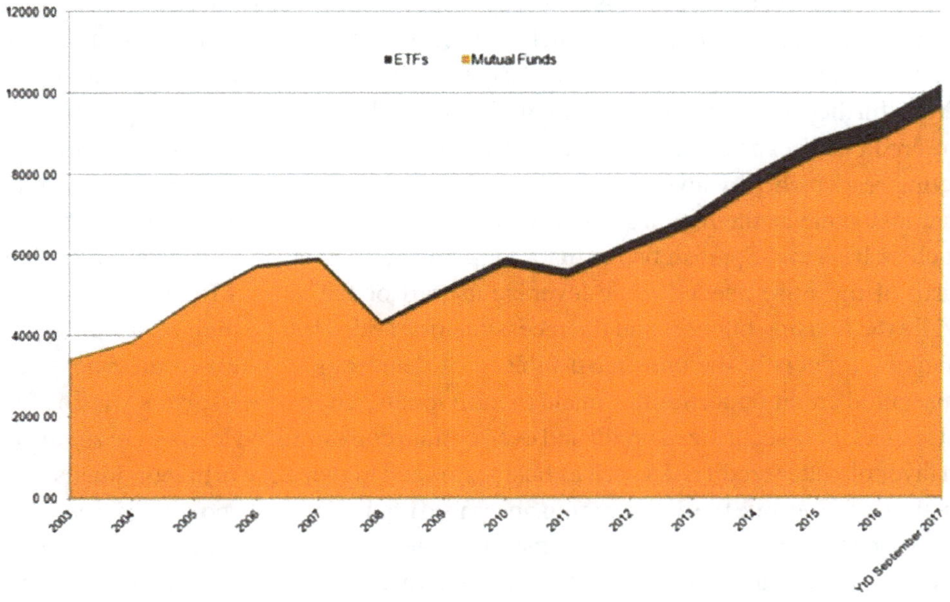

Figure 3. Euro-area household interest earned on deposits

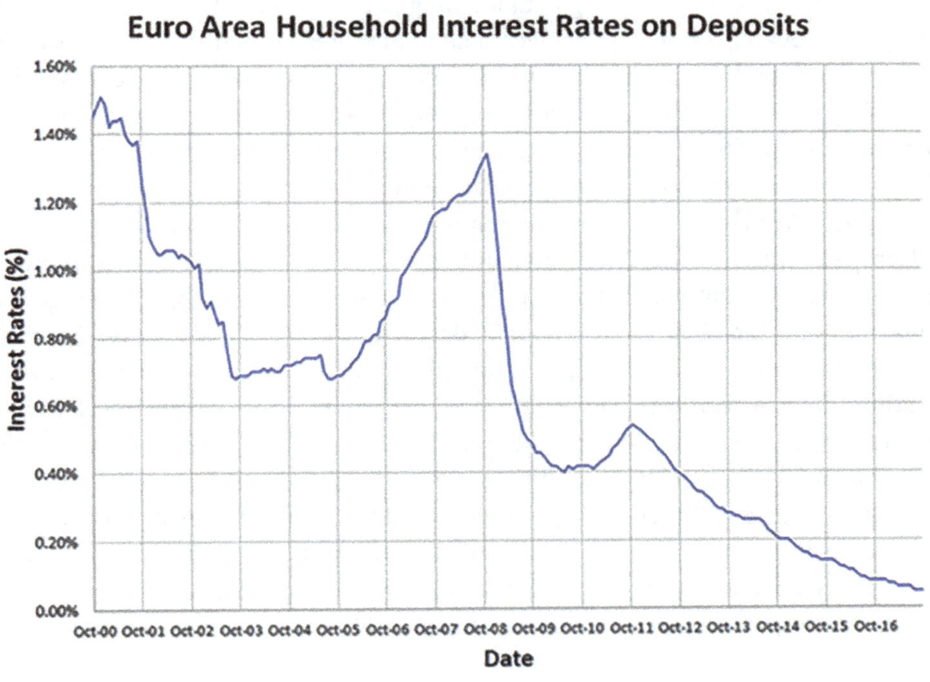

For individuals seeking credit, the emergence of P2P lending Fintech firms gained steam after both crises. Credit is fundamental for a robust, growing, and prosperous economy and credit contractions such as after the 2007-09 Global Financial Crisis made the need for access to cash an acute problem. The ensuing European Debt Crisis further brought along major banking-related events that adversely impacted the credit landscape on the continent. First, national banks in nations such as Greece, acting responsibly by investing assets in sovereign debt, loaded up on Greek bonds that lost value as the nation's true finances were brought to light. Secondly, banks such as Irish and Spanish banks that saw increased leverage immediately after the birth of the Euro needed to de-leverage as part of the terms of the ECB bailout while the housing bubbles and the recession took added toll on their balance sheets. Banks – especially the ones most in need of deleveraging – turned their noses at unsecured credit opportunities made to individuals. To improve solvency ratios for EU banks, non-securitized credit and even certain collateralized loans were shunned in favor of better credit risks such as businesses and higher net worth individuals and institutions. Despite the deleveraging, increased regulation, and economic forces, the need for credit persisted. In earlier times, the role would perhaps fall to other parties such as payday lenders and pawn shops that charged expensive prices and demanded collateral for the risk associating with lending. As technology evolved, Fintech firms began to picture a role as matchmakers that could introduce individuals and small businesses that need capital with those non-institutional investors seeking places to invest capital. The advent of big data learning and artificial intelligence promoted a potential supplement or even replacement to the standard underwriting processes and conventions that had marginalized many in the capital-seeking community. An ability to properly assess a borrower's credit risk using multiple sources of data became key to offering investors returns above other assets and borrowers cheaper alternatives for capital than payday lenders and pawn shops.

The dearth of viable investment alternatives raises the need by household investors in Europe for alternatives. The combination of a search for yield, a contracting credit environment by traditional lenders, and the availability of technology, software and algorithm development, as well as new data sources to qualify and synthesize a picture of a person's creditworthiness brought about the advent of the Peer-to-Peer Lending Fintech platform. This study shall first look at the overall P2P Lending business model and participants in the lending space. Next, we shall look at the stakeholders in the industry including participants in the P2P Lending space. The following section will include opportunities and barriers to the success of the P2P Lending industry including the impact to legacy institutions. Finally, we shall look at the assessment of the future of P2P Lending as an investment asset class.

P2P LENDING BUSINESS MODEL

Peer-to-peer (P2P) lending – sometimes referred to as crowdlending – is an alternative to a bank loan where, instead of borrowing from a single source, companies, organizations, and individuals can borrow directly from one or several individuals prepared to lend. Investors (lenders) often bid for loans by receiving guidance for a rate and opting to offer an interest rate at which they would lend, creating a reverse auctioning process. Borrowers have the option to accept one of the loan offers, typically the one with the lowest interest rate. P2P Lending Fintech firms serve the role of "match-making" borrowers who need capital with investors to lend capital. Due diligence is carried out for each loan request, as crowdfunding platforms have a duty and incentive to protect both businesses and investor interests. P2P lending platforms normally require financial accounts and a trading track record.

P2P lending is cited here as an asset class for investment because, unlike the low interest rates on savings deposits, P2P lending provides investors the means to generate returns in excess of those earned on deposit with legacy banks. These funds are lent with a risk profile that the investor outsources to the Fintech Company to determine. The risk management/risk assessment characteristic P2P lending platform performs is not unlike the job performed by ratings agencies for bonds or brokerage analysts perform toward securities. The Fintech Company – much like banks and even ratings agencies for credit – quantifies and prices the risk based upon the borrower's risk profile. P2P investors lend various-sized amounts directly to other individuals at terms/rates of interest allowed by regulators and set by the Fintech firm. The P2P space consists of two lending mechanisms; first is the direct P2P model with one lender and one borrower. Figure 4 (left) illustrates the relationship between the single P2P Portal and borrowers and consumers while Figure 4 (Right) illustrates the relationship between investors and borrowers in the second mechanism: crowd-funding lending.

Figure 4. (Left) P2P Peer to Peer Individual Lending Model; (Right) P2P Crowd-Lending Model

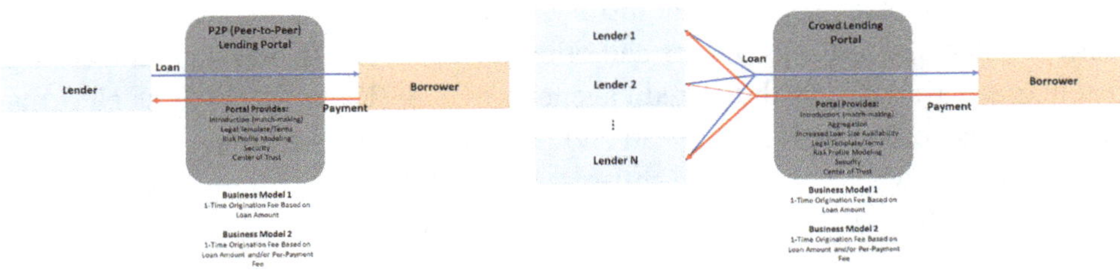

The single-lender model presents the investor the opportunity to capture most of the margins that traditionally went to a legacy bank for a loan (minus fees paid to the Fintech portal). The crowd-sourced loan model provides the opportunity to diversify the lenders' exposure to a borrower's credit risk in exchange for a smaller share of the margins. The crowd-sourced structure provides a lender diversified exposure for the same total amount of capital in the event there is a default on a single loan. The micro-credit model made famous by Nobel Prize-winning Bangladeshi entrepreneur Muhammad Yunus is a version of the P2P direct lending model.

The P2P lending space is not limited only to household borrowers and Fintech companies help extend credit to small businesses. Further, not all loans are uncollateralized: pawn shop P2P lenders also exist. Pawn shop and P2P business lending are considered safer in part because of the amount of borrower credit history and cash flow history available for businesses in addition to the option of collateralization (to back the loan) by using equipment, easy-to-liquidate inventory, or receivables. Of course, pawn shop P2P lenders' utilization of hard assets adds a level of security to loans made through this business model.

The P2P lending Fintech portal seeks a light capital footprint while using economies of scale to decrease transaction fees. The Fintech lending portals' value proposition is twofold: rock-bottom costs to borrowers, and an opportunity to provide several back-office functions for would-be investors at rates they couldn't achieve themselves. Either P2P lending portal provides a Fintech firm multiple options to monetize its involvement in the transactions. Origination fees may be one-time events charged as a percentage of the loan amount. Alternately, the firm may charge a fee for each payment by the borrower that amounts to be larger in the aggregate. However, such an arrangement exposes the Fintech firm to default risk and unpaid transaction fees. P2P lending portals offer several value-added services besides match-making: the firm acts as an intermediary and services payments as a layer of security (or the lender may outsource payments to a third party such as PayPal). Fintech companies also perform risk management by assessing credit risk to borrowers using proprietary algorithms (sometimes in lieu of other credit-determining criteria). Unlike the US with legacy banks' reliance on FICO (Faire Isaac Credit Company) scores to determine credit-worthiness, European legacy banks utilize criteria such as employment history, salary, family situation, existing debts, residence status, court judgements and other factors. The UK has a credit scoring system made available by three credit agencies as well (Kane, 2011). Fintech firms incur various costs: outside overhead and personnel costs (i.e. sales, management and leadership, cybersecurity, and software/IT, etc.). Fintech firms also take on the cost to operate the portal, legal and regulatory compliance, and payment processing.

Some would-be P2P loans would have been made at banks, charged on credit cards or as cash advances, or even made at pawn shops and payday lenders. Some of those loans would've been made outside the formal banking system due to:

- A lack of credit-worthiness on behalf of the borrower in the eyes of the legacy bank/institution
- Uneconomical spreads for the legacy bank, loan amounts that are too small for a bank to issue, or lack of adequate compensation for the risk
- A lack of required background information and history on behalf of the borrower using legacy banks' processes for credit issuance.

In a European financial system that underwent a large concerted initiative to improve capitalization ratios (see Figure 5), lending was suppressed and the unabated need for credit led to the growth of the EU and the UK P2P lending space that became a $5.9 billion market (Leech, 2015).

Figure 5. EU banking capital-to-asset ratio, 2008-2016

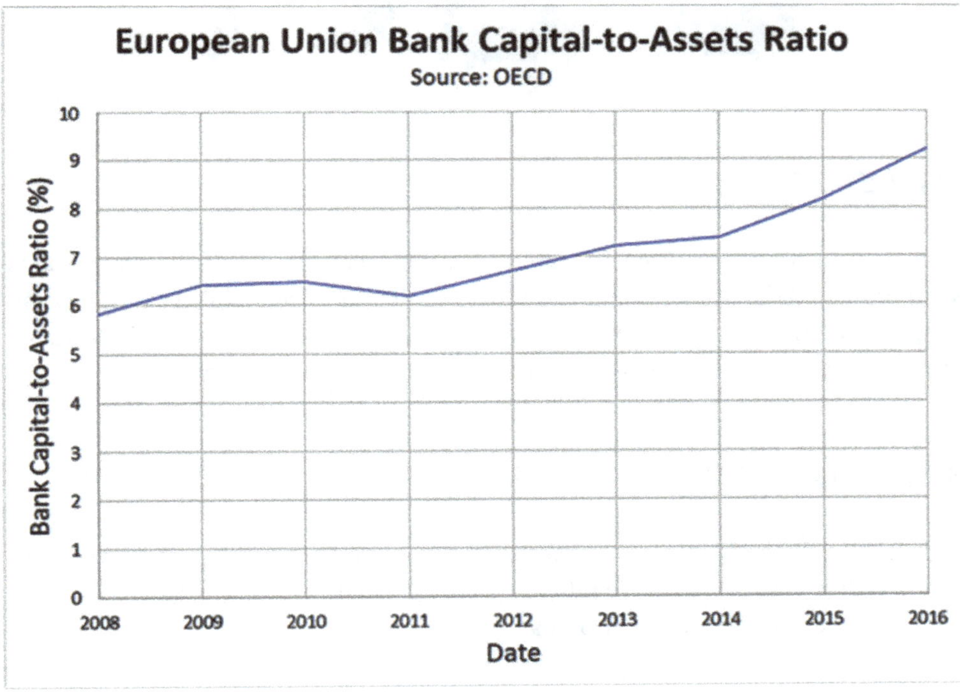

Figure 6. Concentration of P2P lending volume, Source: AltFi

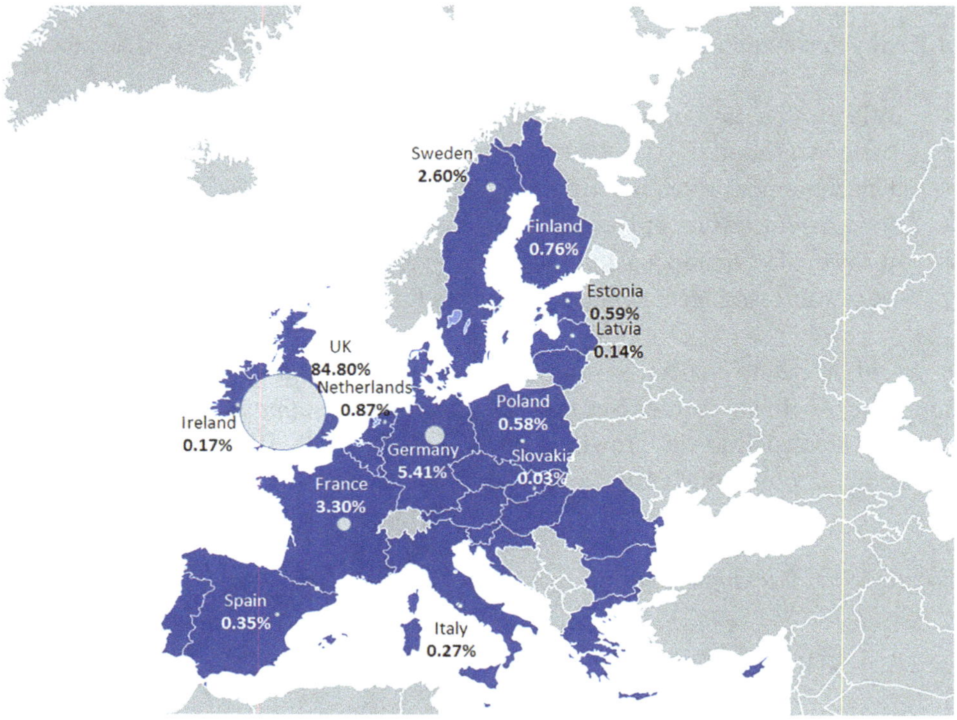

P2P LENDERS

P2P lending originated in the UK in 2005. While the majority of P2P lending remains focused in the UK (accounting for over 84% of the European market), P2P lending is becoming increasingly popular in the rest of the EU (see Figure 6). Table 1 lists some important P2P lenders as well as notable features about each firm:

STAKEHOLDERS

Stakeholders in the P2P lending space are a broad group including the P2P lending firms, borrowers, investors, regulators, and financial institutions such as banks, payday lenders, and even pawn shops:

Table 1. List of notable P2P lenders in the EU and key features

Company	Country of Headquarters	Features
Zopa	UK	An industry pioneer, considered the very first P2P lending platform launched in the world (2005). Focuses on consumer P2P lending, operating in the UK and the US through a partnership with six Credit Unions. Zopa has helped people lend more than £1.54 billion in P2P loans and raised over $56 million in investor funding. Voted 'Most Trusted Personal Loan Provider' in the Moneywise Customer Awards for six years running.
Funding Circle	UK	An online marketplace for business loans, matching businesses with investors. Founded in 2010, considered the first to launch an online platform for P2P business funding in the UK. Helped fund over £1.3 billion worth of loans to SMEs with over 50,000 UK investors and raised over US$270 million in investment funding.
Ratesetter	UK	Founded in 2010, the second largest P2P lending platform in the UK known for having introduced the concept of a "provision fund" into P2P lending. Originated over £595 billion worth of customer loans last year. Has registered over 40,000 lenders and 240,000 borrowers while raising over US$46 million in investment funding.
Auxmoney	Germany	One of Europe's largest P2P lenders, its P2P loan marketplace focuses on private loans and consumer lending, allows private customers to take out personal loans of between €1,000 and €25,000 from investors. Originated over €465 million in consumer loans with €99 million in 2015.
CreditGate24	Switzerland	Launched in 2015, connects borrowers with private and institutional lenders and regulated by the VQF (Verein zur Qualitätssicherung von Finanzdienstleistungen) as a member of the ZEK (Verein zur Führung einer Zentralstelle für Kreditinformationen) and the IKO (Verein zur Führung einer Informationsstelle für Konsumkredit). Offers credit checks using classic credit assessment methods, Big Data tech, the insurance and the solidarity arrangement. Did not have a payment loss in the first year.
Younited Credit	France	Formerly known as Prêt d'Union and founded in 2009, operates a P2P lending platform accredited by the French central bank. The platform allows private individuals and institutional investors to lend money to borrowers directly through a secured bond marketplace. Younited Credit has managed over €347 million in loans and raised US$52.35 million in funding
Bondora	Estonia	Launched in 2009, Bondora invests in European personal loans claims and delivered double-digit returns to over 14,000 investors on the platform, lending over €60 million has been lent through Bondora with over €12 million being paid out to investors as interests, and serves lending clients in Estonia, Finland and Spain.
Geldvoorelkaar	Netherlands	Launched in 2011 as the first P2P lending platform for SMEs, the Dutch Financial Authority (AFM) granted it a credit license and raised over €85 million worth of loans.
Fixura	Finland	Operates a consumer P2P loans platform connecting borrowers with lenders. Has over 70,000 registered users, having lent over €55 million, it is said to be the largest P2P lender in the Nordic countries and the oldest one in Finland.

P2P Investors

The most critical stakeholders in the P2P Lending Space is perhaps the investors (lenders). Investors are primarily small, non-institutional investors offering capital. In considering P2P lending as an alternative investment asset, the P2P lending sector shares many features with other familiar services to put the investor at ease. Consumers/investors have familiarity with intermediaries like online brokerages that don't produce physical certificates for bonds, equities, options, alternative investments like commodities, futures. Customer familiarity with banks offering online banking and money market accounts often don't recognize their money saved in high-yielding money market accounts for what they are: capital from savers/ investors transferred to corporate treasuries through short-term I.O.U.'s known as commercial paper. The biggest leap of faith for investors in P2P lending regards the relationship to the recipients of their capital. Transacting with an intermediary such as the P2P lending portal inspires higher confidence in the prospects the loan will be repaid and the investment is sound. Yet while investors may have varying insight into the actual recipients themselves, the P2P lending portal makes the process as hands off as a money market account (by serving as the credit assessor, receiving/storing investor capital, and processing payments from borrowers). With improving EU macroeconomic metrics and improving non-performing loan statistics (see Figure 7) where loan default rates are lower than historical rates, the current economic expansion provides an opportunity to show the viability of P2P lending as a prudent investment. Figure 8 shows the EU NPL performance of different loan types, showing improvement in loan performance year-over-year from 2014 to 2015. Improved asset-wide NPL performance ups the default rate P2P lending platforms must exceed but indicates that lower default rates that may positively impact P2P lending performance.

Investors must overcome fears of outsourcing the function of judging borrower creditworthiness to the platform – a dependence on the P2P lending platform's competency to generate the adequate algorithms and models to supplement or (in some cases) stand in for a lack of credit history. Once a credit risk is priced, there may be competition among fellow investors that creates an auction platform in addition to a pricing platform. Appraising the compensation required for extending an uncollateralized loan is a potentially difficult task for some to execute, especially cognizant that other reputable, better capitalized, and experienced institutions obviously shied away from such a transaction.

Figure 7. EU and nation NPL cycle

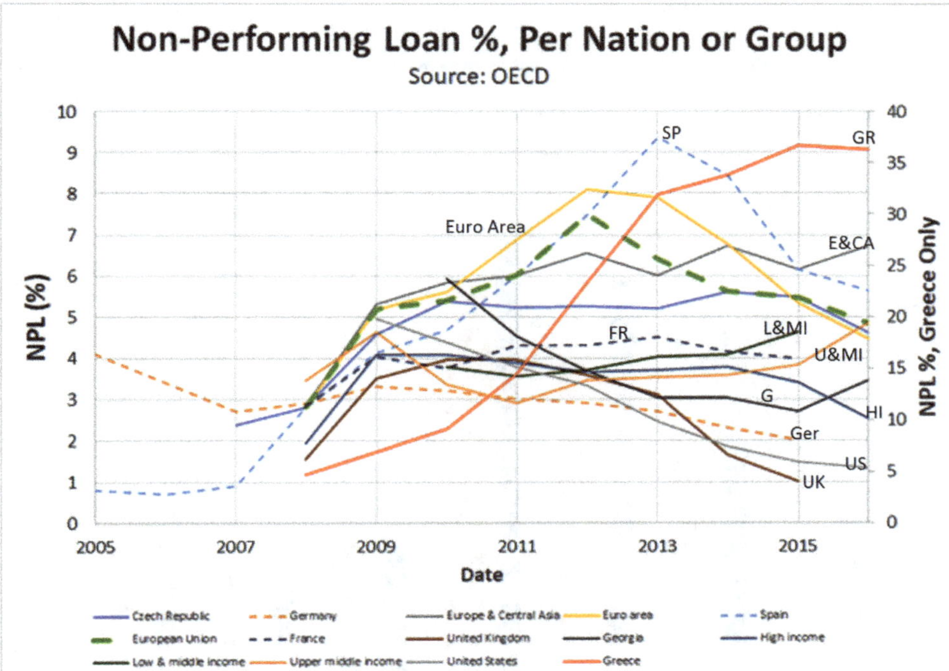

Borrowers

Borrowers are affected by credit availability: non-home purchasing credit to households as well as small business loans offered by banks on the Continent. Since 2014, such credit availability improved as European banks have grown to be better capitalized (see Figure 5) due in large part to central bank intervention. Improved credit availability combined with the advent of P2P lending offers added options and better prices for borrowers. P2P lending offers lower interest rates as investors compete with one another to lend money to borrowers (individuals and businesses) at better offered rates to secure the deal.

According to the ECB, while credit availability is stabilizing/improving (see Figure 9), there is a need for P2P lending since lending issues remain. The ECB's 2016 Q4 Euro Area Bank Lending Survey (European Central bank, 2017) identifies credit tightening issues like the cost of funds, balance sheet constraints, pressure from competition (including other higher-margin products to use balance sheet capital), perception of risk, and banks' risk tolerance. P2P lending platforms lets borrowers get a loan when that borrower is refused by a bank.

Figure 8. Non-performing loan ratios of significant banking groups in the euro area, by sector and loan type

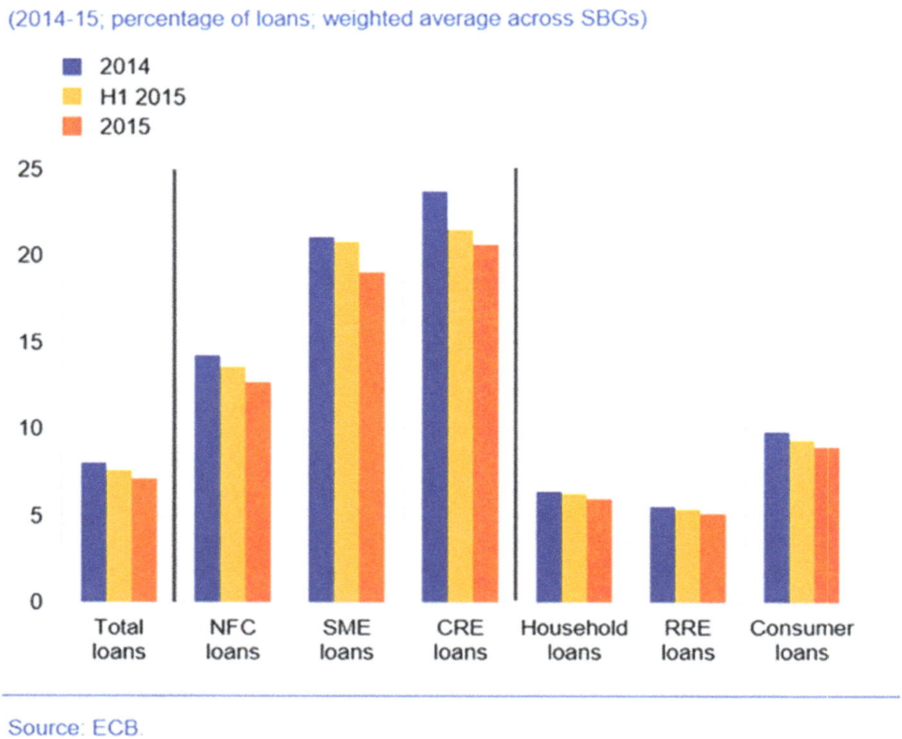

Loan sizes can vary greatly to cater most needs. The minimum loan size may be very small and encourages a wide range of lenders to participate. This feature is important: banks typically shy away from small amounts since the origination cost may outstrip profits from small loans. P2P loans are typically repaid through direct debits to the platform, which in turn distributes repayments out to the lenders. Disclosure requirements are similar to those requirements for a bank. Unlike the bank, however, such disclosures are made public to all investors.

One of the main concerns for regulators like the FCA and EC regards the track record of borrowers that fall victim to over-indebtedness when exposed to easy credit (particularly those with poorer credit histories). An EC-commissioned study of EU nations saw over-indebtedness issues at member states: Latvia, Portugal, Lithuania, Czech Republic, Spain, the UK, France, and Estonia which have seen slower growth in the years since the end of the two crises (iff/ZEW, 2010). Increasingly, regulators seek to walk a fine line: curtail the role of credit-providing entities from payday to

Figure 9. Credit availability as % of GDP in the EU-28. Source: The World Bank

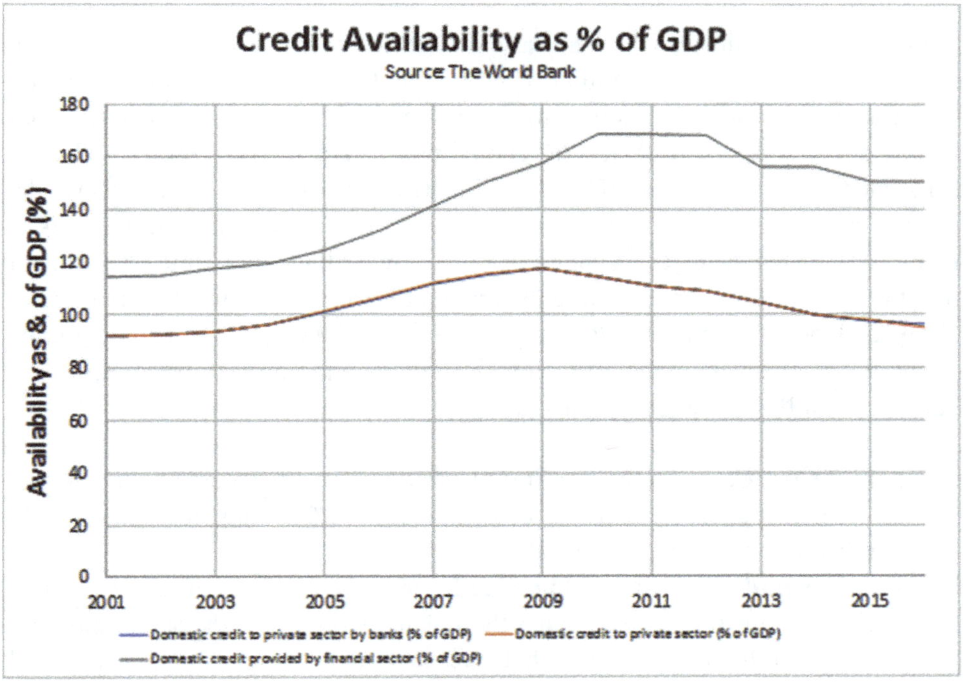

P2P lenders in terms of perpetuating the over-indebtedness cycle while not completely quashing all legitimate and helpful sources of credit.

Equity Investors

P2P lending firms are largely privately-held companies, although the larger players such as LendingTree in the US are public companies with shares traded on major exchanges. Like US and Chinese P2P lenders, European firms – even mature ones like Zopa –undergo late-round financing with eyes towards initial public offerings (IPOs) in the future. Many hedge funds, venture capital (VC) funds, and private equity (PE) firms participated in mid- to late-stage financing rounds with an eye towards 10x returns (1000% returns) on their original investment upon liquidating their stakes. With average holding periods of 5-10 years, it's reasonably assumed that many European P2P offerings will occur within the next few years as of this book's publication. Besides the desire of VC, PE, and other institutional investors to cash out from investments, the capital required to scale these private institutions (even with a capital-light footprint) will increase to the levels typically requiring access to capital markets like equity or debt exchanges.

Regulators

The P2P lending industry is perhaps as close to the payday lending and pawnshop industry as it is to the banking industry. As such, it is subject to certain regulation – not for the purposes of capitalization levels since P2P lending firms do not hold loan assets or deposits. Rather, P2P lending portals are subject to scrutiny by the European Commission in the EU and the FCA (Financial Conduct Authority) in the UK and laws of the member nations where they do business. Under the auspices of the FCA, for instance, recent regulations in 2014 regarding admissible loan amount growth cap over time were admittedly aimed at significantly curbing the number of payday lenders. The FCA's 2014 regulation capped the maximum amount to be repaid by borrowers at 2x the original loan amount and imposed a price cap which includes the following terms (Financial Conflict Authority, 2014):

1. **Initial Cost Cap of 0.8% per Day:** Lowers the cost for most borrowers. For all high-cost short-term credit loans, interest and fees must not exceed 0.8% per day of the amount borrowed.
2. **Fixed Default Fees Capped at £15:** Protects borrowers struggling to repay such that if borrowers do not repay their loans on time, default charges must not exceed £15. Interest on unpaid balances and default charges must not exceed the initial rate.
3. **Total Cost Cap of 100%:** Protects borrowers from escalating debts. Borrowers must never have to pay back more in fees and interest than the amount borrowed.

The 2014 regulatory change took effect on January 2, 2015. Consequently, no borrower would be compelled to pay back more than twice what they borrowed, and someone taking out a loan for 30 days and repaying on time will not pay more than £24 in fees and charges per £100 borrowed.

European regulators are famously cautious when it comes to consumer protections, especially in the financial industry. Caution in this industry is due to notoriety of many legacy institutions, its size after years of consolidation, and the utilization of technology that has not had a chance to be exposed and subjected to adversity (Kharpal, 2017). The ECB, for instance, put on hold the notion of using crypto-currencies until further data is available and further review is conducted (European Central Bank, 2015). As the P2P lending industry has grown over the years, the ECB, EC, and other local and EU-wide regulators have scrutinized both the means of lending activity as well as the rates charged on loans to comply with EU anti-usury laws.

Besides usury-related concerns, Know Your Customer (KYC) concerns are important to the P2P lending sector – especially regarding international investors. Multiple manual verification steps are required for investors and borrowers alike, including photo scans of an ID or passport by the borrower. In the UK, a utility bill is required to verify identity. A few marketplaces within the EU permit video identity verification. In such a case, a user registering for a service would send the required forms and schedule a Skype video call to answer questions and prove identification. Such process steps are often the requirements of individual member nation jurisdictions. In other cases, special software applications may be used to determine photo authenticity when a real-time process is unavailable. Germany allows video identification as an acceptable means of identity verification, in part to replace the traditional PostIdent process – a longtime German process. The video identification process, while integrated in the portal. However, is sometimes performed by a third-party provider. For payments, however, bank accounts are typically required. An alternate form of identification involving multi-party verification is performed in nations like the UK, Sweden, and Germany. Multi-party verification would confirm that a first, middle, last name, gender, phone, address, city, and postal code matches records from other services which typically confirms an identity for KYC purposes within minutes as opposed to 3 days for the form upload process.

Legacy Financial Institutions

Legacy financial institutions are affected by the rise in P2P Lending with the effects spread beyond traditional lending banks. Affected legacy institutions include legacy banks, credit card companies, pawn shops, and payday lenders.

Legacy Banks and Credit Card Companies

Banks involved in household lending are impacted because credit is a crucial to a bank's business. The effect on banks is two-fold: First, banks count on deposits from savers as an inexpensive means of capital for lending. Given the low savings interest rates paid out in the EU, the spread relative to the returns of sovereign debt of a hundred-plus basis points is compelling. Further, the cost of capital via low-cost deposits is vital – not just to employ in higher-yielding assets but to maintain aggressive and significantly higher new capitalization requirements brought upon by the dual financial crises. Second, while many P2P loans are too small for banks to wish to loan to, larger/intermediate sized loans and loans to small businesses is a profitable product line that directly competes with banks' main business. In the case of credit card companies, high yield P2P loans offer investors revenue that credit card companies will directly lose out on (in the case where the borrowers have enough credit history to acquire a loan). The impact will be felt in retailers offering credit cards as well, such as Tesco, Harrods's, Carrefour, and other large retailers in Europe that are currently under enough cost-based pressure to their core business from online retailers.

Borrowers in P2P lending portals often require a bank account to make payments, but the lack of business conducted at banks, or the relegation of a bank account to a back-office operation limits the quality of relationship with the consumer and may adversely impact the ability for the borrower to serve as a source more profitable revenue from cross-selling opportunities in the future.

Pawnbrokers and Payday Lenders

One of the principal competitors to firms in the P2P lending space is the pawnbroker and payday lending industry – both of which are affected by different parts of the economic and even commodities cycle and which may affect P2P lending at contrasting times. In the UK, the payday lending sector is under duress from 2014 consumer protection regulation on payday lenders that kicked in January 2, 2015. The first 5 months of the regulation saw a 35% decrease in loan volume by such lenders, and a decrease of 7% of consumers (estimated at 70,000) that were not in turn able to secure such a loan (Financial Conduct Authority, 2014). EC response to the interest charges in payday lending is more measured but ongoing nonetheless. Cognizance of and data measurement from regulations at the member-nation level is cited in "FSUG (Financial Services User Group) opinion on interest rate restrictions in the EU" documents (Financial Services User Group, 2011).

Ironically, the rules and regulations taken by the FCA to drastically reduce the number of payday lenders may help place the UK's payday lending industry in a better position to compete with P2P lending firms since the 2014 regulation's cap on earnings and rates would make the two sectors' costs similar for the borrowers. EU/EC response, however, has been more measured. Given the cultural and political differentiation among member nations with respect to the high interest rates charged by payday lenders, studies and white papers have been generated for years with no definitive measures similar to those of the Office of Fair Trade (OFT) and FCA in the UK. Yet while payday lending offers credit to those with no credit history (subprime borrowers), P2P lending firm tend to shy away from such borrowers. In a spectrum of consumers, P2P borrowers consists of a significant number of potential payday lender customers, although they do not comprise a significant percentage of their profits.

Pawn shops are an attractive option for many potential P2P lending customers. However, the value of assets held in collateral for loans makes the industry prone to the value of commodities such as gold and silver – commodities that had a higher value during inflationary times, but has been muted since the recessions caused by both crises. Considering the consuming habits of individuals participating in the

P2P lending market, they appear to be asset rich (albeit rich in consumer goods with limited value to pawn shops). High-end pawnbroking, as is the term referred to in the UK, has been on the rise of late while down-market pawnbroking is on the decline amidst higher competition at that end of the spectrum. High end participants in the market have broadened the definition of acceptable collateral to include high end fashion and even art work. Such collateralization is perhaps more attractive to older P2P borrowers with large-ticket assets to post, while perhaps less attractive to younger borrowers without meaningful high-end asset accumulation besides precious metals.

P2P LENDING OPPORTUNITIES AND BARRIERS

Market Opportunities

The P2P Lending space has many potential sources of growth given important demographic, macroeconomic, legal, and regulatory developments over the last decade-plus. Having earlier identified the major P2P Lending stakeholders, we now discuss sources of adversity and support for this Fintech sector.

EU Consumer Finance Demographics

The need for P2P lending is spurred partly by lack of access due to an inadequate credit history/track record. Inadequate history includes the lack of credit history, which is exacerbated through the increase in millennial Generation Y and Z members of the EU-28 population (while an exact definition isn't available, it is generally accepted that Gen Y are men and women born between 1980 and 1996, whereas Gen Z are born after 1996).

As figure 10 denotes, a large portion of the population are aged 34 and below in the EU-28. Gen Y men and women came of age as they obtained access to credit between the 2007-09 Global Financial Crisis and the Euro Debt Crisis – subsequently, many had their credit adversely affected and/or suffered unemployment, and only now are beginning to rebuild their credit records.

Age demographics is a critical indicator to identifying participants in P2P lending given the wide discrepancy between the unemployment rate of those 25 and under in the EU versus those between 25 and 74 (see Figure 11). Younger EU men and women had an unemployment rate 10% higher that of their older compatriots throughout the EU and comprised a demographic not sought for by traditional lenders credit providers.

Figure 10. Demographics of population in EU-28

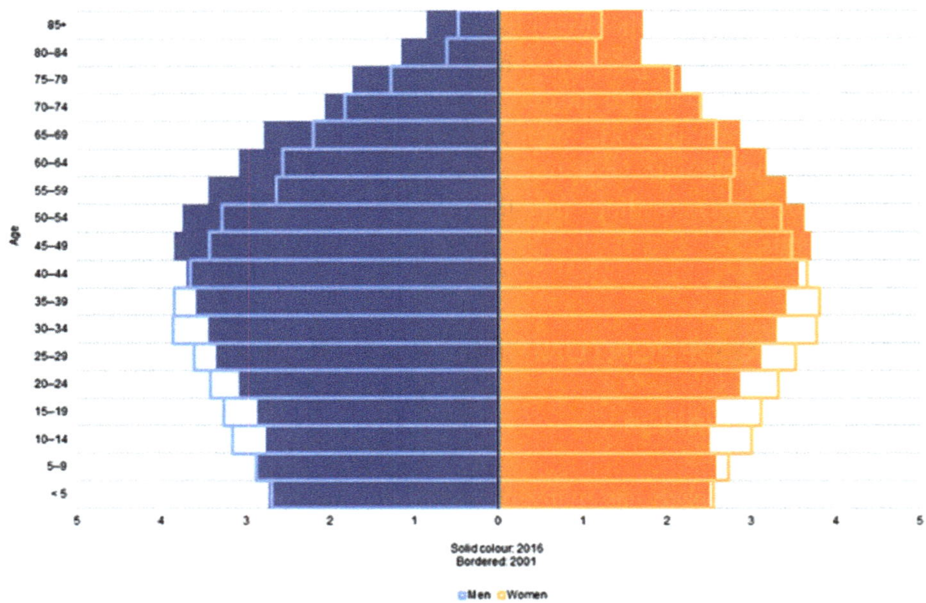

Note: Break in series. 2016: estimate, provisional.
Source: Eurostat (online data code: demo_pjangroup)

EU Macroeconomics

P2P lending's rise in popularity is due to the following conditions:

1. Dovish monetary policy that pushed down short-term rates and an asset-backed security purchase program that bid up the prices (and collapsed the yield) of long-term assets
2. A lack of prudent (non-speculative) investing options
3. Low rates of savings at the bank and sovereign debt issuance level (denoted by the extended negative real rates of return on Swiss and German sovereign debt)
4. Uncomfortably high asset prices

The combination of these factors leaves many savers with a lack of options for meaningful wealth accumulation (i.e. savings, buying homes) and retirement savings. An illustration of the decimation of savers' earnings is seen in the contrast between average interest rates on savings and the lending rates throughout the EU-

Figure 11. EU-28 unemployment by age group from 2000 - 2017

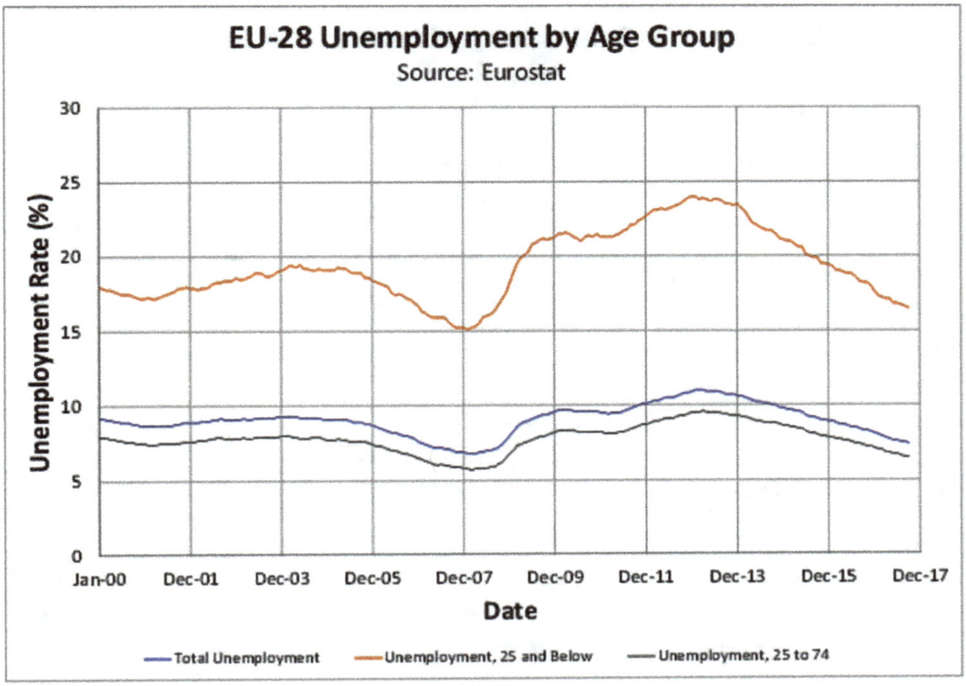

28 nations providing data to the World Bank. Figure 12 shows the average savings account interest rates, lending rates on bank loans, and loan rate spreads as reported by the World Bank for member nations. Average rates for savings interest dropped precipitously over the creation of the Euro from 1999 to 2016 from 5.9% to 0.7% while lending rates almost halved from 18.8% to 9.5% and lending rate premiums dropped from 10.4% to 5.7%.

An FCA report in the UK saw that half of adults (25 million in the UK) were potentially financially vulnerable and would perhaps fall prone to high cost of credit or cash flow shortages (Warick-Ching, 2017). The report saw that 25-34-year-olds have above-average holdings of credit or loan products, personal loans, car finance loans, and regularly switch a credit card balance. Further, the FCA study reported that 12.9 million UK adults were overdrawn in the past 12 months. The FCA's study looked into the wealth gap between UK residents as the impetus for measures such as the 2014 payday lender regulations seeking affordable credit for indebted customers. Given the high costs from overdraft fees and interest rate penalties, the ills befalling younger consumers will likely fester and worsen – more so given moves by the Bank of England to follow the same interest rate increase trajectory as the US' Federal Reserve. Stagnant growth on the continent likely means that young consumers in

Figure 12. EU-28 Savings, lending, and credit premium spreads

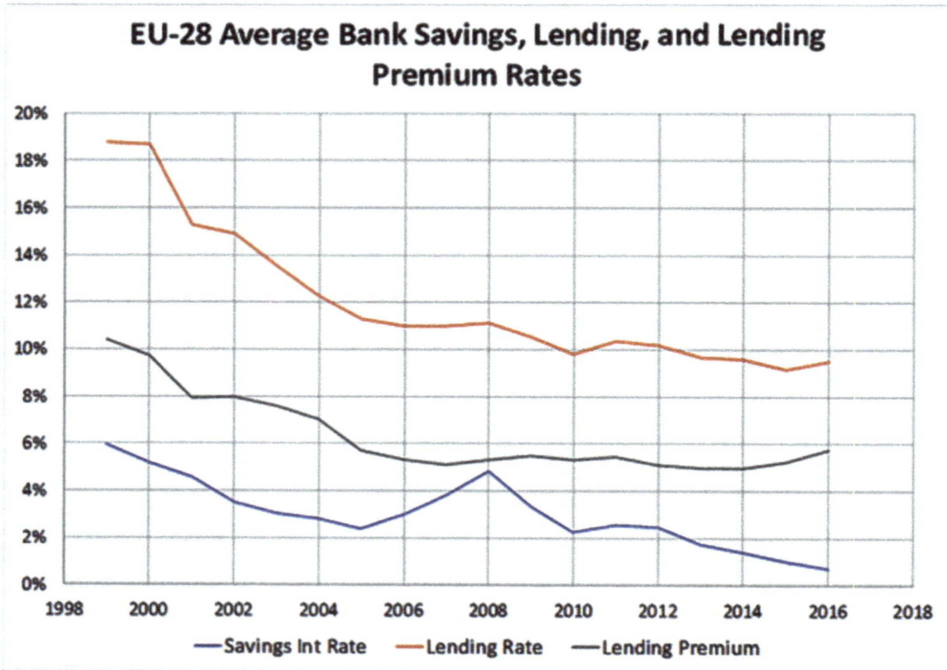

the EU suffer the same reliance on high-cost credit of which P2P lending with its lower interest costs and longer-term loans may help ease.

EU Unbanked and Under-Banked Population

In the decade-plus since Zopa started in 2005, growth in the P2P lending industry expanded from the UK to the rest of Europe. International organizations like the International Monetary Fund (IMF), World Bank, ECB, and EC are dedicated to the digitization of the money supply and the accessibility to the power and benefits provided by the financial system to all consumers to spur growth and prosperity. Reducing (if not eliminating) the unbanked/under-banked population has been a decades-long endeavor for these organizations, and P2P lending is seen as one of many tools that will help make this happen.

The World Bank publishes data on the world-wide unbanked and under-banked population. While many EU economies are developed, a considerable portion of the population are still unbanked or under-banked – especially among the Eastern-most nations in the EU, the young, women (a gender gap of 7% exists between unbanked women to men), the poor and unemployed (81% of those with wage payments have an account while 53% of those without wage payments have an account), and the elderly. While the unbanked population in the EU is relatively low at 13% (see Figure 13), the percentage of adults 15+ amounts to almost 55 million consumers. Surveys from firms such as MasterCard who have roles in payments and money transfers claim the number including unbanked and under-banked people in Europe to number 138 million individuals – including 12% in the UK (Buck, 2016).

Most P2P firms are located in the more developed EU economies with a higher percentage of the population integrated to the formal financial system. As Figure 13 illustrates, the percentage of the population that haven't been induced to the formal financial system by legacy financial institutions range between 1% to 4% of the population for the largest EU economies (i.e. France and Germany). Nations like Italy, Greece, and Eastern EU nations have unbanked populations over 10% in 2014: all are ripe for Fintech firms with novel means of both outreach, access, and engagement.

Other approaches to introduce more people to the formal financial system include education and improving the payments infrastructure (Demirguc-Kunt, Klapper, Sincer, & Van Oudheusden, 2015). Availability of products like P2P lending may

Figure 13. Unbanked population % throughout EU-28; Source: World Bank Group, 2017)

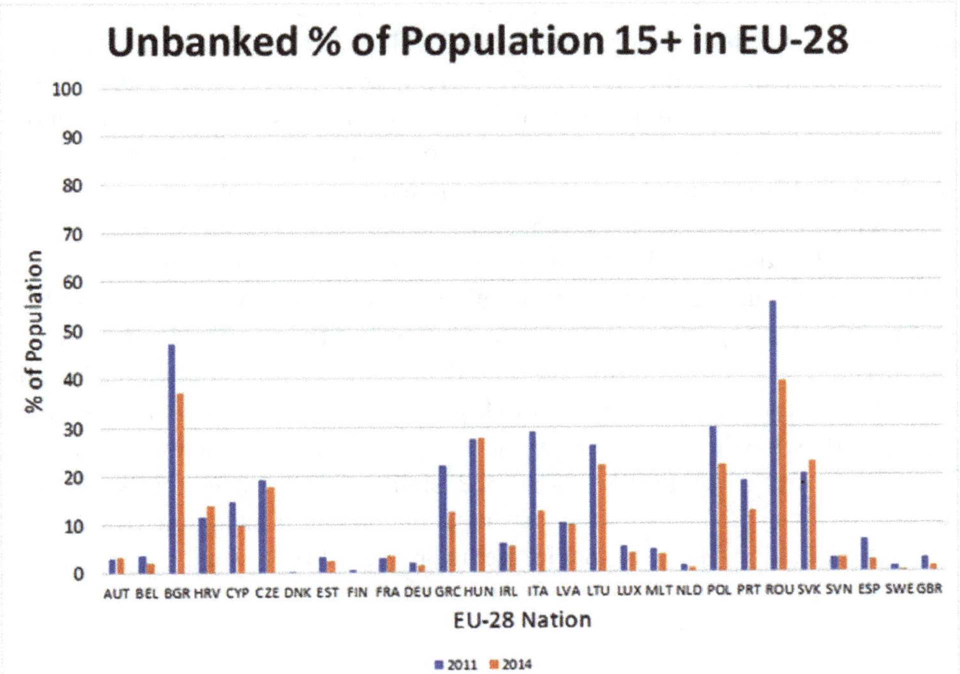

serve to establish an initial link to the formal financial system via an account to use for payments.

In the efforts to bank the unbanked population using Fintech, literacy will factor into which Fintech service will penetrate a given region. Highly literate nations offer high opportunity for P2P lending which requires contractual requirements. In a World Literacy Foundation study of the EU, lack of literacy is credited with costing the EU economy €350 billion annually including for reasons such as higher consumer spending and lost wealth creation opportunities (European Commission, 2015). The World Bank cites the EU literacy rate at 99.1% for people age 15+ in 2010, with women having a slightly lower literacy rate than adult men at 98.9 (although women are more literate than men in Poland) (UNESCO Institute for Statistics, 2017). Overall, the high EU literacy rate is prime opportunity for Fintech firms to find new customers given their levels of transparency, information availability, and ease of use.

Technology

Technology is important to Fintech as a whole, but has increased significance in P2P lending in order to effectively quantify a borrower's credit risk and price the risk to allow the investor optimal returns to compel continued interaction. The two functions are not mutually exclusive as the portal takes a significant stake in insuring success for its own viability as a market-maker while creating the opportunities for it to prosper through fee collection and growing the size of the market. P2P lending portals create many of the underwriting functions of a traditional bank without incurring capital risk itself. In doing so, P2P lending portals surrender the potential to earn the majority of the spread earned by correctly pricing risk. Rather, the firm seeks to grow the volume of activity for its cut in terms of smaller margin functions like origination and processing revenues.

How P2P lending portals are able to quantify risk is one of the biggest demands on the firm's technology. While credit ratings are not exclusively used by all P2P lenders, many lending portals shy away from extending credit to subprime borrowers. The reason is not purely altruistic – high default rates associated with subprime borrowers would not only command risk priced at high levels such as seen with payday lenders, but the default rates would likely burn investors who would shy away from placing credit at high risk of loss. P2P lending was promoted for its ability to generate of a complete credit view of a potential borrower, using metrics such as earnings, obligations, and even social media profiles of borrowers.

EU Net Savings

Net savings underwent significant downturn in the EU due to credit contraction and labor losses from the 2007-09 Global Financial Crisis and the Euro Debt Crisis (see Figure 14). Subsequently, a combination of increased access to capital and lower prices for consumer goods spurred consumption. The negative savings rate in the EU amidst improved consumption is fertile ground for P2P loan activities and for investors seeking opportunities for high returns.

Legacy Firm Complacency and Consolidation

The reduction in major financial institutions in the EU, from 30 in 1960 to 6 major high-street banks shrunk the available providers of credit for EU consumers. Reduction of physical branches (from 232,525 branches in 2009 to 198,744 in 2013) constrains access to credit and other financial products, particularly for the youngest and oldest members of the population and hastens the development of faster

Figure 14. Net EU-28 household savings/borrowings as percent of GDP including individual statistics for Spain, France, UK, and Germany

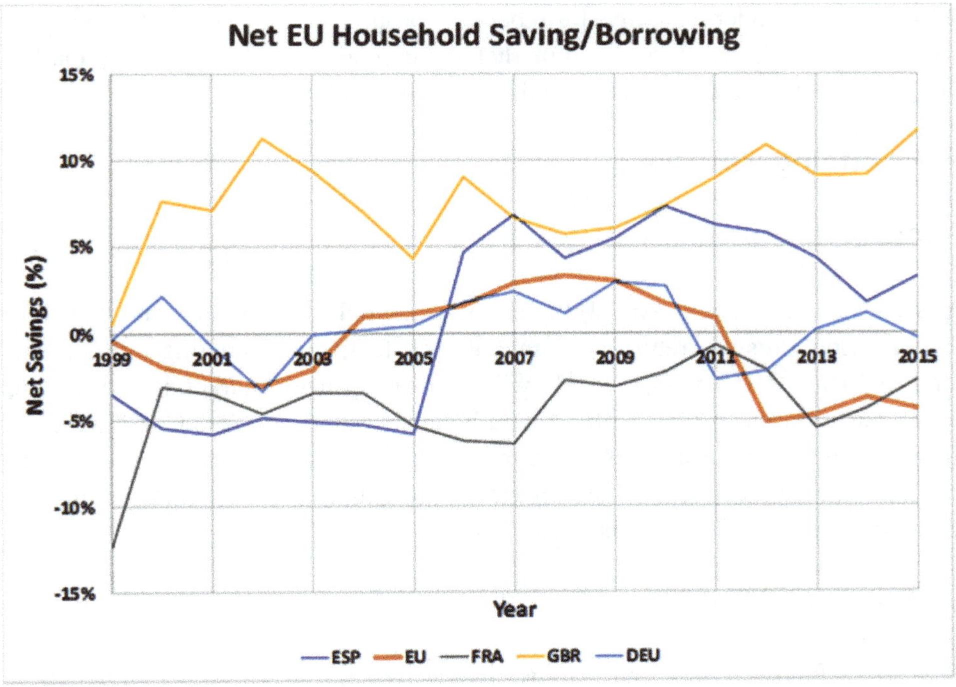

identification verification technologies as well as simplification of the process to open accounts or have access to financial services. Customer dissatisfaction with legacy institutions prompted a search by EU consumers to find a-la-carte solutions to their financial service needs.

EU P2P Lending Operations

Fintech P2P lenders face a challenge foreign to traditional banks in that their awareness must be raised and elevated to investors and borrowers alike while traditional banks leverage their existing and past clients. Perhaps because the activity of depositing funds is so distinct from the own bank's operational activity of performing adequate risk management and lending out deposits, depositors are shielded from their actual role as lenders and the possibility of a default by borrowers. That active fear of loss may (in the short-term) suppress the P2P lending space and reduce it to an operation of investment as opposed to a seemingly safer savings operation. Risk of loss is obscured by the professionalism and reputation of the institutions involved. The question is whether P2P lending Fintech firms can obscure the relationship between lending and borrowing parties sufficiently to remove the semblance of risk in the transaction akin to any other bond while reducing costs for the lending portal and borrowers while improving margins to other investment assets for investors. The ability for P2P lenders to outperform banks' non-performing loan (NPL) cycles is critical to their ability to grow the market and compete with banks/money market firms and other investment equity and credit entities alike.

P2P Lending Barriers

Awareness

Awareness of a product's availability and its potential as an investment asset class that can yield outsized returns is essential for effectively promoting P2P lending Fintech firms. P2P lending portals will rely on up-front marketing, education, communications, as well as word of mouth via consistent outsized performance to grow. Besides mere awareness, however, a track record of success is required to turn P2P lending beyond just a hobby for people willing to participate, and into an asset with a place in investment portfolios.

Security and Data Privacy

Security of funds and personal data is among the more serious issues for Fintech companies. However, a recent issue with funds disappearing from investors' sales of loans via P2P portals Zopa (Neuworth, 2017), for instance, is an issue hastened by discouraging paper records and relying on electronic records that undergo less scrutiny by customers. With time, the impact to scandals will recede – much as scandals involving Uber drivers and passengers do not adversely affect business. However, the Wells Fargo scandal and several other scandals involving mortgage originators leading to the US Housing Crisis show that fraud and abuse exists in brick-and-mortar cases as well.

The EU is on the leading edge of security in payments and banking compared to its US counterparts with very stringent consumer privacy and data protection standards thanks to the EU Data Protection Directive and the successor General Data Protection Regulation (GDPR) in 2016 (EU GDPR, n. d.). The EU GDPR's impact to Fintech is not solely based in the consent they must acquire from customers. A large potential threat to Fintech is the liabilities incurred for lacking adequate data protection under EU GDPR. PSD2 regulations due in 2018 will determine the levels of access Fintech firms will have to consumer data and the one-stop-shop operability they help provide.

Regulation

Regulation affecting P2P lending is unique compared to Fintech sectors due to the company's involvement as a market-maker as opposed to a credit counterparty to borrowers. Because the firm is not explicitly taking deposits to issue out loans, P2P lending firms do not fall explicitly under regulators like the EBA (European Banking Authority) and other deposit insurance schemes in the UK or the broader EU. However, P2P lenders share regulation attention from consumer protection-based regulators at the UK and EU-level such as the Financial Conduct Authority in the UK, the European Commission at the EU-28 level, the ECB because matters of money are involved, and of course individual member-nations who have ultimate jurisdiction over activities within their boundaries. The aforementioned agencies are committed to providing a high level of consumer protection under Article 12 of the Treaty of the Functioning of the EU. (European Commission, n.d.)

Regulation need not limit growth of the P2P lending sector. Rather, the viability of P2P lending Fintech firms *requires* that regulation removes the aura of lawlessness and exudes security and accountability. Regulation legitimizes the industry on par with legacy institutions. P2P lending scandals like the "Lending Club fake loan scandal" must be culled to help firms focus on their role as finance companies with a sober responsibility versus tech companies with interesting but risky characteristics. While regulations and scrutiny by regulators threatens to increase costs for Fintech firms, they may help decrease acquisition costs by helping clear hurdles of credibility and trust.

Curiously, the lack of regulatory harmonization throughout the continent is a drag on the growth of P2P lenders beyond an individual member nation's borders. Studies and white papers by the EC will need to ultimately produce regulation for the industry to protect investors and borrowers alike and enhance the brand of the sector in the process. Harmonization regulation reduces costs like compliance and reporting costs. Identification verification for KYC purposes is an added constraint where different member states have sometimes widely different standards and requirements. Beyond KYC, while regulation may be attained on an EU-wide basis, tax policy and rates to be paid will be a major source for the viability and growth of this industry, particularly in cross-border lending transactions.

LEGACY FINANCIAL INSTITUTION BUSINESS MODELS AT RISK AND IMPACT TO LEGACY FINANCIAL INSTITUTIONS

The biggest threat posed by P2P lenders to affected institutional stakeholders will ultimately come from a downward pressure on credit spreads to consumers through the introduction of enhanced analytics that not only takes into account exiting metrics of consumer creditworthiness, but adds other metrics and data previously unavailable. The magnitude of the pressure by P2P lenders on legacy banks will depend on the P2P lending sector's ability to:

1. Keep loan non-performance sustainably below that of traditional lenders
2. Evenly grow overall lending volume and the number of willing and committed investors for the P2P lending portal
3. P2P lending portals' ability to monetize involvement and enhance value-added services for investors and borrowers alike

Business Models at Risk

The Consumer Banking Business Model: Lending

Licenses and regulatory requirements provide formidable barriers to entry for lucrative services. In the case of P2P lending, Fintech provides the opportunity to bypass loss-leading sticky services that are heavily regulated and restricted to offer other profitable services. Advantages abound for Fintech firms playing in the arena of established financial institutions including:

Figure 15. Consumer banking business model

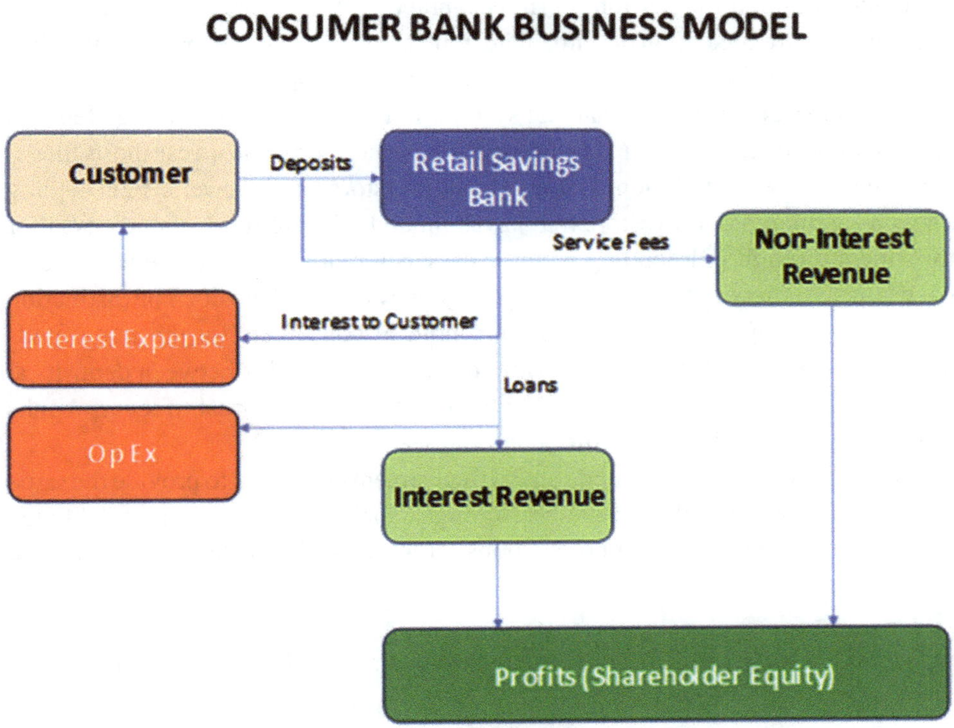

1. **Regulations:** P2P lending Fintech firms enjoy the fact that they do not have many of the regulatory oversight of traditional legacy firms/banks, nor do they have the capitalization requirements associated with traditional lenders. Besides KYC (Know Your Customer) requirements, consumer protection regulations in terms of lending rate setting are the only EU-wide regulations these firms must adhere to.
2. **Low Rate Environment Options:** While banks use ECB rate-setting as a guide for the rates on deposits they offer, P2P lenders allow investors/lenders to by-pass the intermediary role of banks in lending to capture more of the margins.

The P2P lending space impacts legacy financial institutions in terms of higher top-line spreads and containing bottom-line operational expenditures through lower infrastructure costs.

While liquidity may be injected to the overall financial system through P2P lending, the exact impact is not completely knowable, but actual levels of added liquidity will not be as pronounced as many proclaim at the consumer level. Liquidity is provided through the net long-term acceptance of risk where often short-term acceptance with expectation of immediate transfer is mistaken as providing liquidity. For instance, securitization provides liquidity as it raises capital while the acceptance of risk still exists within the financial system. In the case of P2P lending, deposits that would normally go to banks (where banks would lend out/invest those funds) would instead be directly lent (invested) by investors to borrowers. P2P lending serves as another asset investors may invest in and allocate funds. Funds used for investment through P2P lending is not necessarily mutually exclusive from funds that would've otherwise served as savings or money market deposits. The distinction would be made by assessing the risk threshold of consumers for the same funds (see Figure 16). Funds where clients are indifferent to the risk between deposits at a bank and funds invested through P2P loans would pose a threat to legacy banks seeking low-cost capital for credit.

The advent of provisional funds on P2P investments is crucial to provide the same manner of protection bank deposits earn through deposit insurance or through Money Market Fund insurance. Investor protections combined with broader moves in deposit

Figure 16. Deposit and P2P lending industry population Bayesian illustration

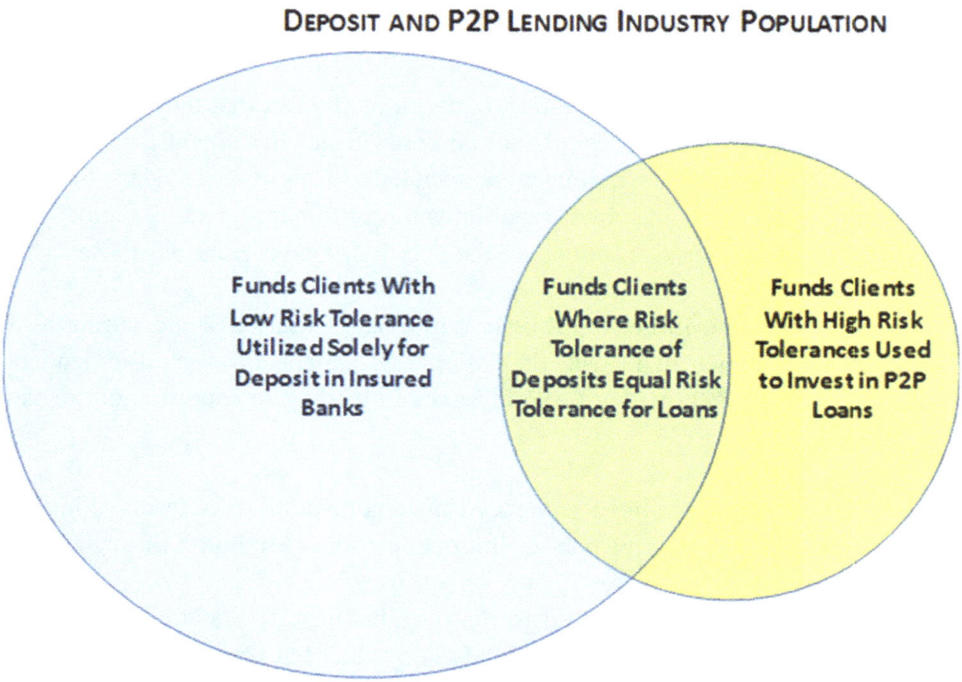

interest rates and the growth within the current boom in the NPL (non-performing loan) cycle will ultimately determine the first major plateau in the market's size.

Impact to Legacy Banks' Business

After earlier listing P2P lending stakeholders, this section seeks to quantify the impact using individual case studies including a review of historical business line business and market shares. While the ECB lists over 200 banks as significantly important or large institutions under their oversight, this study will analyze the consumer banking space for the following 6 "high-street" banks: Barclays (UK), HSBC (UK), Banco Santander (Spain), Deutsche Bank (Germany), ING (Netherlands), and BBVA (France). Not only do these institutions offer diversified products, they are also publicly traded corporations with audited SEC filings.

Lending and Savings: Barclays, HSBC, Banco Santander, Deutsche Bank, ING, BBVA Banco Frances

Financial statements for Barclays, HSBC, Banco Santander, Deutsche Bank, ING, and BBVA Banco Frances were obtained from their 2016 annual reports to observe the amount of deposits on hand from household consumers. Also obtained was the level of unsecured lending they engaged in to compare with P2P lending which is also unsecured. Conversion rates to dollars at a rate of 1.18 dollar per euro and 1.28 dollars per pound were used. Figure 17 shows the comparisons between deposits and credit cards or personal loans. As a general practice, such personal loans are unsecured and are lower weighted assets when tallied up to determine a bank's Tier I ratio or Basel III capitalization level. Personal and credit card loans aren't valued as high as other loans like mortgages because they are not collateralized debt instruments, regardless of the interest rate they earn.

Figure 18 shows the deposits per capita of each EU member nation according to EBA (European Banking Authority) plotted against the household deposit growth rate. The figure shows that the more affluent nations host the greatest number of Fintech firms operating. The greatest opportunity for a competitive edge for a Fintech company is where the costs are prohibitively large for legacy banks to compete (in the lower household deposit per capita regions). Yet much of the funding opportunities are geared toward banking in larger nations with higher deposit prospects such as the UK, France, the Netherlands, Belgium, and Germany.

Figure 17. Customer deposits in 6 major EU banks

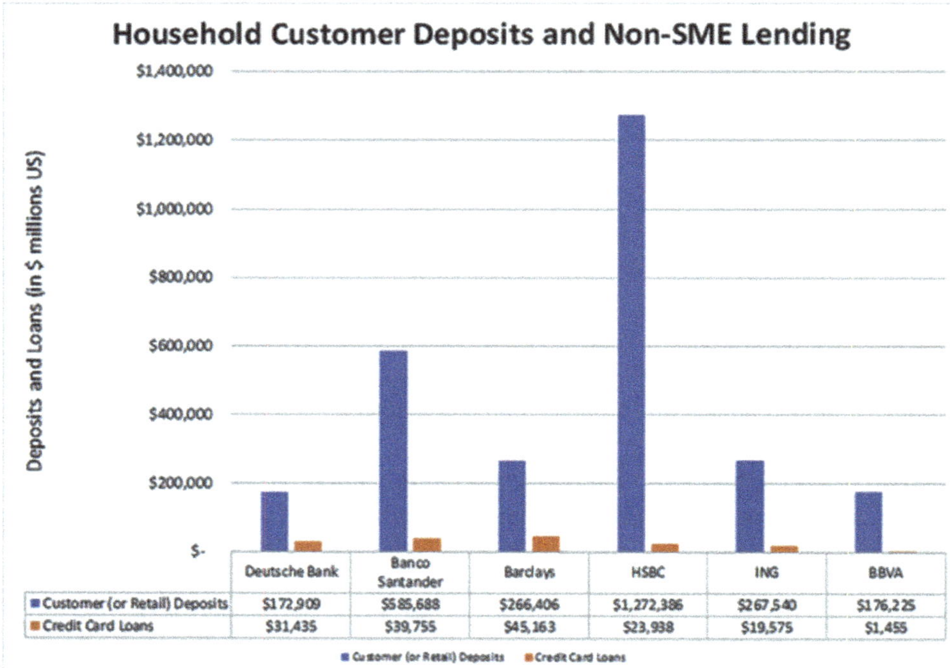

Figure 18. Deposits per capita and deposits growth rate, 2013-2014. Source: EBA

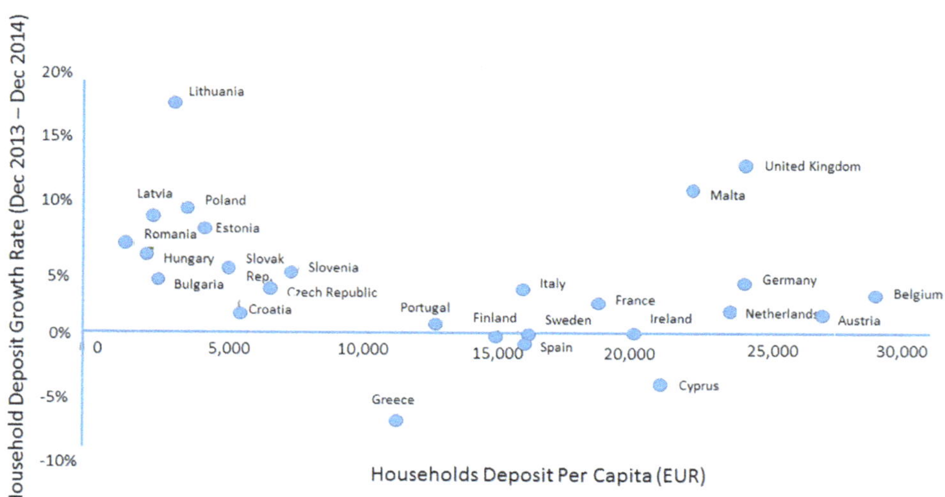

The decrease in consumption growth and disposable household income and savings (see Figure 19) increases the need for access to credit to fill household needs for those in need and the prospect for higher returns to those with capital to offer.

The competition for customers may not have been led by the legacy banks but they are engaged having seen the progress made by Fintech startups in P2P lending. Many of the concepts initiated by Fintech firms such as API-based savings platforms and P2P lending using alternate criteria to standard credit assessment measures appear obvious today but were considered major gambles long ago. From Alta Vista and Blackberry (ceding way to Google and Apple, respectively), business history is littered with examples where first-mover advantage proved to be no advantage at all.

The P2P lending space is a highly competitive battleground between Fintech and legacy banks for a number of key reasons. First is the matter of trends in the market of consumer non-secured lending (see Figure 20) which does not include products like mortgages, car loans, or home equity lines of credit – each backed by some type of collateral.

Despite post-crisis ECB injections totaling €1.7 trillion of liquidity (with the total projected to be €2.2 trillion by the end of 2017) (Speciale, 2016), banks are still reluctant to meaningfully increase consumer lending. From 2007 to 2014, non-secured or "subordinated" credit to households waned. A possible factor to that phenomena is the impact of the Basel III Capital Adequacy Ratio calculations and the risks associated with subordinated debt. First, some definitions of different Tiers of capital under Basel III:

Figure 19. Growth in consumption, disposable income, and savings rate, 013-14. Source: EBA

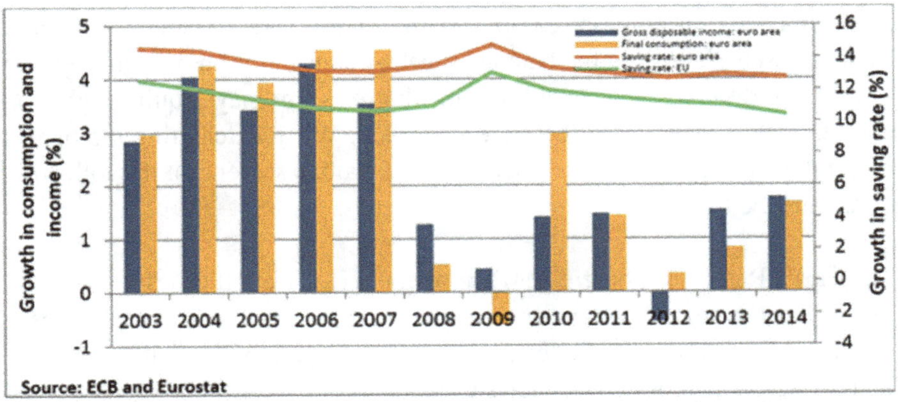

[14] Growth of nominal consumption and nominal gross disposable income of households (annual percentage rates) and the saving rate (percentage of gross disposable income).

Figure 20.Trends of consumer non-secured lending

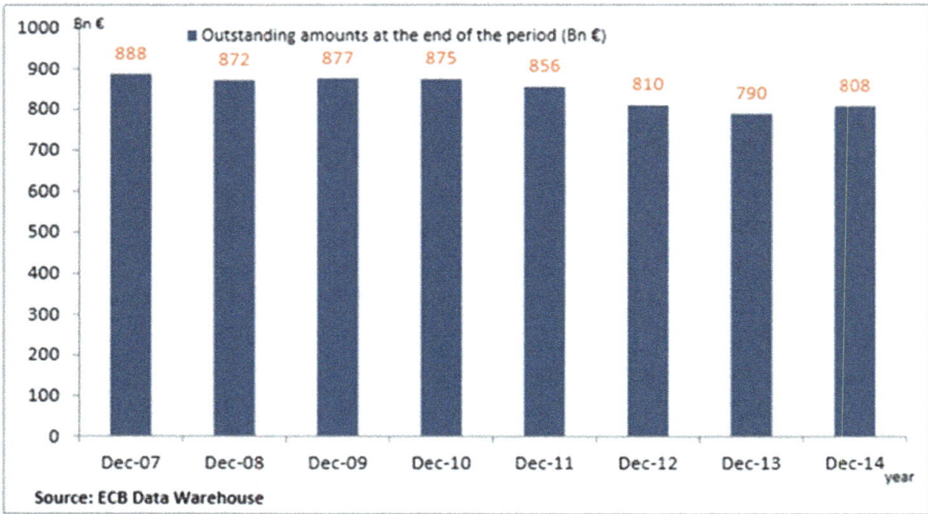

- **Tier 1 Common Capital** = Common Shareholders' Equity – Goodwill – Non-MSR Intangibles
- **Tier 1 Capital** = Total Shareholders' Equity – Goodwill – Non-MSR Intangibles + Qualifying Hybrid Securities and Noncontrolling Interests
- **Tier 2 Capital** = Subordinated Debt + Non-Qualifying Hybrid Securities + Qualifying Allowance for Loan Losses
- **Total Capital** = Tier 1 Capital + Tier 2 Capital

Under the Basel criteria, consumer credit such as personal loans and credit card loans, such capital is considered Tier 2 capital which must have an allowance for loan losses removed from common equity. A bank's liabilities and shareholder equity must balance assets (loans) on the balance sheet. A concentration of high-yielding loans such as credit cards may be a boon for the bank but may require high loan loss reserves which adversely affects Tier 1 capital. When calculating the bank's CAR (capital adequacy ratio, which determines a bank's solvency), the following conditions must be true:

- The Tier 1 Ratio must be greater than or equal to 4% at all times
- The Tier 1 Common Ratio must be greater than or equal to 2% at all times
- The Total Capital Ratio must be greater than or equal to 8% at all times
- Tier 2 Capital cannot exceed Tier 1 Capital

Consequently, it becomes very costly for a bank to have a large subordinated debt portfolio since it means the firm must take other measures to shore up its Tier 1 capital and exposing itself to downturns in the broader non-performing loan cycle. As P2P lending portals aren't banks and lenders are investors, they have the potential for a competitive advantage over legacy banks in the consumer loan market. 2014 consumer credit outstanding is almost 10% lower than 2007 levels. While that is a big gap for Fintech firms to make up ground on, it is not insurmountable. However, regulation is a wild card: if regulators key in or impose restrictions on P2P lenders, legacy banks buy time to re-establish lending levels to consumers. The second wildcard is the response to investors in an uptick in non-performing loans and whether that eradicates the liquidity these investors provide and the ultimate effect on the brand of P2P lenders.

CONCLUSION: THE FUTURE OF P2P LENDING

Overview

By reducing interest rates, central banks like the Bank of England spur growth and inflation while leaving investors with few options to earn any meaningful yields on funds. P2P lending firms have seen an increase in lending funds (Staff, 2017) due to the dearth of yield-returning options available for savers, compelling them to absorb some additional risk in the name of seeking higher potential returns. Savers' deposits are welcome by institutions seeking to recapitalize amidst capital shortfalls. Only meeting solvency requirements by ECB and Basel III will those institutions ace stress tests and obtain permission for increased transfer of profits to shareholders. Eventually, adequate capitalization activity will reach a nadir when the banks will have more than enough funds to meet capital adequacy ratios and additional deposits will be unwelcomed – manifest by a persistence of inadequate returns on deposits (such as the case seen today). This indicates the near zero yields on deposits will persist for longer and hint at an increase in P2P lending activity.

The fate of the P2P lending sector will be determined by economic cycles. The end of the Financial Crisis and Euro Debt Crisis began one of the most sustained periods of growth– especially in Europe. In periods of economic expansion, it is difficult to assess the viability of a business model when it hasn't undergone a full loan performance business cycle. An increase in non-performing loans is inevitable and will be joined by an increase in interest rates which will close the credit premium gap between investing in P2P borrowers and depositing money in a bank. It remains to be seen what the response will be for P2P lending amidst an NPL increase. Firms that supplement the existing credit agency data will likely fare better and undergo lower forbearance and NPL rates than legacy banks.

It's worth saying that for all the research in risk assessment models by Fintech P2P firms, the models have yet to prove their endurance. Models are accepted as valid until events occur requiring modification. Risk profiles are valid until they reach a crisis that undermines the assumptions generating that profile. For instance, since 1997, the Securities and Exchange Commission (SEC) accepted the Value at Risk model as an adequate quantifiable measure to incorporate into financial statements (U.S. Securities and Exchange Commission, 1997), largely for players with derivatives and other mark-to-market assets by marking to model. As the JP Morgan "London Whale" conference call on May 10, 2012, the value at risk (VaR) model for the firm used to be $67 billion prior to the conference call but reverted to VaR value of $129 billion when the hedge was proven invalid.

A proxy for anticipated performance amidst a downturn in the economy may be seen in the market for the ABS (asset-backed securities) market (see Figure 21 and Figure 22).

The European ABS issuance figure is more relevant, particularly with respect to similar consumer and credit card loans which are more cyclical and volatile. It is unknown what individual investors to P2P loans would do. In individual EU nations, lending cycles follow similar patterns to macroeconomic cycles.

Another consideration is the introduction of loan loss provisions by Fintech firms that don't offer the feature as a means of insurance on investments. Ed Norton's character in the movie "Fight Club" said: "on a long enough timeline, everybody's

Figure 21. US ABS issuance volume, 1985 to 2016. (SIFMA, 2017a)

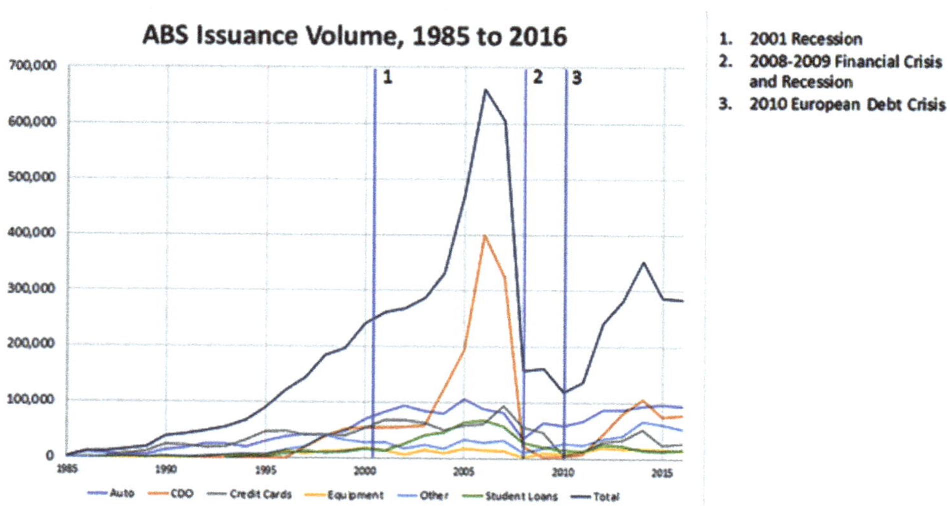

Figure 22. European ABS market from 1985 to 2016 (SIFMA, 2017b)

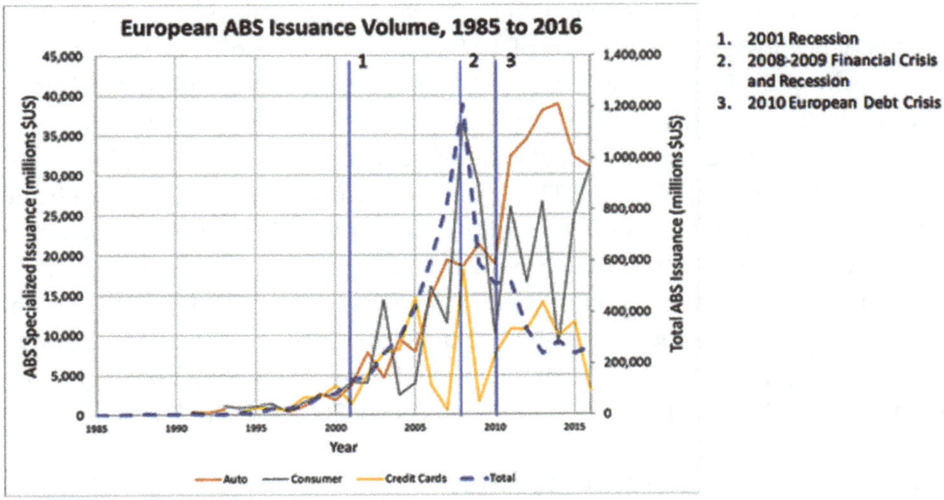

probability of survival goes to zero." While some firms claim to never having undergone a loss to any investors, such a record can't expect to continue forever. RateSetter has seen over £2 billion invested in its platform while seeing no net losses by any investors (RateSetter, 2017). The record was largely accomplished through the firm's use of a provisional fund in addition to its service fees for investors – akin to loan loss provisions used by legacy banks. The fund's contribution by each loan is a function of loan size and the credit risk assessment by the firm. Coverage by a provisional fund is assessed in terms of both interest coverage and projected losses on loans outstanding using their credit risk methodology. Seeing RateSetter's activity as an example, one sees that the rate of new investor growth had increased at a faster clip (CAGR of 123%) than the relative loss percentage of loans outstanding (CAGR of 19.5%) while the provisional fund maintains adequacy in protecting versus future losses. On the basis of depletion of the funds and actual loan performance, however, the trends follow similar trends to broader indicators of loan activity and EU NPL performance.

Since a Fintech firm is a business, after all, a major trend taking form regards the service payment for the P2P lending portals. Firms such as RateSetter are eschewing the model for a 1-time fee to originate a loan and opting instead for a stream of fees off each monthly payment. The change in business model and adoption of provisional funds will increasingly chisel away at the spreads investors will enjoy from such a service. It remains to be seen what investor supply will look like with lower yields and/or borrower demand amidst higher interest rates required on loans.

Despite the doubts, P2P lenders should be considered an investment asset class akin to a real estate investment fund (REIT). Currently, companies such as RateSetter still experience losses such as a loss of £4.9 million versus profits of £0.5 million the year prior (P2P Finance News Staff, 2016). Still these firms will find a sizable niche amidst the reassessment of risk by legacy firms. The degree of service as well as the Fintech firms' ability to model and evolve their risk models will play a role in their growth potential. To-date, P2P lending has grown significantly (albeit from a low baseline). As Figure 23 shows, both within and outside the UK, P2P lending growth increased at an 83% CAGR from 2011 to 2015e while EU minus UK P2P lending increased at a 206% CAGR over that same timeframe. The projected growth in the P2P lending volume for non-SME (small and medium enterprise) in both the UK (see Figure 24) and EU minus UK (see Figure 25) is expected to grow to approximately $40 and $150 billion, respectively, by 2025 (Leech, 2015).

Trends in P2P Lending Companies

As the P2P lending industry matures and legacy institutions strengthen again, the niche filled by these lenders evolves as well. While the spreads are attractive assuming no undue risks are posed to invested capital, capturing the spread between the interest issued and the cost of capital to the investor is easy to exceed the returns on safer investments such as investment-grade bonds. However, the more successful P2P lending firms with the lowest rates of default (lower than rates seen on the NPL cycle curve) would require loan loss provisions, similar to those employed by traditional banks. P2P lending firms will need to "thread the needle" by enlisting capital to be used for loan loss reserve provisions (thus, P2P lending portal firms that earn a

Figure 23. UK Online P2P lending (Left); EU Minus UK online P2P Lending (Right)

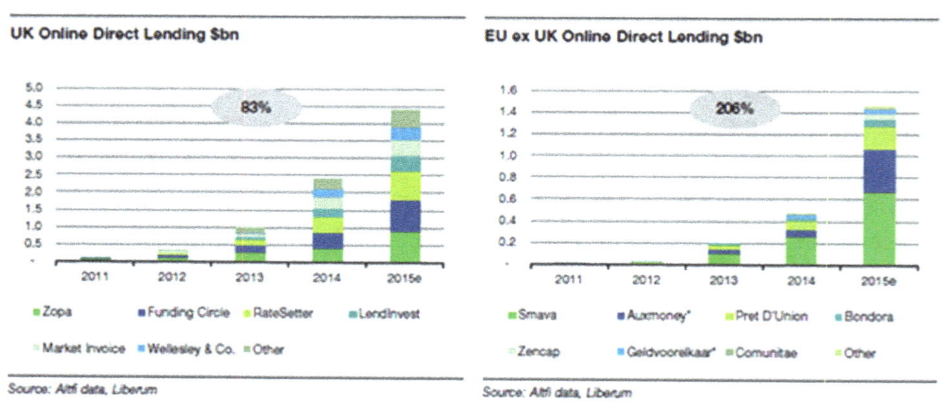

Figure 24. Forecast UK direct lending volume by segment. Source: Liberum

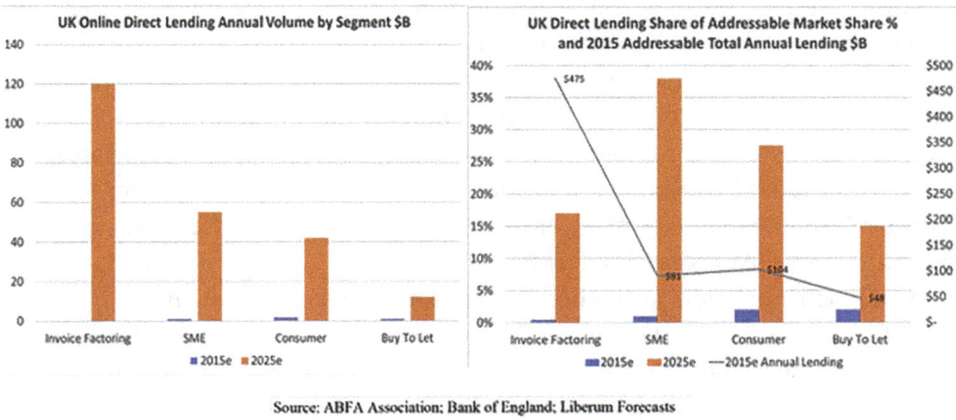

Figure 25. Forecast EU minus UK direct lending volume by segment. Source: Liberum

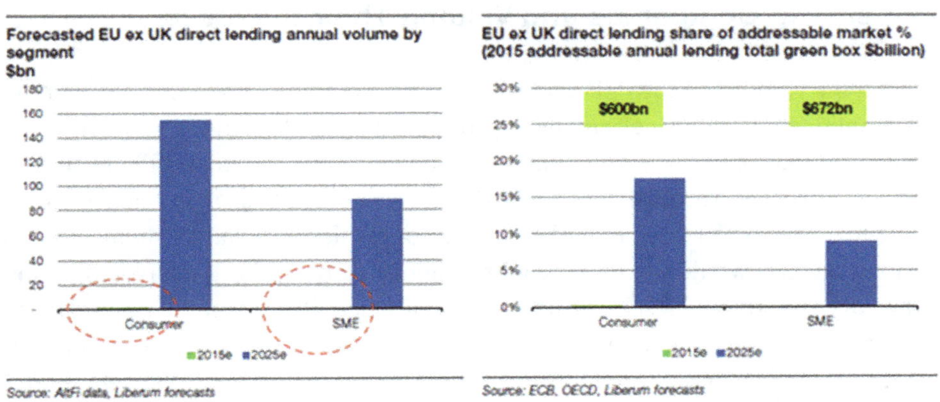

regular fee over the life of a loan will preserve their revenue stream longer), while not chipping away too much at profits to investors who may find the cost-adjusted risk unacceptable and would be tempted to switch to another investment class.

A trend being observed in P2P lenders like Zopa is the switch from a pure P2P lending platform to seek a banking license (Dunkley, 2016). To some, such a move potentially undermines the growth potential prospects of P2P lending, or at least in its pure form. The transition will allow for additional revenue streams for the firm; either in terms of providing a cheap source for loan loss provisions, the utilization of the firm's risk-assessment algorithms for use of its own account, or loans and other products to clients. The Financial Times noted that Zopa may also offer credit cards and overdrafts with a license as well as protection by the Financial Services Compensation Scheme (a service like the Federal Deposit Insurance Corporation in the United States). Such a service isn't available to P2P lenders. Regardless, the move is a nod to the stickiness of retail deposits as the firm – and others with prospects to make similar moves in the future – looks to offer consumers and investors enhanced stability. The move followed a partnership with Metro Bank in 2015 to offer services to the bank similar to money market accounts which earn profits for savers by using funds for commercial paper investments.

As a practical matter, such a license acquisition makes sense depending on the type of loan to be issued. Given the variety of lending options (individuals, small and medium enterprises (SMEs), etc.), there are vastly different capital requirements involved and licenses would be more suitable for individual lending purposes. Ultimately, the move signals a willingness for P2P lenders to evolve their business model to suit the needs of consumers and investors alike rather than perpetuating a still-nascent business model.

P2P Lending Companies and Valuing Their Business

To assess the viability of the P2P lending businesses requires an assessment over the fundamentals to their business models: profits and costs. Mainly, there is the costs for operations and cost for acquisition of new clients. With the story still unfinished regarding client churn behaviors of P2P lending companies, it is reasonable to assume that successful track records of these companies will endear themselves to investors (much as would was the case with online brokerages that tout diversification to stave off investment losses). Offering loan products that are small nominally to permit diversification, the reduction of costs through automation, stickiness through rewards or other incentives including promotions, and the promotion of the portal to would-be borrowers to increase aggregate demand for the product will go a long way toward increasing its attractiveness to investors.

In terms of valuing the enterprise for equity investors into the portal companies, the move from one-time fees on new loans for P2P lenders to one where fees are assessed with each loan in repayment makes for a better and more even estimate of future performance in use for a discounted cash flow analysis. From then, the growth in business will largely be a function of both market size and projections on interest rate moves and assessment of the macroeconomic cycle and the loan performance cycle.

Summary

The consumer loan line of business for legacy banks is a profitable business when conditions permit its expansion. Economies of scale in terms of focusing the firm's attention on the smaller-sized loan issues as opposed to a diversified credit line set will aid P2P lenders to reduce costs.

To say that P2P lending firms have an inherent advantage over legacy banks in any business line in large part due to a lack of baggage or of legacy infrastructure that is often hard to modify – let alone overhaul – is a gross oversimplification. The main advantages of P2P lending firms is the customer-centric nature of these new businesses, for the truly successful Fintech firms – like any truly successful startup – is that the firm begins with the customer and her or his needs.

P2P lending provides many new opportunities for banking and finance. Through specialization and customization, cost reductions may be feasible which makes the unbanked population eligible to ultimately be banked, at least at higher than current levels. By providing a means to provide investors opportunities once available only to legacy banks, and consumers the access to the credit current banks are unable or unwilling to extend, the P2P lending industry already succeeds where decades of legacy banking failed.

REFERENCES

Buck, G. (2016, December 13). MasterCard: Millions of Europeans still unbanked. *GTNews*. Retrieved from https://www.gtnews.com/2016/12/13/mastercard-millions-of-europeans-still-unbanked/

Demirguc-Kunt, A., Klapper, L., Singer, D., & Van Oudheusden, P. (2015, April). *The Global Findex Database 2014 Bringing the 2 Billion Unbanked into the Financial System*. The World Bank Group. Retrieved from http://pubdocs.worldbank.org/en/113791483565360488/N2UnbankedV5.pdf

Dunkley, E. (2016, November 16). Peer-to-peer lending pioneer seeks banking license. *Financial Times*. Retrieved from https://www.ft.com/content/53aa3dec-ac07-11e6-9cb3-bb8207902122

EU GDPR. (n.d.). *GDPR Key Changes: An overview of the main changes under GPDR and how they differ from the previous directive*. EU GPDR. Retrieved from http://www.eugdpr.org/key-changes.html

European Central Bank. (2015). *Virtual currency schemes – a further analysis*. Retrieved from https://www.ecb.europa.eu/pub/pdf/other/virtualcurrencyschemesen.pdf

European Central Bank. (2017, January). *The euro area bank lending survey: Fourth quarter of 2016*. Frankfurt, Germany: ECB. Retrieved from https://www.ecb.europa.eu/stats/pdf/blssurvey_201701.pdf?6c44eff3bac4b858969b9cb71bd4a8fa

European Commission. (2015). *Literacy in Europe: Facts and Figures*. European Commission. Retrieved from http://www.eli-net.eu/fileadmin/ELINET/Redaktion/Factsheet-Literacy_in_Europe-A4.pdf

European Commission. (n.d.). *Consumer Rights and Laws*. EC. Retrieved from http://ec.europa.eu/consumers/consumer_rights/index_en.htm

Eurostat. (2017, July 20). Population pyramids, EU-28, 2001 and 2016 (% of the total population). *Eurostat*. Retrieved from http://ec.europa.eu/eurostat/statistics-explained/index.php/File:Population_pyramids,_EU-28,_2001_and_2016_(%25_of_the_total_population).png

Financial Conduct Authority. (2014, November 17). *FCA confirms price cap rules for payday lenders*. Financial Conduct Authority. Retrieved from https://www.fca.org.uk/news/press-releases/fca-confirms-price-cap-rules-for-payday-lenders

Financial Conduct Authority. (2014, November 17). *FCA confirms price cap rules for payday lenders*. Financial Conduct Authority. Retrieved from https://www.fca.org.uk/news/press-releases/fca-confirms-price-cap-rules-for-payday-lenders

Financial Services User Group. (2011, March 22). *Financial Services User Group's (FSUG) opinion on interest rate restrictions in the EU*. FSUG. Retrieved from https://ec.europa.eu/info/sites/info/files/file_import/irr-2011_03_22_en_0.pdf

iff/ZEW. (2010). *Study on interest rate restrictions in the EU, Final Report for the EU Commission DG Internal Market and Services*. Project No. ETD/2009/IM/H3/87. Retrieved from http://ec.europa.eu/internal_market/finservices-retail/docs/credit/irr_report_en.pdf

Kane, C. (2011, February 9). Do Credit Scores Matter Outside the US? *CNBC*. Retrieved from https://www.cnbc.com/id/41498097

Kharpal, A. (2017, October 19). Cryptocurrencies like bitcoin are not 'mature' enough to regulate, ECB chief Mario Draghi says. *CNBC*. Retrieved from https://www.cnbc.com/2017/10/19/cryptocurrencies-are-not-mature-enough-ecb-chief-mario-draghi.html

Leech, C. (2015, October 20). Liberum LendIt Presentation. *Liberum*. Retrieved from http://www.liberum.com/media/69233/Liberum-LendIt-Presentation.pdf

Leech, C. (2015, October 20). Direct Lending: Finding Value/Minimizing Risk. *Liberum*. Retrieved from https://www.google.com/url?sa=t&rct=j&q=&esrc=s&source=web&cd=1&cad=rja&uact=8&ved=0ahUKEwj6uKXJ_tHYAhVY8WMKHVngAT8QFggnMAA&url=https%3A%2F%2Flendingwell.com%2Fresearch-posts%2Fliberum-direct-lending-finding-value-minimising-risk%2F&usg=AOvVaw01QH7xpk-J2FNNoBGrOZB_

Neuworth, S. (2017, August 11). *Zopa provides update on 'disappearing money' IT glitch*. P2P Finance News. Retrieved from http://www.p2pfinancenews.co.uk/2017/08/11/zopa-update-it-glitch/

P2P Finance News Staff. (2016, September 20). Ratesetter blames £4.9m loss on new fee structure. *P2P Finance News*. Retrieved from http://www.p2pfinancenews.co.uk/2016/09/20/ratesetter-loss/

RateSetter. (2017, August 12). Retrieved from https://www.ratesetter.com/

SIFMA. (2017a, July 14). *US ABS Issuance and Outstanding*. Retrieved from https://www.sifma.org/resources/research/us-abs-issuance-and-outstanding/

SIFMA. (2017b, April 13). *European Structured Finance Issuance and Outstanding*. Retrieved from https://www.sifma.org/resources/research/europe-structured-finance-issuance-and-outstanding/

Speciale, A. (2016, December 8). ECB Expands Stimulus to $2.4 Trillion as Monthly Purchases Slow. *Bloomberg*. Retrieved from https://www.bloomberg.com/news/articles/2016-12-08/ecb-extends-bond-buying-at-reduced-pace-until-end-of-next-year

Staff. (2017, August 13). Ratesetter doubles new investor accounts since base rate cut. *Peer2Peer Finance News*. Retrieved from http://www.p2pfinancenews.co.uk/2016/09/16/ratesetter-doubles-investors-rate-cut/

The World Bank Group. (2017). *Global Financial Inclusion Database*. The World Bank Group. Retrieved from http://databank.worldbank.org/data/reports.aspx?source=1228#

United Nations Educational, Scientific, and Cultural Organization (UNESCO) Institute for Statistics. (2017). *Literacy rate, adult total (% of people ages 15 and above)*. World Bank Database. World Bank. Retrieved from http://data.worldbank.org/indicator/SE.ADT.LITR.FE.ZS?locations=EU

U.S. Securities and Exchange Commission. (1997, July 31). *Questions and Answers About the New "Market Risk" Disclosure Rules.* U.S. S.E.C. Retrieved from https://www.sec.gov/divisions/corpfin/guidance/derivfaq.htm

Warick-Ching, L. (2017, October 20). Young People Rely on Parents and Credit Cards to Cover Costs. *Financial Times*. Retrieved from https://www.ft.com/content/d3dddbcc-b4df-11e7-aa26-bb002965bce8

Chapter 4
Fintech Impact on EU Retail Savings and Investment

Manuel Gutierrez
Loyola Marymount University, USA

ABSTRACT

Years after the 2007-2009 financial crisis and European debt crisis, the European Union's (EU) banking system sustained persistent strain due to those two shocks, austerity and economic contraction, political events, poor banking operations, enhanced regulation, and litigation. The European Central Bank's response was significant: short-term interest rates collapsed and markets were flooded with money via quantitative easing programs. Consequently, investors fled risky assets for the safety of government debt. Yet as banks recovered, savers sacrificed asymmetrically: yields on bank deposits and bonds were decimated. As macroeconomic challenges subsided, Fintech increasingly threatened legacy financial institutions' business models to the benefit of the EU savings public. This study assesses the impact by Fintech companies on legacy banks in the EU with respect to savings, lending, and wealth management. This study also assesses and makes recommendations on a strategy by Fintech to benefit savers, and the measures legacy institutions must take to survive amidst this new competitive landscape.

DOI: 10.4018/978-1-5225-4148-6.ch004

INTRODUCTION

The EU financial system emerged weakened after the Euro Debt and 2007-2009 Global Financial Crisis: institutions failed, new capitalization buffers emerged, and a still-evolving regulatory framework bemoaned the banking power among fewer large institutions. The end of the Euro Debt and US Financial Crises coincided with the rise of the Fintech industry. Fintech (a combination of the words "finance" and "technology") refers to the use of technology hardware and software to perform banking and financial services such as money transfers, payments, banking transactions, wealth management, and insurance. For diversified European financial institutions, traditional banking functions (i.e. wealth management, payments, savings, deposits) serve as steady revenue sources to dampen highly volatile business lines like trading and investment banking. Despite lower returns on equity for traditional banks versus investment banks, traditional banking functions are increasingly important in the aftermath of the Global Financial Crisis. Even investment banks, such as Goldman Sachs, shifted resources toward retail lending to offset the volatility of trading and non-interest income. This project investigates the vulnerability of major legacy (existing) financial institutions to Fintech software companies, barriers for Fintech, and the effects of Fintech on EU savers and investors.

Over 5000 businesses worldwide are categorized as Fintech firms (Dietz, Khanna, Olanrewaju, & Rajgopal, 2015), this project focuses on savings and wealth management firms in the EU-28 (Austria, Belgium, Bulgaria, Croatia, Cyprus, Czech Republic, Denmark, Estonia, Finland, France, Germany, Greece, Hungary, Ireland, Italy, Latvia, Lithuania, Luxembourg, Malta, Netherlands, Poland, Portugal, Romania, Slovakia, Slovenia, Spain, Sweden, and the United Kingdom). The UK is included despite uncertainty regarding the United Kingdom's Brexit procedure under Article 50 as of this publication. Fintech companies provide compelling product offerings for savers, promote financial service imitation, and squeeze margins for legacy banks and wealth managers. Success by Fintech firms will be a double-edged sword: while some may carve out viable and profitable businesses, M&A (mergers & acquisition) will be spurred by early Fintech investors looking to cash out, the need to achieve scale to effectively rival legacy banks, and transactions by banks who seek to purchase services more effectively rather than develop such services organically. This study proposes that effects on EU retail customers will be somewhat muted due to M&A and evolving regulations at the EU- and individual member-state level. Fintech will impact savings and wealth management by making clients align the value of services and features towards performance.

This study is comprised of five other sections: (1) Background including an overview of savings and wealth management Fintech products; (2) An outline of stakeholders involved in Fintech's rise and proliferation; (3) Description of the EU's market, its growth, opportunities, and challenges including cultural, and demographic features of the EU pertaining to the movement of people and money; (4) Identification of the main threats to existing/legacy financial institutions including major EU institutions such as HSBC, Banco Santander, Deutsche Bank, Lloyds, ING; (6) A summary of why Fintech may not completely upend the banking industry but make positive contributions to retail savers and investors.

BACKGROUND: FINTECH EVOLUTION AND BUSINESS LINES

Recent Banking History and the Need for Fintech in EU Savings

The surge in Fintech investment in developed economies (i.e. the US and EU) increased awareness of the industry through reach and word-of-mouth. Fintech has existed for decades in some shape or form: (1) The creation of charge plates by department stores to issue credit in the 1920s (Tsosie, 2017); (2) The ATM's creation in 1967 by John Shepherd-Barron (History, 2010); (3) Mobile phone-based money transfer service "Smart" in the Philippines on 2000 (Runde, 2015). Fintech provides new opportunities for consumers while threatening legacy businesses models. While Fintech includes hardware developments, software is the largest concentration of new Fintech services and investment – thus the focus of this project.

Legacy financial institutions' practices, processes, and regulations were built off centuries of actions that answered regulatory, political, societal, and economic needs. The financial industry primarily transfers capital from those who have capital to those who need it: a depositor opens an account to deposit money for savings with the stated promise that capital is available on demand (plus some modest accrued interest). Under the fractional banking system, a fraction of deposits (the reserve requirement ratio) is held for on-demand claims while "excess" funds are lent out and reinvested. Historically, banks held loans as assets on their balance sheets which counter-balanced deposits counted as liabilities and profits (equity). In addition to default risks, administrative costs, and opportunity costs, interest rate risk is priced into the interest rates charged by lenders for mortgages and other credit. Since increases to interest rates may wipe out loan income, the interest rate was sizeable for long-dated loans to compensate for that risk. The advent of securitization in the 1980s expanded mortgage amortization periods while down payments and interest rates charged decreased. EU banks were slow to implement securitization: only at the turn of the millennium did securitizing mortgages become widely-accepted practice.

Central banks serve as lenders of last resort – an important tool in fighting the threat of bank runs. The 2007-2009 Global Financial Crisis and European Debt Crisis shook the confidence in trusted financial institutions and hastened the rise of Fintech thanks to:

1. Unprecedented nation-wide depressed home values in nations like the US, Ireland, and Spain
2. The moment the "buck was broken" on September 16, 2008 when the Reserve Primary Fund money market fund held losses from Lehman commercial paper (Gullapalli, Anand, & Maxey, 2008)
3. In a sight not seen in the UK since the 1866 Overend, Gurney, and Co. bank run, the Northern Rock bank run (see Figure 1) shook depositors' confidence of their banks' solvency (Wall Street Journal Staff, 2007).

Figure 1. People queuing outside a branch in Golders Green, London, on September 14, 2007, to withdraw their savings due to fallout from the U.S. subprime crisis; source: Cunningham, 2007

This image was originally posted to Flickr by Alex Gunningham at http://flickr.com/photos/89319548@N00/1378965141. It was reviewed on 15 September 2007 by the FlickreviewR robot and was confirmed to be licensed under the terms of the cc-by-2.0.

Customers and tax-payers in Europe witnessed events unfold that undermined confidence in their governments and the institutions those governments were tasked with supervising. At the height of the 2007-2009 Global Financial Crisis, Ireland backstopped its state banks while exposing taxpayers to liabilities equaling multiples of its GDP. To become a Euro member, Greece altered its economic and budgetary statistics to comply with EU guidelines. The reckoning from their perilous economic state precipitated the climax in the Euro Debt Crisis which unleashed an IMF bailout and harsh austerity measures. A buying spree by Icelandic banks led to the nation's bankruptcy. Several large banks participated in the 2008 London Interbank Offered Rate (LIBOR) manipulation scandal. In the US, the 2016 Wells Fargo cross-selling scandal disclosed millions of fraudulent accounts opened on behalf of unsuspecting clients.

During the 2007-2009 Global Financial Crisis, moral hazard was in full view: governments stood by their banking systems using taxpayer money while taxpayers were conversely allowed to undergo foreclosures, austerity, and decreased incomes on fixed-income investments and banking deposits. The erosion of confidence in the financial system stirred a desire for eliminating intermediaries and institutions whose interests ran counter to those of their clients' best interests. The rise of Fintech was not the creation of a new financial product per se or even a new channel of distribution for banking products. The rise of Fintech in the EU is the realization by the savings public that the prudent and responsible vehicles to accomplishing financial goals were broken and a new transparent approach was required.

Fintech Business Lines

With exceptions like artificial intelligence and "big data learning" (the analysis of large data sets to determine patterns, habits, and networks to influence human interactions), much Fintech innovation involves speed, increased functionality, and 24-hour world-wide access to traditional financial services. Speed and accessibility is important for customers in EU urban centers: The EU consists of many high-density cities such as Paris, London, and Frankfurt where the top 40 cities have a Tomtom traffic index average of 30% (traffic travel time is on average 30% higher than regular travel times) and cities like Bucharest possessing a 50% traffic index (Tomtom staff, 2017). Meanwhile, the migration of many large institutions away from rural areas toward urban centers makes accessibility increasingly important to rural areas that lost bank branches.

Legacy banks' response to improving customer service focuses on accessibility through apps, automation, and online distribution of existing banking services for convenience. Fintech firms, meanwhile, propose to enhance consumer choice, provide liquidity and customization in a more cost-effective manner with potential returns to benefit consumers. Fintech businesses emerged in areas like consumer banking and wealth management.

Banking: Savings and Checking

New Fintech banks typically have an extremely limited (or non-existent) brick-and-mortar footprint with the same Financial Services Compensation Scheme protection as legacy banks – (exception challenger banks, discussed later). Combined with a mix of robust services and easy/intuitive interfaces, Fintech firms threaten legacy banks' discrete business lines and markets. Findings from the recent Competition and Markets Authority (CMA) review of financial institutions disclosed a need for more competition in the financial industry via "Open Banking" (discussed in a latter section).

Savings and checking are some of the most basic yet most capital and regulation-intensive operations (besides remittances). In Europe, the Deposit Guarantee Schemes (DGS) was developed to provide deposit insurance across all non-UK EU member financial institutions while the UK's Financial Services Compensation Scheme (FSCS) performs the same deposit insurance functions and coverage levels in the UK. The need for harmonization emerged in the 2008-09 Financial Crisis as savers sought secure institutions and different EU members initially undercut the security of their partners. After the Lehman Brothers collapse, Ireland instigated an EU-wide flight of capital when the government stood behind all depositors in its 6 Irish-owned banks to the tune of €440 billion (The Telegraph Staff, 2011), compelling other EU nations to hastily follow suit and undercut other partners in the process. In addition to these operating requirements, KYC (Know Your Customer) regulations seek to prevent money laundering and corruption. Three main Fintech banking categories emerged:

Neobanks

Neobanks (see Figure 2) are financial service providers that do not hold a banking license; rather, they align with licensed banks that provide access to their license, infrastructure (predominantly payments and money transfers) and some back-office operations in exchange for compensation in the form of small transaction/payment fees. Neobanks' offerings include convenient features such as:

1. **Ease of Opening Accounts:** Verifying identity requires a combination of a video call, scans of identification credentials like a passport or ID card, an email, and downloading the Neobank's mobile phone application

Figure 2. Neobank Model

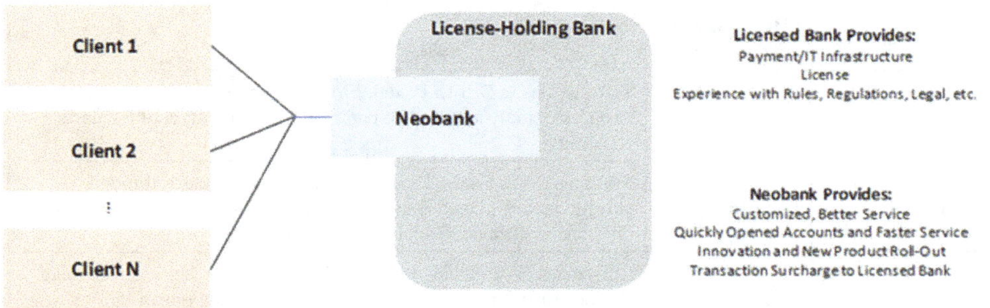

2. **Quick Time to Open Account:** Some like N26 promote that an account may be opened in 8 minutes
3. Integration with the Existing Payment System: Issuing credit cards with partners such as Europay, Mastercard, and Visa (EMV)
4. **Low Cost:** Free account maintenance in many cases

The ease provided by mobile banking, mobile payments, and credit cards further the efforts of organizations such as the World Bank, ECB, and EBA (European Bank association) to digitize money in circulation and push the economy toward a "cashless society." Neo and Challenger Banks offer several sources of value for consumers: service, accessibility, transparency. Fintech firms passed on most savings to consumers by way of lower transaction costs, higher interest on savings, or some combination of the two. Neobanks and other Fintech companies often provide leading-edge functionality and capabilities (like budgeting and social media functions) to establish a more intimate, cooperative, and friendly relationship with customers. Such novel features include traditional payments, transfers to the smartphone, e-mail, Twitter, micro-loans in a single click, currency exchange, purchase of precious metals, and social brokering (copying the behavior of other brokers). A practice that seems taboo to many older consumers (openly discussing financial matters like income and payments) is embraced by Millennials. In an American Public Media "Marketplace" interview with PayPal CEO Dan Schulman (owners of Fintech services like Venmo) cited that 90% of Venmo users tag their payment with a comment present in some of their timeline feeds – completely optional features unique (Rysdal, 2017). Fintech businesses will continue to take chances and experiment with such features in an industry where new product wrinkles traditionally develop glacially.

Table 1 lists some notable EU Neobanks.

Table 1. Notable EU Neobanks

Company	Headquarter Country	Features
Atom Bank	UK	Transactions in British Pound. First Neobank to receive BOE license to operate solely as a mobile bank but has no credit card issuance capability
Number26	Germany	Transactions in Euros. Issues MasterCard credit cards through partner bank Wirecard Bank. Withdraws can take place in shops and pharmacies with confirmation in SMS text
Nemea	Malta	Currencies used include the Euro, Dollar, Pound Sterling, Canadian Dollar, Australian Dollar, and Swiss Franc. Rated one of the easiest processes by which to open account and holds banking and investment licenses. Requires a €50 transfer from another account and a 2 day delay. Supports 31 countries.
Fidor Bank	Germany	Utilizes PostIdent feature requiring ID verification at a post office. Issues MasterCard credit card. Offers "smart account" with 25 functions including but limited features in UK as well as social media space to share financial advice and information with others.
Sberbank Direct and VTB Direct	Germany	Holds banking license in Germany. Requires bank account and mobile number with online application and video call to verify identity. Direct Pay systems can do direct transfers to Russia but available only to German residents.

Source: Banks, 2016)

Challenger Banks

Unlike Neobanks that pick and choose distinct banking lines to operate, challenger banks compete directly with legacy banks like the major "high-street" banks (i.e. Barclays, HSBC, Lloyds, Nationwide, Santander, and RBS) (Lyons, 2017). Challenger Banks were ushered into being after main regulatory authorities sought new competition into the banking landscape after the Competition and Markets Authority (CMA) investigation.

Challenger Banks offer traditional financial products without the baggage of legacy institutions. Their value to consumers comes by way of superior service and better terms on loans and savings products. Challenger banks focus on enhancing customers' banking experience through valuable easy-to-use surfaces and global, round-the-clock access from a mobile phone or web site. Table 2 shows Challenger Banks throughout the EU as well as available features and services.

As Tesco Bank, Harrods Bank, and Virgin Money illustrate, Challenger Banks are started by companies with a built-in customer base from other services or products. Tesco and Harrods, for instance, are retailers with prior experience with consumer credit, while Virgin incorporates several consumer businesses including

Table 2. Challenger banks and unique features

Company	Headquarter Country	Features
Charter Savings Bank	UK	Offers only savings products
Metro Bank	UK	Launched in 2010 with 36 branches in London and over 360,000 accounts
Tesco Bank	UK	Pays interest on accounts up to £3,600. Account linked to Clubcard reward system
Carrefour Banque	France	C-Zam accounts activated within 10 minutes online with €1 monthly fee
Virgin Money	UK	Virgin Money stores are available for deposit as well as post offices with different deposit limits, and other bank accounts.
Monzo	UK	Mobile-only digital bank with Mastercard pre-paid card partnership. Granted license by Prudential Regulation Authority and the Financial Conduct Authority.

Source: BI Intelligence Staff, 2017

telecommunications and airlines. Another difference between challenger banks and Neobanks regards distribution where, unlike Neobanks' online-centric presence, Challenger banks have the advantages of physical point of services (i.e. retailers' store fronts). Many current Challenger Banks are focused in the UK (see Figure 3).

iBanks

iBanks are a unique breed of Fintech banks with a large internet-centered distribution channel approach. Unlike a traditional dependency on credit spreads between deposits and loans and interchange fees, iBanks use unorthodox revenue sources such as data monetization (including geo-location and geo-nudging) and advertising. The lower cost for iBanks' services may seem an acceptable trade-off for the lack of privacy for consumers by this practice. Smaller in number than other Fintech banks, a few examples of iBanks exist such as the UK's Loot. Focused on students, Loot provides budgeting tools and holds a credit card agreement with Mastercard. SeccoBank's SeccoAura iBank uses tokens to purchase products and services using social media for everyday activities.

Wealth Management

One of the most compelling and fastest-growing Fintech advances in consumer finance is wealth management by Robo-advisor services. Robo-Advising makes

Figure 3. Challenger Banks by Country, 2016 [1]

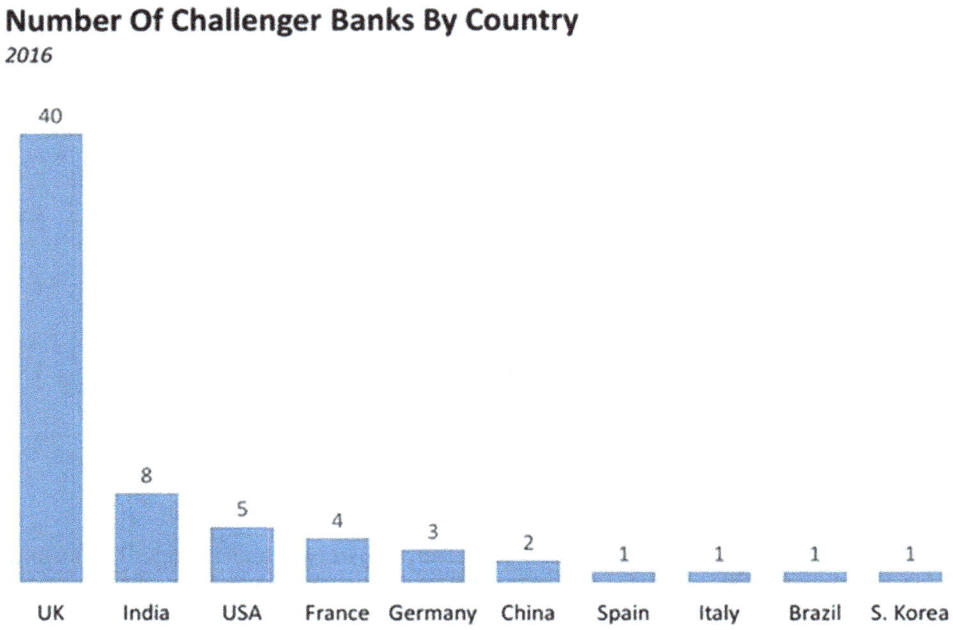

wealth management available for non-affluent customers at a fraction of the cost of traditional services from fund managers (typically 1-2% per annum) without large up-front costs thanks to multiple technological advances:

- Algorithm-based trading techniques and analytics incorporating: (1) a wide history of varied trading strategies across multiple asset classes (e.g. bonds, securities, commodities, and futures); (2) clients' detailed profiles (including need-dates, return objectives, and risk tolerances), and; (3) macroeconomic trends and business cycle data.
- Specialized features such as: (1) aggregated view of multiple accounts and assets; (2) the ability to emulate other clients' investment strategies/portfolios; (3) internet trading via desktop and mobile, and; (4) trading around the clock for all exchanges.
- Low-cost assets such as ETFs, stocks, bonds, and no-load mutual funds.
- Transaction cost reductions from over $0.25 per share in 1980 to $0.03 per share in 2003 (Pisani, 2010). The drop in transaction costs is due to in large part to technology and policy including: (1) the decimalization of securities trading in the US in April 9, 2001; (2) the advent of the DOT (Designated

Order Turnaround) and SuperDOT systems in 1976 and 1984, respectively, which bypassed floor traders in routing orders; (3) the SOES (Small Order Execution System) expansion by NASDAQ after the 1987 crash.

Most wealth management services develop algorithms managing client portfolios based at least in part on Modern Portfolio Theory, made famous by Yale University's David Swenson. The need for alternatives to traditional pensions and asset managers for long-term savings is not just performance-based. Rather, it is due to factors including management fees/commissions, asset allocation, investment manager choices, asset performance, financial repression, and taxes (Better Finance, 2016). By some estimates, the Fintech wealth management industry is projected to manage 10% of global wealth over time (Rubini, 2017). However, the true figure may well be less due to the share of major institutions/investors such as pension funds, governments/sovereign wealth funds, and private companies and corporations who pursue active and alternative investment options with maximum returns beyond the realm of Robo-Advisors. What remains is the individual retail investment sector, of which 10% may be achievable long-term.

There are two primary markets for Robo-advisement. First, in the business-to-business (B2B) space, the Robo-advisor service interacts with an investment manager, advisor, or planner. Second, in the business to consumer (B2C) space, the Robo-advisor reports and/or transacts directly for the investor – the focus of our study. Within the B2C space, Fintech provides two essential services for clients: manual Robo-Advisement and automated Robo-Advisement (see Figure 4).

In the two aforementioned markets, there are four overall stages in Fintech wealth management (see Figure 5). Cost-reducing measures by Robo-Advisors is possible by utilization of complex algorithms and the cost of the products available such as no-load mutual funds and ETFs.

Figure 4. (Left) Manual Robo-Advisor Advisement Service; (Right) Automated Robo-Advisement Service

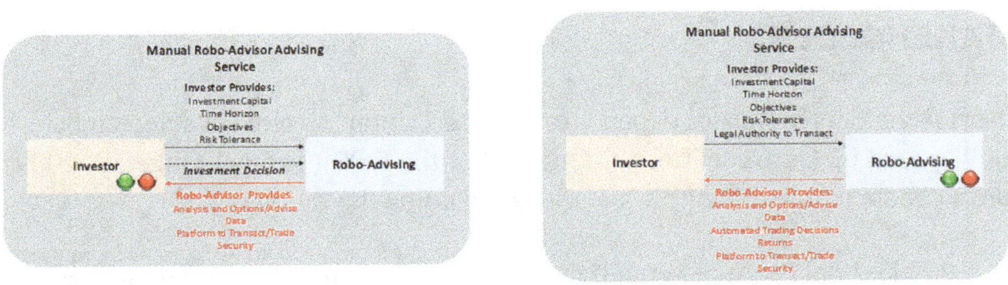

Figure 5. Four Evolutionary Steps to Full Automation Fintech Wealth Management

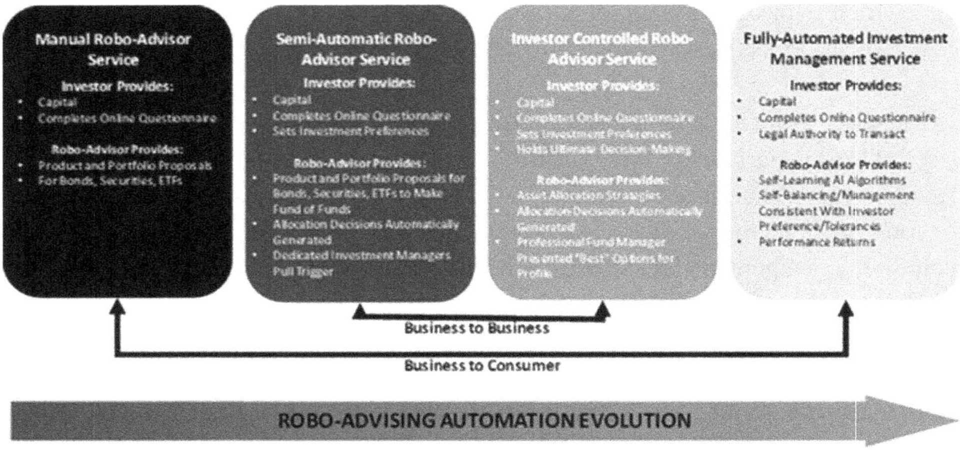

Much of the Fintech focus in EU wealth management is on the B2C space, especially in the UK and Germany (hosts of the biggest exchanges in the EU). Most Fintech wealth management firms charge very small fees based on assets under management (AUM) compared to legacy asset managers and eschew commission-based charges (between 0 and 0.8% AUM). Additionally, they lack fund-creating divisions legacy competitors use as an income stream via cross-selling. New entrants into the German wealth management space (corporate-connected), however, have higher fees (0.9 – 1.4%) than other EU counterparts – perhaps due to overhead from their parent companies, or perhaps due to the realization that the fees may betray the notion that higher costs are necessary.

In the US, protests over proposed US Department of Labor Fiduciary Duty standards for financial advisors resulted in a wide shift to charge on an AUM basis rather than on a commission basis. Despite initial protests, the shift has proven more beneficial to asset managers than previously expected. Table 3 shows a list of some Robo-Advisors and some EU notable features.

STAKEHOLDERS

The stakeholders in finance depends on the transaction. As cited in a September 21, 2009 speech by Gertrude Tumpel-Gugerell of the ECB (Tumpel-Gugerall, 2009), banks conduct a broad range of activities including the following:

1. Provide liquidity (money) and payment services through deposit-taking business.

Table 3. EU Robo-Advisors

Company	Country of Headquarters	Features
Wealth Horizon	UK	Requires £1,000 to open an account
Vaamo	Germany	Only for German residents with future expansion plans. Goal-alignment and projection services provided.
Nutmeg	UK	Helps clients invest towards goals (including weddings, retirements, homes charging low robo-advisor fees. £500 minimum to open account
Stockspot	Austria	Founded in 2013 with low account opening barriers. Offers diversified assets to create dynamic portfolios.
MoneyFarm	Italy	Advises clients to invest in ETFs.
OwlHub	Germany	Recently acquired by LIQID Asset Management. Focuses on small accounts, with deposits made at regular intervals from clients' paychecks
Zen Assets	UK	Targeting large investors with assets seeking to earn competitive returns but paying smaller fees than actively-managed funds
Scalable Capital	Germany	€300 million assets under management with clients based in Germany, Austria, and the UK.

2. Transform assets in terms of denomination, quality and maturity, and manage the associated risks.
3. Process information and monitor borrowers using specialized technologies; thus establishing long-term relationships with clients to mitigate the negative impacts of adverse selection and moral hazard involved in resource allocation

The following are the main stakeholders for the savings and wealth management:

Financial Service Clients

Perhaps the most important stakeholders in the Fintech consumer banking services sector are the consumers whose faith and trust comprise the backbone for the financial system. Despite the loss of trust due to actions leading to the US Financial and Euro Debt Crises by legacy institutions, savers may question trust to firms not that didn't participate in those episodes due to lacking a physical presence or track record.

Characteristics of the modern-day bank branch seek to gain a sense of trust and exude professionalism to clients: stoic attire from employees and formal attire from management; a façade of brick and solid material enveloping the branch; a vault with thick steel doors within view (even if the vast majority of bank holdings are now digitized); lines that snake inside the branch so as to not extend outside the doors of the branch and incite a panic. The question for today when many in the population utilize direct deposit and go without collecting a physical check for his or her wages, or seldom step foot inside a bank branch, is whether these images are still necessary to reinforce the perception of security to consumers. Achieving client trust may be easier for younger generations who largely grew up without dependence on interaction with physical infrastructure and even physical money versus older generations for whom are still evolving to the new financial system. In wealth management, investors may either entrust fully automated Robo-Advisors to make judgement and decisions for them, or to entrust the research of hybrid Robo-Advisors to provide sound and prudent analysis the investors may act upon.

Financial service companies must earn market share from other competitors and reach new segments of the EU population – segments that either lacked the ability to pay for financial services or whose acquisition costs for legacy firms were too high.

Fintech Service Providers and Investors

Fintech companies themselves are major stakeholders. Many of these firms are private companies as of the date of this publication and face unique headwinds including EU-wide and local regulatory compliance requirements. In the consumer banking space, Challenger and iBank stakeholders include partner firms such as Virgin and Tesco with a widespread brick-and-mortar presence and capital lending experience.

Fintech firms are largely startups that undergo multiple investment rounds by a combination of founders (who bootstrapped investments early on), early investors, corporations, and venture capital (VC) – the latter seeking investment returns on the order of 1000% (10x investment). VC investors typically seek exits to their investment through sale to another investor or an initial public offering. VC obtains capital from: (1) affluent individual investors or limited liability partners (LLPs); (2) asset management funds; (3) private equity funds; (4) corporate players in the financial industry such as large banks such as Deutsche Bank, Banco Santander, and other high-street EU banks.

Legal and Regulatory

The financial services industry is one of the most heavily regulated industries due to the importance of functioning capital markets in economies. As far back as 33 A.D. in the Roman Empire when Emperor Tiberius responded to a massive bank run with a 3-year interest-free loan to shore up the nations' banking system (Calomiris, 2010), governments have quickly responded to stand by their nations' financial institutions. Unsurprisingly, the ECB reacted swiftly and overwhelmingly to save its financial system after the Global Financial and Euro Debt Crises. The EU regulatory framework includes regulators like the ECB (which regulates the largest banks in Europe principally), EBA (European Banking Authority), EC (European Commission), and various local agencies within EU member states that regulate the non-major high street banks. The payments system must conform to MiFID2, PSD2, and TARGET2 requirements while broad savings-based regulations such as KYC (Know Your Customer) and anti-money laundering regulations impose compliance rules and processes for all institutions. EU regulators seek harmonization via common processes to reduce aggregate costs in the overall system and provide requisite protections for firms and consumers alike.

The ECB works closely with the largest, most systemically important EU financial institutions. The ECB grants and withdraws licenses, conducts supervisory reviews and on-site inspections, performs investigations, and sanctions firms that violate rules. The ECB assesses the acquisition and disposition of bank holdings and ensures harmonization of EU rules and processes and confirms compliance with Basel III and other capitalization requirements. Overall, the ECB directly supervises 124 significant banks in the EU (all together these 124 banks hold 82% of banking assets in the Euro area) (European Central Bank, 2017). Besides the ECB, the SSM (Single Supervisory Mechanism), is based out of Frankfurt, Germany and led by the ECB. The SSM aids in the stability of the EU banking system and the broader economy, and works closely with national supervisors of countries using the Euro. EU-28 nations who don't embrace the Euro may subject themselves to work with the SSM at their discretion. The SSM provides day-to-day supervision, ensures regulated banks comply with harmonized EU banking rules, helps tackle problems early on, and ultimately determines whether bank is fit to operate (European Central Bank, 2017).

The financial system favors entry of new competition in the industry given Competition and Markets Authority (CMA) findings over the heavy concentration of assets in 1/5th the number of institutions that existed 50 years ago – concentration exacerbated by financial crises affecting the EU. BBVA, for instance, was product of the collapse of several Spanish *cajas* merging into the largest bank in Spain; Lloyds was rescued by the BOE through capital injections but exited the crisis healthier and more competitive than firms that didn't require such intervention. Ultimately to combat the hyper-consolidation in the banking sector, 17 new bank charters were issued in the UK alone from 2004 to 2017.

While much focus is on banking, EU regulators pay increased scrutiny to Fintech in wealth management as well (Binham, 2015). A joint EBA-ESMA (European Securities and Markets Authorities) discussion paper stated "potential risks include the possibility that consumers could misunderstand advice provided to them without the benefit of a human adviser to support the advice process, and the potential for limitations or errors in automated tools that may not be easily identifiable for consumers of financial institutions" (EBA, ESMA, and EIOPA Joint Committee, 2015). EU regulators are continue to compile public comment even after codifying measures such as MiFID 2 affecting securities and broad areas of banking.

Legacy Financial Institutions

Legacy financial institutions are aware of the rise of Fintech in finance, especially with respect to the long-term impact of technology on the industry. Jamie Dimon's 2016 JP Morgan Chase shareholder letter referred to JP Morgan Chase as not just a bank but a technology company, having spent $6.5 billion on technology in FY 2016 - $3 billion dedicated to new initiatives and $600 million allocated to fintech and partnerships with Fintech businesses and organizations. However, given the diverse business lines in a major legacy bank, much of banks' IT expenditures regard bolstering existing infrastructure and providing incremental improvements in functionality. Legacy firms' capex is nudged by Fintech to new areas in savings, payments, and even blockchain technologies. Due in large part to the length of the RFC (request for comment) and review process to adjudicate and develop new regulations, legacy financial institutions focus on influencing regulation, compliance, and lobbying efforts. The resulting ECB's wait-and-see approach regarding blockchain and delays in roll-out of the MiFID-2 rules and TARGET2 payment system illustrates the effectiveness of regulation lobbying to hold Fintech threats at bay.

Legacy financial institutions traditionally produce innovations more slowly than other industries. Introduction of new technologies such as the credit card, ATMs, and internet banking/trading had long roll-out periods. The decision by Schwab to incorporate mobile trading in addition to its financial advisors in the 1990s was made after much hand-wringing and research after the rise in internet trading companies (McFarlan & Tempest, 1999/2001). In finance, a wait-and-see approach is often preferred over establishing a first-mover position, especially since technology and coding is often easily replicable. In fact, legacy financial institutions outsource innovation by providing investment capital to Fintech firms.

Mobile Carriers and Major Retailers

Mobile carriers are became EU Fintech providers thanks to Sir Richard Branson's Company: Virgin. Mobile service providers have advantages that could extend to banking services:

1. **Existing Client Base:** Mobile carriers have a large existing client base. One of the most difficult and costly tasks for Fintech firms is new client acquisition and retention.

2. **Trust:** A difficulty Fintech firms encounter is establishing trust to initiate an intimate customer-service relationship. Mobile carriers already established credibility to provide customers the service of communication and customers have generally accepted the mobile carriers' ability to charge a fee for service; many reliably with auto-pay.
3. **Data:** Mobile carriers have long-demonstrated a capacity to use big data to their advantage. Carriers have links to customer bank accounts for payment, knowledge of data usage, and geo-location data at their disposal – an important feature for monetizing data.

Retailers possess advantages for banking including multiple points of sale and brand awareness that pure-digital Fintech companies don't have. Retailers' market position and instant rewards programs present an opportunity to induce clients to use the service and increase the volume of activity. Ultimately, retailers pose a significant threat to legacy financial institutions in the EU savings and lending spaces.

FINTECH OPPORTUNITIES AND BARRIERS

Market Opportunities

The Fintech industry holds opportunities to benefit consumer savings and investment. Having earlier identified the major stakeholders, we shall discuss market and demographics opportunities affecting the savings and wealth management spaces.

EU Consumer Finance Demographics

European population and consumer demographics provides Fintech opportunities to grow a clientele base. The EU savings financial services market it is a tale of two EUs – the more aged population (controlling the greater share of wealth) and the millennial population which is driving much of the current gains and consumer growth for Fintech firms. Fintech adoption by Millennials is driven by politics as much as history and an intimate ease and familiarity technology. Those born during the 1990s did so after the fall of the Berlin Wall, the rise of democracy, globalization, and freer communication throughout the farther reaches of post-Soviet Eastern Europe. This generation has not felt first-hand price instability of past generations. In fact, the only peoples of the EU traumatized by hyperinflation are the Germans for whom the images of Germans carting around piles of Deutsche Marks in wheel-barrows or Germans wallpapering their homes with worthless paper currency are ghosts: vivid, haunting illusions. This Millennial generation comes of age with the rise of the internet and the blurring of European national borders. However, as deficits force austerity in countries like Ireland, Greece, and Portugal after the European Debt Crisis, Millennials will bear the cost to the excesses and concessions enjoyed by prior generations. Polls in like the United Kingdom leading into and after the Brexit vote of 2016 showed the disparity in posture toward the European Union for the millennial generation versus older generations. Millennials sought out tighter EU integration, freer travel, and freer flow of information versus more protectionist postures by older generations. Time Magazine polls prior to the Brexit vote reported 19% of voters from 18 to 24 supported leaving the Eurozone (72% favored remaining). Meanwhile, 59% of pensioners favored "Leave" (34% favored remaining) (Shoster, 2016).

Adoption for Fintech is arguably as much for the ease and convenience as it is for the inherent cost-based advantages towards savings and retirement investment. The appeal to Fintech firms is due to conditions on the ground: dovish monetary policy and low bank and sovereign bond yields (denoted by the extended negative real rates of return on Swiss and German sovereign debt) eroding the effectiveness of prudent, traditional income generation. Figure 6 shows the average savings account interest rates, lending rates on bank loans, and loan rate spreads as reported by the World Bank for member nations. Average rates for savings interest dropped precipitously since the Euro's creation from 5.9% in 1999 to 0.7% in 2016 while lending rates dropped from 18.8% to 9.5% and lending rate premiums dropped from 10.4% to 5.7% over that time.

In wealth management, Fintech provides a valuable channel of distribution for high-liquidity/low-cost investment products like exchange-traded funds (ETFs) that track broader indices and require little management and administrative burden. As John Vogel and Warren Buffett demonstrated during interviews in March 2017, fees charged by some mutual funds demonstrated an annual loss of wealth for investors of over $100 billion. Non-value-added costs transfer wealth from savers/investors to intermediaries in the system, paving the way for low-cost alternatives like Robo-Advisors.

Immigration

Immigration throughout the EU is encouraged by uniform passport and travel policies within the Eurozone. Uniformity in economic metrics like growth, inflation, and deficits harmonize prosperity and the need for economic stimulus. Immigration is largely due to pursuits of education, family-related reasons, or labor disequilibrium due to downturns in economic cycles. A vibrant economy depends on an influx of young immigrants or a sufficiently high birth rate to positively influence growth, market size, and innovation. Besides the economic prosperity and geo-political stability of

Figure 6. Average Savings, Lending, and Lending Premium Rates

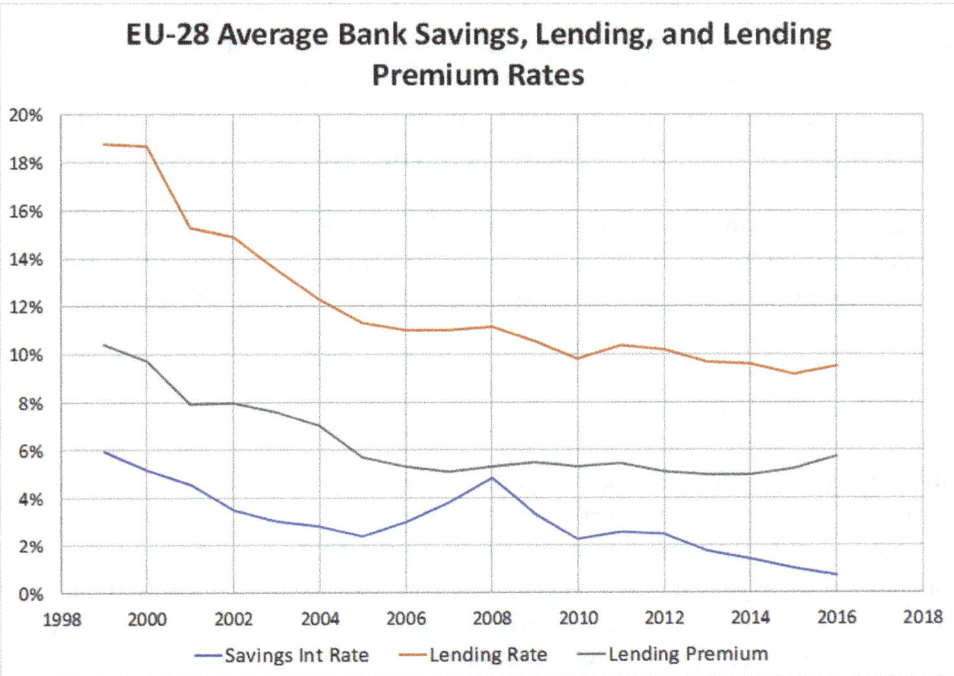

the EU since its inception, the proximity to other politically instable regions such as the Middle East and N. Africa led to its designation as a place of opportunity and refuge from war and strife. The number of political asylum applicants increased from 2008 to 2015, although a decline was seen from 2015 to 2016. The largest nations of origin for asylum-seekers have come from Syria, Iraq, and Afghanistan (Eurostat, 2017). In January 1, 2016, 20.7 million EU-28 residents were born in another EU-28 country while 35.1 million EU-28 residents were born outside the EU-28. The largest concentration of these "non-nationals" were in Germany (8.6 million), the UK (5.6 million), Italy (5 million), Spain, and France (4.4 million each) while more than 5% of the populations of Luxembourg, Cyprus, Estonia, Latvia, Austria, Ireland, Belgium, and Germany consisted of non-nationals. A majority of EU immigrants were under 40 years of age with a higher concentration of young immigrants coming from non-nationals (Eurostat, 2017).

Immigration impacts financial for two main reasons. Increased population from immigration provides both a larger market size and a source for the need for money transfers to nations of origin. Where immigrants comprise roughly 8% of the overall EU population, the non-national portion of the immigrant population presents a need for savings, payments, and money transfers to require currency conversion to local fiat currency. It's worthwhile to note the reasons immigrants site for migration: 27% cited economics and employment, while family and education comprises 29% and 20% of reasons cited on applications, respectively. The "economic" justification increased from 24% in 2014 (Eurostat, 2017). The prosperity of the EU versus their home economies provided the economic justification motive accounting for 28% of migration to the EU in 2008-2010 (the timeframe of the Financial and Euro Debt Crisis).

Static brick-and mortar bank branches offer limited value to immigrants. Movement across borders presents a need to store value with mobile access for payments and access to funds. Given migration statistics, the EU poses an ideal market for Fintech companies given service portability.

Religion and Banking

Shifting demographics in the EU with regards to religion presents another benefit and opportunity to Fintech firms. The Pew Research center projects a declining population in Europe overall between 2010 and 2050. While the Christian and Jewish populations are expected to be in decline, other religious groups within the region are expected to grow due to higher fertility, younger populations, and immigration. The Muslim population in Europe is projected to grow 63%, from 43 million in 2010 to 71 million in 2050. The median age of Muslims in Europe is 32, where 67% of the religious group is between age 15-59 and another 22% are younger than 14. A high fertility rate (2.1 children per woman) insures the need for greater income and savings for such needs as education, home ownership, and other items related to family formation (The Pew Research Center, 2015). The World Bank Findex Survey cites that 25% of the population without an account in predominantly Muslim nations such as Niger, Uzbekistan, and Turkey cite religious concerns (regarding interest) as reason for not having an account (Demirguc-kunt, Klapper, Singer & Van OUdheusen, 2015). Muslim immigrants to the EU may hold religion-based aversion to bank accounts for that reason. Fintech banks are malleable-enough to provide Sharia Law-compliant features without a major disruption to overall operations by facilitating special savings accounts designed to steer clear of both interest and its utilization for interest-bearing purposes. Besides outreach, accessibility, and customization, Challenger banks offers the opportunity to help introduce a large demographic to banking and provide benefits such as wealth through savings and home-ownership by utilizing rewards instead of interest for savings and purchase activity.

EU Unbanked and Underbanked Population

The World Bank and institutions like the ECB, European Commission (EC) and the International Monetary Fund (IMF) are concerned with the world's unbanked and underbanked population. Research concludes access to banking and credit improves the lives of individuals and helps raise people out of poverty (Zhuang 2009 (Akanji, 200201 Bossone & Sarr 2002, Khan 2013). In fact, the advent of micro-credit helped illustrate the benefits of access to credit and financial services toward improving one's opportunities for economic ascent (Akanji 2008) (Akanji, 2002).

Despite the maturity of many EU economies, a considerable population is unbanked or underbanked – especially the young, those in Eastern EU nations, women (a gender gap of 7% exists between unbanked women to men), the poor, unemployed (81% of those earning wages have an account while 53% of the unemployed have an account), and the elderly. As Figure 7 shows, the unbanked portion of the EU is relatively low at 13%, but nominally totals almost 55 million potential consumers. Firms like MasterCard (with roles in payments and money transfers) publish figures for the unbanked and underbanked people in Europe of 138 million individuals – including 12% in the UK[2].

Figure 7. Unbanked Population % Throughout EU-28[3]

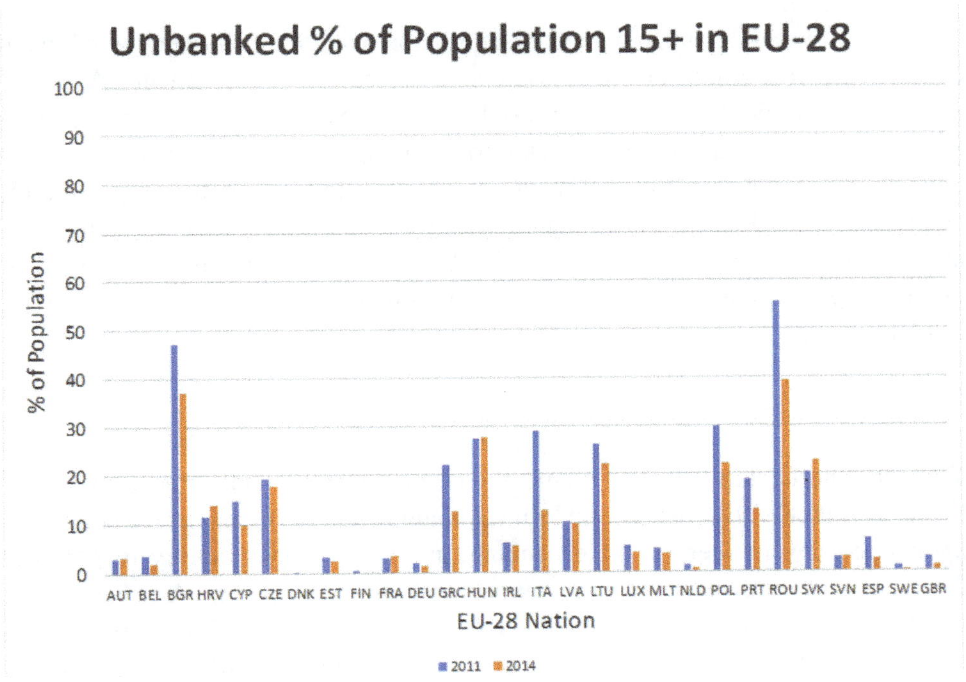

The World Bank Global Findex Survey reported that, globally, 59% of those surveyed who were "unbanked" cited "a lack of money" as a reason why they do not open an account while 19% cited "a lack of money" as the only reason. 30% of survey-takers cited "no need for an account" and that "another family member already has an account" as reasons. In Europe and central Asia particularly, 30% of respondents cited lack of trust as the primary reason and overall while 5% cited "religious reasons."

The Findex Survey Report credits digitization of payments/wages, agriculture payments, and remittances as a primary means of banking the unbanked population that can be accomplished through initiatives at the government level. Given the high levels of the European population already in the formal banking system, gains by introducing new participants to the formalized financial system would be small percentage-wise. Other approaches to introduce unbanked people to the formal financial system include education and improving the payments infrastructure (Demirguc-kunt, Klapper, Singer, & Van Oudheusden, 2015).

Most Fintech firms in savings and wealth management focus on the more developed EU economies. As Figure 7 illustrates, the percentage of the population that haven't been induced by the legacy financial institutions to-date range between 1% to 4% of the population for the largest EU economies such as France and Germany, while nations like Italy, Greece, and Eastern EU nations have unbanked populations over 10% in 2014 – all ripe for Fintech firms with novel means of both outreach, access, and engagement.

In the efforts to bank the unbanked population using Fintech, literacy will play an important role amidst new disclosure and transparency requirements. A World Literacy Foundation's study cites lack of literacy for costing the EU economy €350 billion annually for reasons such as higher consumer spending and lost wealth creation opportunities (Elinet, 2015) despite an EU literacy rate at 99.1% for people age 15 and above in 2010 where women have a slightly lower literacy rate than adult men at 98.9% (World Bank, 2017). The high EU literacy rate provides Fintech firms opportunity to acquire customers and provide added transparency and information availability.

Technology

While internet banking has existed since the 1990s, the recent boom in access was due to the advent of APIs (Application Programming Interfaces), mobile and computer application development, mobile banking, and smartphones' ability to perform functions that were once solely the domain of desktop computing. Another source of growth potential for Fintech is higher-speed 4G LTE connections in the EU. A 2015 GSM (Global System Mobile Association) study reported 78.5% of Europeans are unique subscribers with 20% of those comprised of LTE connections in 2015. Penetration is projected to be 81.4% in 2020 with 93% of those 4G LTE connections. Data usage is projected to grow from 0.9 GB per person per month in 2014 to 5.7 GB per person per month in 2019 while smartphone percentage of mobile phones is projected to grow from 17% in 2010 to 76% by 2020 (GSMA, 2015).

Many EU member nations have relatively low costs of communications expenditures as percent of GDP, on par with the rest of the developed and mobile-heavy nations worldwide (see Figure 8). Low costs provide added opportunity for customer outreach and usage.

API is critical to the development of Fintech applications integrated toward payments, money transfers, investments, and open banking. As Fintech companies pick and choose services that offer greater margins, there is a benefit to seamlessly integrating their services with those of other firms like legacy banks. Doing so lends Fintech companies credibility and helps the Fintech company appear as value-added “complements” as opposed to “either or” consumer choices. A compelling interface will promote ease of use and likely increased transaction activity. Given the desire to provide integration of finance among nations with disparate local regulations requirements and the fragmentation between mobile phone operating systems, API-based design is critical.

Fintech technology is not solely concentrated on interfaces or even through streamlining operations and back-end functional development. Artificial intelligence (AI) provides potential competitive advantages to savings and wealth management alike. Combined with big data learning, AI provides additional security measures for mobile banking features like payments. Analysis of consumer habits and data can be monetized as an additional source of revenue, as cited in an APM (American Public Media) Marketplace interview of PayPal CEO Dan Schulman (Ryssdal, 2017).

Figure 8. Communications Expenditures as Percent of GDP

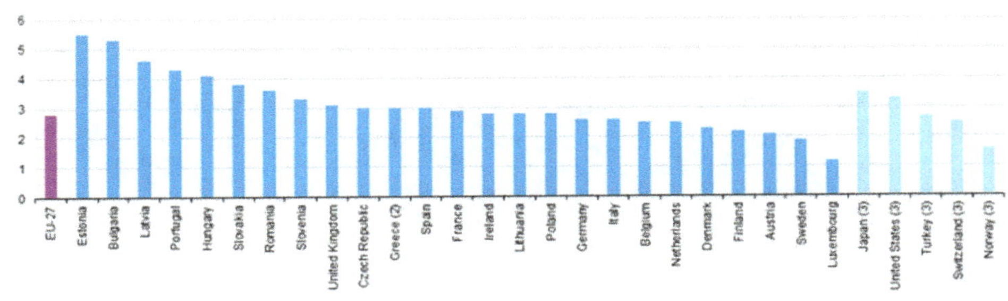

(¹) Croatia, Cyprus and Malta: not available.
(²) Provisional.
(³) 2008.
Source: Eurostat (online data code: isoc_tc_ite), European Information Technology Observatory (EITO)

Neobanks, essentially, provide a test bed on financial service innovation for legacy firms. Yet one of the adverse effects to the demonstrated ability to wield software to produce additional services effectively is enhance an easy-to-replicate toolset with diminishing durability of competitive advantages for legacy banks play catch-up. Artificial intelligence and big data learning is increasingly used by legacy firms such as Santander, HSBC, and Barclays who have explicitly called out the need to develop more organic Fintech capabilities in annual reports.

EU Net Savings

In recent years, net savings in the EU has decreased – largely as major EU economies fought economic slowdowns and nations with net negative household savings joined the EU (see Figure 9). As the OECD net EU household savings/borrowing data shows, savings rates in the EU improved with the introduction of the Euro ushered economic prosperity. However, the 2008-2009 Financial crisis and the European Debt Crisis in 2010 prompted a reversal in net savings rates EU-wide. As EU economic conditions improve due to ECB intervention, however, added opportunities exist for Fintech companies in savings and investment.

Figure 9. Net EU Household Saving/Borrowing as Percent of GDP

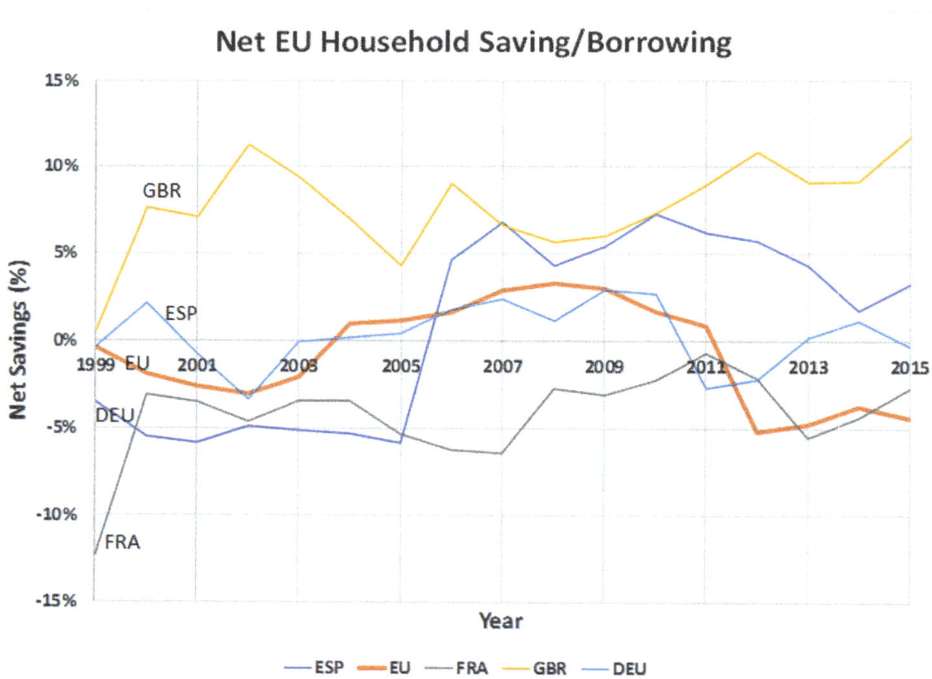

Legacy Firm Complacency and Consolidation

The Fintech industry benefits from consolidation and concentration in the EU's financial industry despite the consumer market share and regulatory influence power legacy institutions wield. In 1960, there were 30 major banks throughout the EU. Today, that number dwindled to 6 major firms. The steady decrease in the number of MFI (monetary financial institutions) over the years restricts option and accessibility to everyday consumers; particularly the youngest and oldest members of the population (the latter comprising the largest demographic). The strength and size of legacy institutions spurred regulators to encourage competition by issuing licenses at an accelerated pace to Neobanks, Challenger banks, and iBanks that can demonstrate the compliance, infrastructure, and profitability acumen to successfully usher in desperately-needed new competition to the space.

Customer dissatisfaction with legacy banks is a major source of opportunity for EU Fintech banks. The largest UK's bank has an NPS (Net Promoter Score rating by Medallia determining the likelihood of promoting a product or service to others) of -24 while the largest German banks have NPS of -8 and -22. Due to dissatisfaction and initiatives after the CMA findings, 18 digital-only banks opened in the UK since 2015 (e.g. Atom, Monzo, Starling, Tandem, and Tide). Meanwhile, the ECB reports steady decreases in the number of bank branch units within the EU; 232,525 branches existed in 2009, dropping to 198,744 by 2014 (Nowodazkij, 2014). Surviving branch locations are increasingly concentrated in major urban centers as assets of credit institutions decreased from €3.041 trillion to €2.289 trillion over that time frame (European Central Bank, 2014).

Loyalty and Consumer Perceptions and Use of Fintech

An Innofact AG customer survey commissioned by Mastercard reported on digital banking services users (Wolff, 2017). Mastercard's role as the partner for 30 digitized banks (including N26, Monzo, and Holvi) throughout Europe offers insight about transactions and mobile banking users. "Convenience" and "ease of use" were polled to be one of the most significant advantages of mobile banking (claimed 70% and 59% of over 11,000 consumers surveyed) with 90% having a positive view of the role of digitization for their banking solutions. In that survey, an important theme was the perception of banking as a commoditized service/product. 49% of responders considered switching to a digital bank in the future. German customers are particularly disloyal to current banks with 27% considering switching in the next 12 months – the highest in the EU. Netherlands customers, conversely, were most loyal to their existing banks with only 7% considering changing their banks and 52% preferring to remain with their current banks. It was noteworthy to consider the variance in transaction frequency by country. For EU customers who have bank accounts, 85% of consumers bank digitally (online or mobile) with 38% transacting or accessing accounts at least weekly or daily. France, at 62%, had the lowest digital banking usage rate while Spain at 92% had the highest. 42% of banking customers cited availability and geographical accessibility, 27% said flexibility/ time convenience, and 24% cited security and real-time information as reasons for bank service selection considerations.

Loyalty and trust are important to Fintech firm for client acquisition and retention. Research by Frederick F. Reichheld in 1993 showed the effects of customer retention on profits, using the MBNA credit card as an example where increasing customer retention by 5% had an outsized 60% increase in profits (Reichheld, 1993). Reichheld's 1993 paper discussed investments, differentiating between potentially loyal customers to those the business considered "lost." Big data analysis performed by Fintech firms include metrics like account access frequency, levels of investment and assets under management, and deposits for Robo-Advisor wealth management. Measures of customer loyalty by Fintech firms according to Reichheld (1993) include understanding the cause-and-effect relationships to deliver superior value to customers. Customer loyalty is best quantified by retention rate, market share, or both. Customer loyalty second-order effects include revenue growth due to customer sales and word-of-mouth referrals. Positive engagement leads to declining costs for customer acquisition and retention while decreasing churn and reducing idleness of accounts that increases per-client costs.

Legacy banks and Fintech firms recognize evolution in consumer perception of digital-only/digital-dominant banking and the opportunities mobile banking provides. The value proposition of those benefits to consumers is measured often through consumer surveys from companies such as PwC. In a PwC UK survey, 10% of consumers believed Fintech firms' offerings are more attractive than their existing banks, 29% believe the big banks' offerings are more attractive, 34% thought the offerings of both class of banks were equal, and 27% were unsure.

Challenger Banks with ties to retail companies have strong, trusted brands, and serve the needs of customers loyal to the parent group as a whole. Examples include Tesco bank; Sainsbury's bank, and Virgin Money (Lyons, 2017).

A MasterCard survey on digital banking use (Thorpe & Keal, 2017) showed that 88% of UK consumers use digital banking – the highest percentage in the EU – compared to 62% in France, 54% in the UK customers use the service daily, and 28% in Italy. 70% of EU survey responders cited time savings as the primary advantage for digital banking, while 59% cited ease of use. In the UK, 56% cited safety as a primary reason to opt for digital banking compared to 21% of Dutch responders. Overall, almost 60% of survey takers welcome mobile banking and believed the benefits were essential. While 14% of UK residents would switch to a new Fintech bank, 70% would elect to remain with their current banks and wouldn't entertain a new bank. A TopCompare survey showed Belgian customers don't take digital tools into account when selecting a bank, instead basing selection on proximity and historical/traditional preferences. A UK survey by PwC and YouGov (see Figure 10) determined 65% of consumers were not likely to switch to Neobanks while 18% were at least fairly likely and 17% were unsure.

A PwC study on EU Challenger banks[4] discovered 68% of consumers banked at 1 institution while 32% banked at multiple institutions whereby institutions are sought for specific purposes as opposed to replacing other institutions. Of those consumers willing to bank at multiple locations, 54% of respondents would choose a bank for the features and products they desire, 30% would do so for a one-stop-shop capability, and 16% were unsure. Of UK consumers, however, only 3% chose to utilize the Current Account Switch Service to switch banks while 85% of those who did elected to move to another high-street bank. Consumers did business with additional banks for various reasons: 59% did so for mortgages, 53% for personal loans, 53% for credit cards, 42% for cash ISA accounts, and 30% for savings accounts. Multiple surveys from various organizations throughout the EU indicate that challenger banks faced difficult prospects meaningfully eradicating legacy banks'

Figure 10. PwC and YouGov UK Survey on Switching to Neobanks

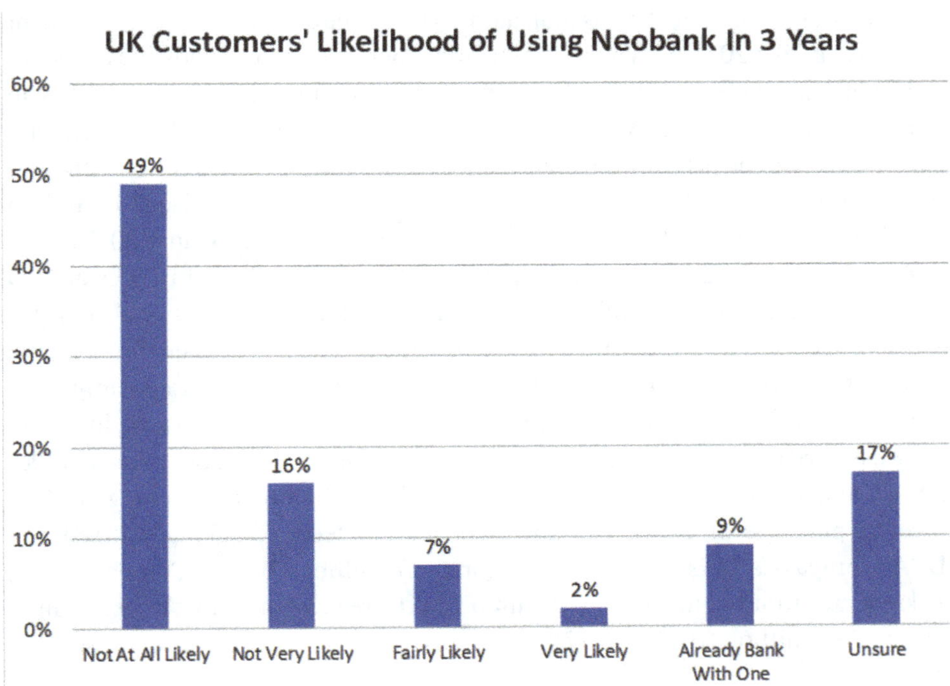

market share outright. Growth in the Fintech banking sector requires "growing the pie" by providing new services and features that customers need but cannot acquire from their legacy service providers.

Fintech Investment

While other metrics are better indicators of the effectiveness and viability of the Fintech's wealth management and retail banking sectors, equity raising, market valuations, and M&A activity are still valuable metrics. Typically, new rounds of investment help grow the business, achieve other value-added objectives to the company (albeit typically at losses in the short run), and provide several months of "runway" (breathing space) for businesses to operate. Another important metric is the source of new investment capital. In many cases, EU banks and firms like department stores and mobile phone operators provide capital for Fintech firms where returns on investment expand beyond revenue sharing to include a front-row view into marketing and operations. While the actual customer data may be proprietary to the Fintech firms, an investment provides a deterrent to would-be acquirers – especially if the investment is large enough that control would be difficult (such as 40%-70% equity stakes).

The European Fintech space in 2016 remained robust despite year-over-year declines in both volume of deals and invested capital from 2015 (see Figure 11). Years with a large number of deals indicate a high volume of smaller-sized investment rounds. Years like 2014 with an outsized large volume of capital indicate a smaller number of deals often result from major companies taking large equity stakes in latter rounds or engaging in M&A. Among the 2014 deals was the BBVA Compass acquisition of Simple (a European mobile banking platform). SoFi (Social Financing) bought Zenbanx for $100 million in February 1, 2017 after having raised $10 million from China's Tencent Holdings and Recruit Strategic Partners (Wang, 2017).

Many Fintech firms seek an initial public offering (IPO). Not only does an IPO offer liquidity and an opportunity for investors to cash out, the additional equity is important serves as a buffer against loan losses and as a means to scale the business. Many Fintech firms in the EU indicated IPOs were pending. UK-based Atom bank announced a $125 million equity investment in early 2017 from existing investors after raising $160 million from hedge fund Toscafund and majority shareholder BBVA Compass, with plans to raise another $125 million before seeking an IPO. However, purchase of Challenger and Neobanks by legacy banks are mixed: the BBVA Compass acquisition of Simple for $117 million (BBVA Compass Staff, 2014) equated to an acquisition cost of almost $1100 per customer that's been written down by $90 million (Barba, 2017).

Figure 11. Fintech Investment in Europe, 2010 to 2016, Source KPMG International

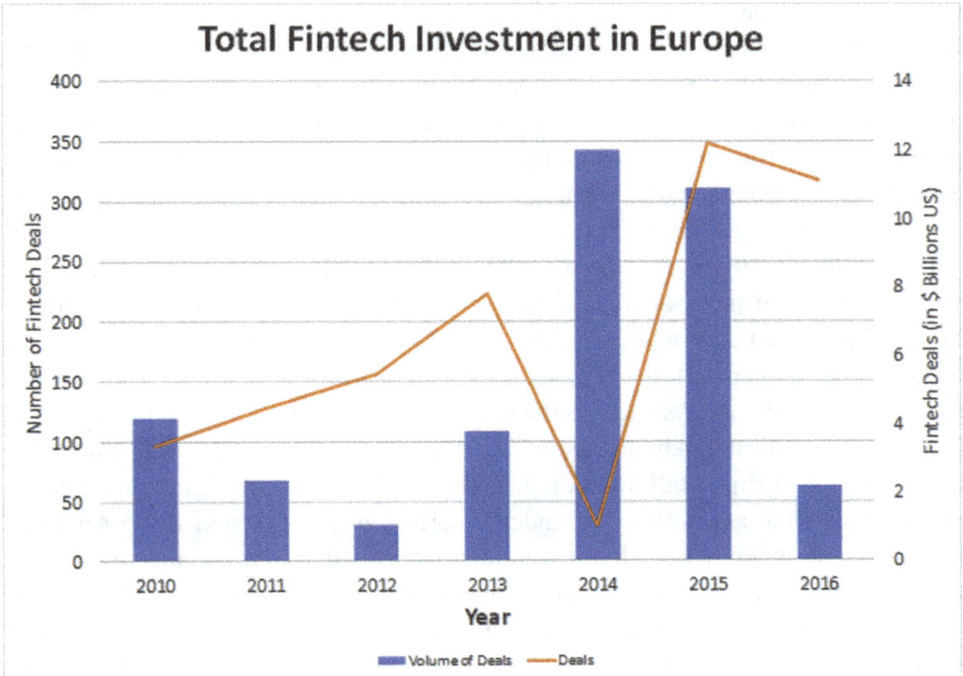

Fintech Operations

The Fintech savings space will be influenced by new regulation in the EU such as MIFID 2 on January 3, 2018 which impacts a diverse range of banking and securities operations. An important consideration for Fintech firms is the acquisition of new clients, spurring engagement, and inducing client activity. A client with limited numbers of deposits and schedule of set of payments typically doesn't require a much engagement with the bank. If the savings feature is supplemented by budgeting, cross-account solvency level data, and payments, the need to engage with the Fintech interface increases. Providers of content to Netflix fought to have their studio logos appear at the corner of the screen since, absent of this feature, viewers would associate Netflix – not the studio – with the content. Absent of reminders to who manages customer accounts and transactions, Fintech firms relegate the bank to back-office status while lending the banks' security and protections to the Fintech firm.

Robo-Advisors see a path to gaining traction by communicating its lower costs to potential consumers. Robo-advisors' makes wealth management affordable. Formerly, wealth management was traditionally a service that was previously the domain of only the affluent. Of course, performance is critical to influence word of mouth and comfort among users – not only for participation in a hybrid B2C model, but eventually to adoption of the automated service model. Comfort level will likely increase through an increase in the presence of artificial intelligence in consumers' daily lives. However, comparative underperformance among other users of hybrid Robo-Advisement service versus automated Robo-Advisors may lead to a preference for full-automation. Past performance by many retail investors may point to such an outcome: as famed investor Benjamin Graham argued, temperament is just as important as technical acumen in outperforming one's investments. Graham's definition of investment is: "the practice upon which thorough analysis promises safety of principle and a satisfactory return; all other practices failing that criteria is by definition speculation and should be avoided." (Hagstrom, 2013). Robo-Advisors, utilizing a customer profile defining satisfactory returns and utilizing big data learning to perform security analysis leaves the task of transacting to the client in the hybrid B2C model. Unfortunately, as Jason Zweig of the Wall Street Journal references in his writings, many retail investors follow a herd mentality in market downturns and upturns alike and ultimately underperform. Fully-automated Robo-Advisors lack the emotional hindrances and reaction to stimuli that may end up consistently outperforming their hybrid B2C counterparts to further wider acceptance of full automation. As illustrated in prior research, individual investor trading results in systematic and economically large losses, due in part to the data provided by brokers (Barber, Lee, Liu, Odean 2008). In fact, the UC Berkeley research of Taiwanese exchange data disclosed: "that aggregate portfolio of individuals suffers an annual performance penalty of 3.8%. The individual investor losses were equivalent to 2.2% of Taiwan's gross domestic product or 2.8% of the total personal income – virtually all individual trading losses can be traced to their aggressive orders. In contrast, institutions enjoy an annual performance boost of 1.5 percentage points, and both the aggressive and passive trades of institutions are profitable." Research implied that retail investors, lacking the proper temperament and restraint to hold to their investments instead of trading, will transact themselves into losses. Thus, automated Robo-Advisors stand to outperform their hybrid counterparts and induce higher assets under management over time.

Fintech Barriers

Awareness

A problem unique to Fintech banks regards awareness: brick-and-mortar institutions have a constant presence in the travels, if not the lives, of consumers. Marketing and awareness has to be a major expense for Fintech firms to establish and reinforce trust and endurance. A Fintech firm's cost and performance advantages in the world are rendered meaningless without awareness. In a Price Waterhouse Coopers consumer survey, responders were asked their awareness of digital banks: 91% were unaware of their existence and the remaining 9% cognizant of at least 1 digital bank (Sullivan, Garvey, Alcocer, & Eldridge, 2014). Digital-only banks will rely heavily on an up-front marketing, education, communications, as well as some physical presence such as third-party distribution channels, call center expense, and regional offices. Challenger banks aligned with established companies such as Virgin and Tesco, however, have a logistical and cross-selling advantage over digital-only Neobanks. Fintech firms in wealth management face the same awareness conundrum as Fintech banks.

Security and Data Privacy

Security of funds is among the more serious issues for online Fintech players, yet fraud and abuse also exists in brick-and-mortar cases as well as online. The EU is rigorous on matters of security in payments, banking, and privacy, spearheading the use of pin-and-chip credit cards, possessing a long history regulating mobile banking and payments, and implementing very stringent consumer privacy and data protection standards thanks to the EU Data Protection Directive and the successor General Data Protection Regulation (GDPR) in 2016 (EU GDPR, 2017). The EU GDPR's impact to Fintech expands beyond consent firms must acquire from customers. A large potential cost for Fintech concerns liabilities incurred from data protection requirements under EU GDPR. PSD2 regulations due in 2018 will determine the levels of access Fintech firms will have to consumer data and their ability to provide one-stop-shop operability for consumers.

Regulation

The regulatory environment affecting Fintech and their role and place within the overall financial system is an area that is still in flux. Regulators like the ECB and EBA (European Banking Authority) continue to solicit comment and render decisions on issues including payments, compliance, and new banking licenses issuances throughout the EU. Lacking the money and experience, Fintech firms expressed concern over the impact of legacy banks' lobbying efforts on EU's PSD2 (second payments services directive) "open banking" legislation rolled out January 2018. In the regulatory offing are proposals to improve regulatory proportionality, reduce capital treatment disparity, increase access to payment systems and transparency in order to improve customer understanding of product value and client trust.

Figure 12. Pre-PSD2 and Proposed Post-PSD2 Account Access Flow Diagrams

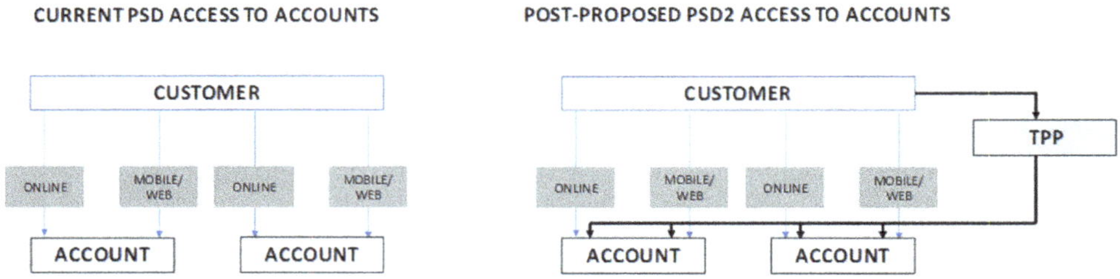

PSD2 regulations seek to boost competition by opening access to customer data (who authorize its use) to third party providers (TPPs) such as Fintech firms (see Figure 12). Under such an arrangement, theTPP may either act as an account information service provider (AIS), access customers' accounts, provide details, and help budget, or the TPP can be a payment initiation service provider (PISP) to engage in payments – thus serve as a one-stop shop or aggregator. In exchange for such access, Fintech firms would be subjected to increased regulation. For savings and even wealth management, the PSD2 regulation is especially crucial since providing a more appealing portal with accesses to accounts from different banks lets the Fintech firm break the monopoly legacy banks had on providing account information and facilitating transactions. Fintech firms could glean over valuable customer data and identify opportunities and receptiveness to other services. Legacy banks' concerns stem from the fact that by losing that connection with customers, they lose the opportunity to cross-sell more profitable products while losing access to valuable customer data. Further, their services become commoditized, forcing fees and terms to equal those of other competitors given the transparency Fintech firms could provide. Budget limits, account balance warnings and budget management tools Fintech firms provide would reduce the possibility of incurring fees like overdraft fees – important sources of banks' non-interest income. The PSD2 rules are billed as a vital source of protection to enhance client confidence, perception of legitimacy, and improve service reliability (Arnold & Brunsden, 2017). However, Fintech firms argue the real concern is a desire to limit inquiries about accounts and access to information that allow Fintech firms an ability to execute basic banking service-based operations. In fact, a CEO for a major legacy bank in a January 2017 Financial Times Article cited the fear customers will believe Fintech firms may offer more services for free than their legacy bank (Arnold & Brunsden, 2017).

Ironically, the viability of Fintech firms depends on requires that regulation's ability to legitimize the industry by establishing regulatory parity with legacy institutions – and the projected security attributed to such regulation. Attention of regulators to Fintech scandals such as the Lending Club "fake loan scandal" brings scrutiny to the sector, yet helps flush out bad actor Fintech firms and focus on their role as finance companies with a sober responsibility. Regulations and scrutiny by regulators may increase compliance costs for Fintech firms but decrease customer acquisition costs by establishing guidelines for credibility and trust.

The Fintech wealth management sector is subject to securities regulators and regulations like the ESMA (European Securities and Market Authority) and the MiFID 2 (Markets in Financial Industry Directive) directive starting January 3, 2018, the FCA (Financial Conduct Authority), and FSA (Financial Service Authority) in the UK. However, the regulatory framework is late crafting responses to the rise in Robo-Advising and the advice provided to clients who utilize advisement and automated investment services. MiFID 2, for instance, is 4 years in development. Despite the concerns from the financial advisement industry, the FCS's Financial Advice Market Review (FAMR) determined as many as 16 million citizens in the UK may be caught in a financial advice gap where would-be clients seek financial advice but can't afford it – hastened by a 2013 measure where advisors were prohibited from offering free advice in exchange for commissions on gains (McDermott & Roxburgh, 2016). The result is a new business model where transactional costs can be replaced by asset under management (AUM) fees. FAMR surveys cited that 69% of advisors turned down potential clients who can't afford advisement fees under a fees-based model. The number of advisors has slowly dwindled over time (see

Figure 13. Decline in Advisor Numbers 2011-2014

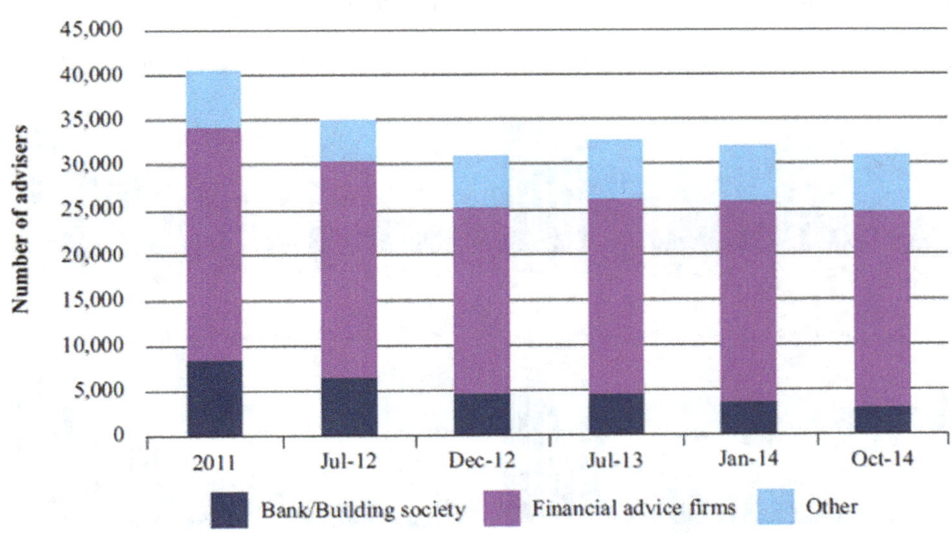

Sources: RS Consulting (2011) and FCA (2012 onwards)

Figure 13 (McDermott & Roxburgh, 2016). Yet while the trend and economics appears to favor introduction of more low-cost advisors, many organizations such as the European Financial Planning Association (EFPA) place the Robo-Advising industry as more of a supplemental solution to expand the market versus a true competition for its industry's legacy members.

Ease of Doing Business

Potential barriers to Fintech firms regard the ease of doing business and legal protections offered by host countries. Data available from the World Bank Doing Business World Rankings was evaluated for: overall doing business scores, ease of starting a business, getting credit, enforcing contracts, and resolving insolvency. Nations hosting most Fintech firms currently such as the UK and Germany have high global rankings overall (7th and 17th globally) and some of the top rankings in the EU-28 (See Figure 14). In matters of resolving insolvency, Germany is ranked 3rd while the UK is ranked 13th globally. While both nations rank highly in credit access (Germany ranked 32nd and the UK 20th), nations with high GDPs and low rankings in ability to get credit pose opportunities for Fintech firms (e.g. France, ranked 82nd in the world, Italy and Portugal at 101 each, and Luxemburg at 170).

Figure 14. Global Rankings of Doing Business in EU Nations. Clockwise from Upper Left: Overall Global Rankins; Getting Credit Global Rankins; Resolving Insolvency Global Rankings; Enforcing Contracts Global Rankings

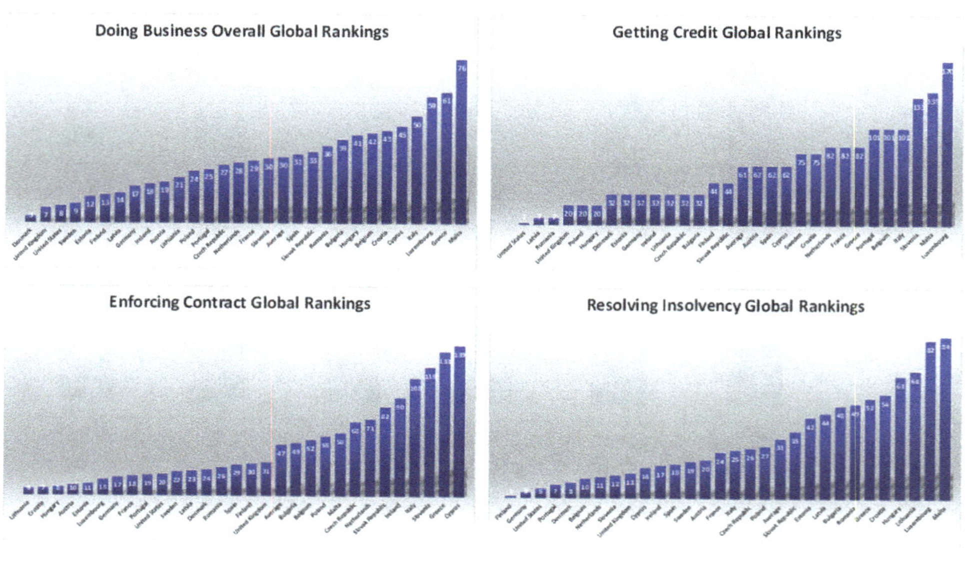

LEGACY FINANCIAL INSTITUTION BUSINESS MODELS AT RISK AND IMPACT TO LEGACY FINANCIAL INSTITUTIONS

The biggest threat posed by Fintech firms is their ability to provide financial services using new channels in distribution. It is this study's perspective that ultimately, margins on many services will be squeezed in the savings and wealth management spaces –the former being one of the more cost- and labor-intensive sectors in finance.

Business Models at Risk

The Consumer Banking Business Model for Savings

The consumer banking business model offers banks many potential revenue sources (see Figure 15, left). Diversified banks depend on cross-selling of high-margin products to customers for much of their net income and shareholder equity. Once trust is established through "sticky" loss-leading services such as checking and savings accounts, losses are offset by services like insurance, wealth management, trading, payment, and remittances. A TopCompare website survey reported how few customers in Belgium took into account digital services in their banking, focusing instead on more historical or traditional factors such as past banking traditional preferences and branch proximity.

ATM's made labor and real estate cost reductions possible. Since 2000, while banks' overall geographical footprints have increased, real estate square footage per consumers and real estate square footage per assets have shrunk in the EU. Promoted in their annual reports, major international banks like JP Morgan Chase, Barclays, and Banco Santander launched mobile based transaction capabilities via their mobile

Figure 15. Banking Business Model (Left) and Wealth Management Business Model (Right)

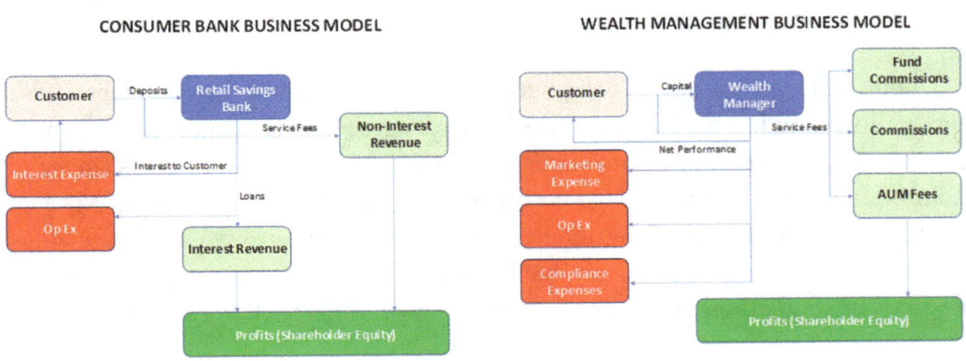

apps (applications) and reduced teller count despite the geographic footprint increase in that time. Checking and savings accounts continue to be an important gateway leading to profitability and establishing trust. Due to Fintech competition, established financial institutions have eagerly sought to squeeze costs from operations. Fintech firms see advantages versus legacy institutions from the following:

1. **Regulations:** The EU regulatory framework does not allow financial institutions to cherry-pick regulators based on favorable criteria as in the US, basing regulators on transaction-based regulators and operation locations. In cases where regulations do not explicitly exist, Fintech firms may influence rule-making through favorable engagement with customers.
2. **Profitable Business Selection:** Given levels of equity capital required up-front, Fintech firms may judiciously select business lines where leadership provide the greatest levels of expertise and avoid loss-leading business lines when possible. Further, retail banking overhead is a cost advantage over established brick-and-mortar firms.
3. **Open Architecture:** Fintech firms and legacy financial firms, by using open-API's, standardize the architecture for transactions to open competition for servicing clients to provide convenience, capability, and ease.
4. **Engagement:** While much is made of the Fintech-provided technological features, the true power of Fintech is on its ability to relate to and engage with customers and their lives. Non-interest earning-power of banks is determined by engagement which leads to cross-selling opportunities of higher margin products. Customer engagement with Fintech firms removes that potential for legacy banks.
5. **Low Rate Environment Options:** Banks use ECB rate-setting as the basis for the rates on deposits offered. Fintech savings Neobanks, Challenger Banks, and iBanks can offer higher rates to lure in more customers as well as provide alternative rewards programs to induce transactions, engage further with customers, learn from transactional habits, and remove the clients' direct engagement with their banks.

Fintech's influence by way of higher efficiencies is essential on a continent whose international borders are blurred by the EU zone agreement and where movement of money is as important as the movement of people: consolidation of one's financial holdings is not only attractive but necessary since tethering consumers to physical branch networks stems the flow of movement and economic growth. Fintech savings banks attracts clients that legacy banks are unable to accommodate. Yet much depends on regulations such as MiFID 2 and new PSD2 regulations: restrictions to account access and customer financial data by Fintech firms may hinder Fintech's ability to grow beyond a low-single digit percent of deposits.

Wealth Management

Like the retail banking business model, the wealth management business model (see Figure 15) depends on multiple sources of revenue such as trading commissions and fees based on assets under management (AUM) basis. Unlike the current US Department of Labor criteria, asset managers in the EU are not imposed a fiduciary standard. Fees are on the rise in the US, though largely due to the scheme of charging account holders based on fees – a benefit that is easy to replicate in the EU.

The EU wealth management industry turns away approximately 69% of potential clients due to an inability of prospective clients to meet minimum investment requirements and thus surpass costs – such prospective clients are prime candidates for Robo-Advisor wealth managers. This market is not a concern to the legacy industry who rely on commissions and transactions for revenues and require higher initial investment capital from clients. Those that can meet minimum investment thresholds

Figure 16. Wealth Management Bayesian Illustration

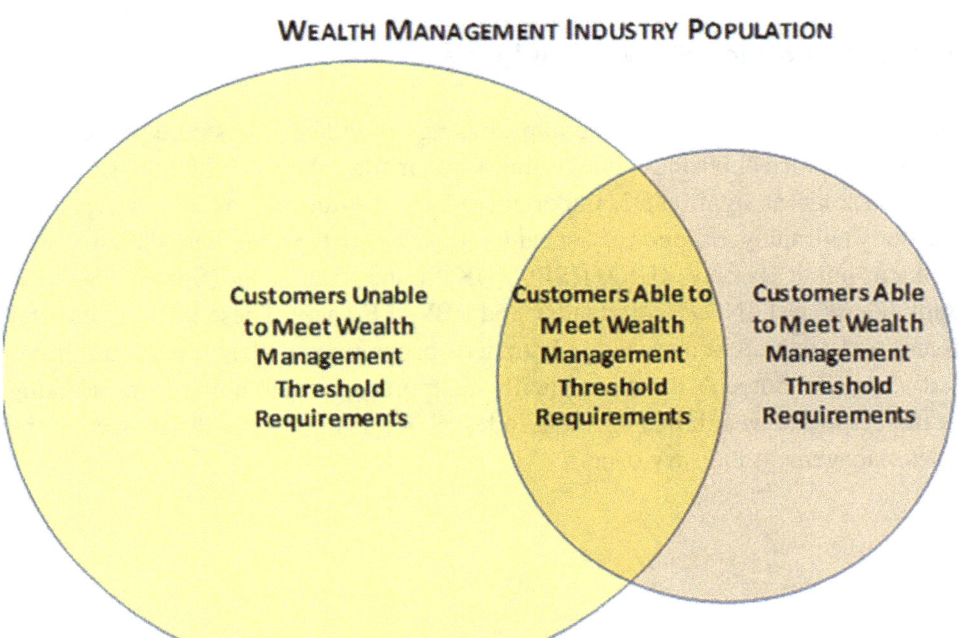

of legacy firms but seek to achieve (theoretically) higher investment returns comprise the prime market at stake for the legacy wealth management industry. The threat Fintech Robo-Advisement firms pose to legacy wealth managers comprises the Bayesian intersection – the investment public that can meet the required thresholds but seek alternatives on a cost or perceived performance basis (see Figure 16). The wealth management industry in 2016, for instance, saw an outflow of $340 billion from actively managed funds while passive index funds saw inflows of $500 billion (Malkiel, 2017). Wealth management is a concern for millennial savers. An EFAMA (European Fund and Asset Management Association) study of the Pan-European Pension Product (European Fund and Asses Management Association, 2016) observed funding and entitlement discussions brought to light from the European Debt Crisis. Among the issues uncovered was a fragmented pension and savings industry with no portability and limited choices for savers yielding little in terms of returns. It was determined that the EU pension system has a €2 trillion shortfall for retirees between 2017 and 2057, with €7.6 trillion held in bank accounts comprising 41% of overall household savings. Solutions embodied the dissolution of cross-border distribution barriers, portability, and competition for added choice and cost reductions. To that end, the Robo-Advising and Fintech wealth management industry appears to address many of the problems identified. From a European Financial Congress posting on Encouraging Long Term savings, nations like Poland were deemed to suffer from insufficient long-term savings.

Impact to Legacy Banks' Business

This section seeks to quantify the impact using individual case studies including a review of historical business line business and market shares. While the ECB lists over 200 banks as significantly important or large institutions under their oversight, this study will analyze the consumer banking space with the following 6 "high-street" banks in mind: Barclays (UK), HSBC (UK), Banco Santander (Spain), Deutsche Bank (Germany), ING (Netherlands), and BBVA (France). These institutions offer diversified products and are publicly traded corporations with audited SEC filings that often cite Fintech's threats explicitly in reporting to shareholders. In assessing an impact to the wealth management industry, we shall observe the impact on the asset management industry overall.

Savings: Barclays, HSBC, Banco Santander, Deutsche Bank, ING, BBVA Banco Frances

Financial statements for Barclays, HSBC, Banco Santander, Deutsche Bank, ING, and BBVA Banco Frances were used to observe household consumers' deposits on hand (conversion rates used: 1.18 Dollars per Euro; 1.28 Dollars per Pound Sterling).

Figure 17 shows each EU member nation's deposits per capita according to the EBA (European Banking Authority) plotted against the household deposit growth rate. Not coincidentally, more affluent nations (i.e. the UK, France, the Netherlands, Belgium, and Germany) host the greatest number of Fintech firms and hold higher deposits per capita. The greatest opportunity for Fintech companies to attain a competitive edge exists where legacy banks' costs are prohibitively large (typically where lower household deposit per capita exist).

Despite high growth, Fintech savings firms cannot yet compete with large legacy financial institutions. Fintech companies' operations is generally restricted to a single country due to local regulatory requirements. A decrease in consumption growth and disposable household income and savings as reported by the EBA (see Figure 18) makes it increasingly difficult to attract funds to compete with larger legacy institutions and encourage high levels of engagement and transaction.

Legacy banks were long hesitant to engage in technological innovations undertaken by Fintech startups. While past incidences of high IT expenditures harken to the past fruitlessness of expenditures such as for the Y2K bug, most hesitation is a combination of three factors:

Figure 17. Deposits Per Capita and Deposits Growth Rate, 2013-2014, Source EBA

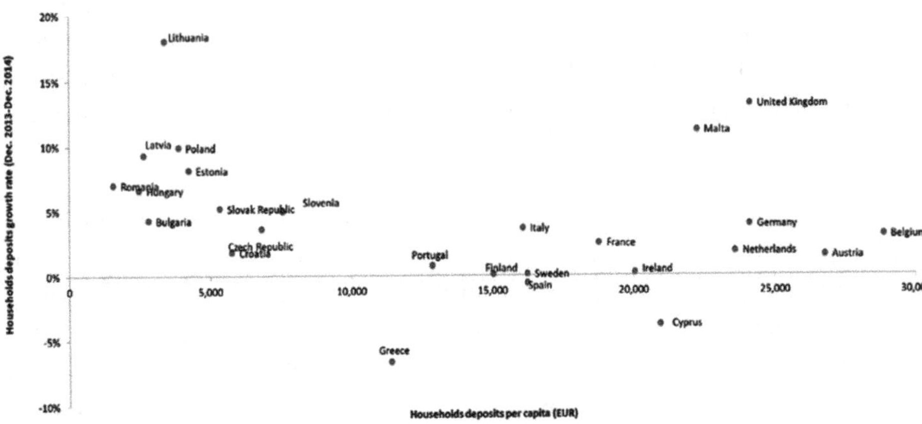

Figure 18. Growth in Consumption, Disposable Income and Savings Rate 2003-14, Source: EBA

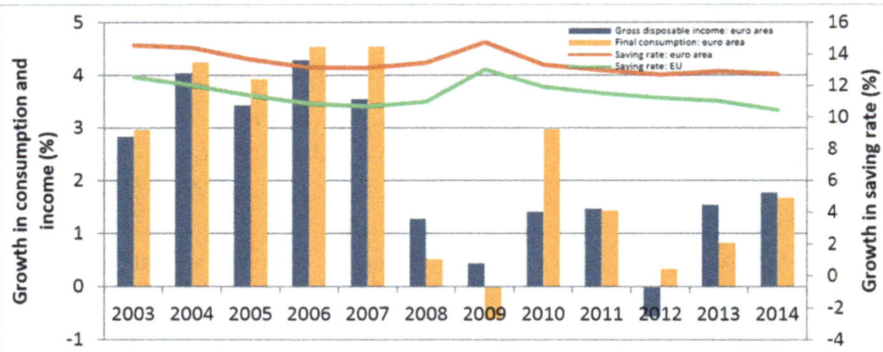

Source: ECB and Eurostat

[14] Growth of nominal consumption and nominal gross disposable income of households (annual percentage rates) and the saving rate (percentage of gross disposable income).

1. The cost and effort to make any meaningful changes to enterprise software and the firm's IT architecture with transactions and processes
2. The sheer cost involved competing with companies that are focused on customer base growth and aren't yet responsible for profits. Indeed, many Fintech firms have yet to turn a profit despite customer counts that number in the hundreds of thousands. The rise of Fintech firms has been on the cusp of a rise of several startup "unicorns" from Snap, to Blue Apron where select technical talent and business valuations sky-rocketed. Such excitement emerged amidst a flooding of cheap money from central banks that and a dearth of options for employment. Legacy banks' experiences recognize the effects of measures that ultimately raise asset prices versus shifting aggregate demand.
3. Many of the concepts initiated by Fintech firms such as API-based savings platforms, while logical today, equated to major gambles at their inception. From Alta Vista and Google to Blackberry and Apple, business history is littered with examples that first-mover advantage proved to be no advantage at all.

As value-added ideas emerged from Fintech firms, legacy banks (especially large ones like BBVA, Banco Santander, and Deutsche Bank) have taken it upon themselves to ramp up offerings of digital financial services. In its 2016 annual report, Banco Santander reported its "Santander Innoventures" effort collaborates with Fintech startups including Openbank (a digital bank in Spain slated to earn mention in 2017 10-Q filings during the first half of the year). Openbank, run by only 100 employees, has over a million clients in Spain and is already profitable. Deutsche Bank – in its filings – reported mobile banking apps that allows the same cross-bank account access capability as some Fintech firms. HSBC cited its investment in partnerships and joint ventures with Fintech firms, while most of its internal investment focused on security measures via biometrics and artificial intelligence. In its 2016 annual report, ING reported investment in 65 Fintech firms including areas such as instant lending.

Legacy banks spent (and continue to spend) billions upgrading their capabilities, developing new technologies, and investing in startups with promising features and capabilities. Such investments are typically made with intent for eventual pursuit of control: a sizable investment dissuades would-be future investors from pursuing control cognizant that an acquisition price would be too dear. The investment beachhead, meanwhile, clears an easier path for eventual control. For instance, a firm that takes a 40% equity stake in a startup requires slightly more than a 10% block to gain control of the firm. The premium attached to that small additional stake is minor relative to the overall firm's valuation. However, any other potential acquirer would need to acquire a large stake in the firm from less available shares where the premium would prove much more daunting.

Technology infrastructure budgets in annual reports often do not call out specific allocations of Fintech spending, infrastructure upgrades and modifications, or new initiatives. However, reports do call out the threat posed by Fintech firms. Much of the credit is attributed to the level of customer service and new features that make it, as the ING annual report calls, "more challenging to compete… against the backdrop of persistent low rates that are putting pressure on bank margins" (ING Group, 2016).

Figure 19. (Left) EU Assets Under Management Over Time and; (Right) 2016 Capital Sources, Source: EFAMA

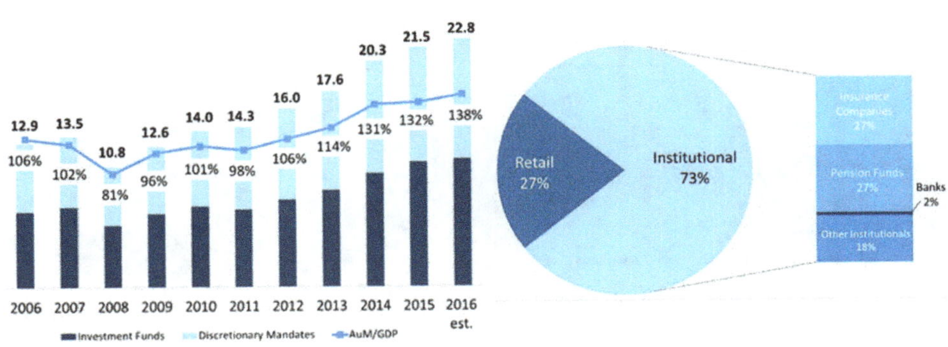

Wealth Management

An appropriate target market Robo-Advisors is retail investment (rather than pensions). In an EFAMA Report on Asset Management in Europe (2017) (Delbecque & Carrol, 2017), of the €22.8 trillion in EU assets under management (AUM), 27% or €6.2 trillion are retail clients (see Figure 19). The growth rate of retail investing AUM is on the wane (see Figure 20), perhaps due to lower savings and increased consumptions after the Euro Debt Crisis.

In spite of the growth in Robo-Advising is significant, the sector still does not equate to 1% of all assets under management of the overall wealth management industry (with over 4000 EU asset managers). Projections in some publications that 10% of retail assets under management may utilize Robo-Advisors would amount to €620 billion – a staggering amount from today's levels, but a figure consistent with the Joint Committee of the European Supervisory Authorities estimate of €450 billion by 2020 from a level of €20 billion in 2015 (EBA, ESMA, and EIOPA Joint Committee, 2015). The industry currently undergoes a shift from a commission-based system to one dependent on fees off AUM (typically 1-2% per annum). Realization of a 10% of AUM from actively managed retail funds towards Robo-Advisors at a 1% AUM fee equates to a minimum of €6 billion in lost fees per year. That figure, however, doesn't capture the pie-growing potential of the Robo-Advising industry. Indeed, one of the advantages to Robo-Advisement is the minimum investment required to open an account and active managers require. Regardless, EFAMA has staked a position desiring the advancement of a Capital Markets Union (CMU) that streamlines the asset management industry by relying less on third parties and bank intermediaries to directly engage with the retail investor.

Figure 20. (Left) Asset Managers' Client Base from 2007-2015; (Right) AUM Growth Rates by Client Type, Source: EFAMA

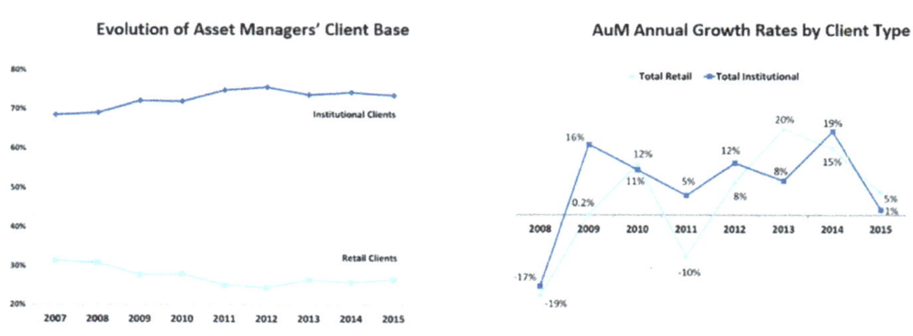

Robo-advisors typically compete on a cost-bases since most of the asset monitoring, rebalancing, and market monitoring functions are performed via algorithms rather than a human analyst. The hybrid Robo-advising model produces the requisite data and analysis for an investor to judge and allocate assets. In contrast, the fully-automated Robo-Advisor company makes ultimate buy, sell, and hold decisions for the investor. The split between hybrid and fully-automated use of Robo-Advisors over time is indefinite, as recent surveys indicated that most investors possess the preference of maintaining ultimate control over their portfolios while target-dated funds which allocate capital among resources provides the clearest initial path toward full-automation (yet still is subject to human investor decisions). The Joint Committee of European Supervisory Authorities cited fully automated Robo-Advisory services as limited by permission to operate in a few EU jurisdictions. Further, concerns exist regarding liability to the advice if the worst should happen. However, the posture of the joint committee is that Robo-Advisory services, especially fully-automated services, will grow client acceptance over time. The bias towards human-based control may evolve over time given continued proliferation of artificial intelligence in everyday lives, including autonomous vehicles, home devices like Amazon's Alexa and Apple's Siri. Wider acceptance of AI may be likely followed by fully automated use of Robo-Advisors in time.

CONCLUSION: THE FUTURE OF FINTECH

Overview

A combination of the desire to take control one's finances/wealth and the lack of confidence in the financial system combined with the rise of Fintech. Some areas of Fintech provide the ability to drastically alter the competitive financial landscape and accomplish major goals. Fintech will have an impact on reducing the unbanked population in the EU. Fintech firms will enable savings a reality for those who do not have the up-front capital to make such a commitment under existing guidelines. Fintech will alter wealth management by providing more options and competitive platforms to conduct transaction of bonds, securities, mutual funds, and ETFs.

What's unclear, however, is whether the success of Fintech firms will translate to success of standalone Fintech firms. Instagram had tremendous success in a photo sharing space that had seen many unsuccessful ventures, only to be acquired by Facebook for $1 billion in 2012. In the current low-interest credit financial environment, purchasing the capabilities a company lacks is often cheaper than developing that capability organically. Initial investments in Fintech companies by major legacy firms, even at current "unicorn" $1 billion+ valuations might be advantageous if the firms offer compelling value to the acquiring firm. Fortunately for legacy banks, the open API many contribute to provides such a major advance in the integration of Fintech firms. Investments and joint ventures give legacy banks an insight into operations, marketing, and strategy, as well as an insight into valuation and handling future growth.

While Fintech firms' lack of profitability to-date is a concern, there are many items to consider. First, many companies take years to reach profitability – especially given the need to grow the business in early rounds of fund-raising. Startups need to show that the firm can achieve milestones involving acquiring new clients with low churn rates. Amazon, for instance, didn't turn a full-year profit from its founding in 1994 until fiscal year (FY) 2004 (Wingfield, 2004). It is tempting to view the success of Fintech through investment levels, valuations, investment rounds, and initial public offerings (IPOs). However, not every tech firm do what Amazon did and have investors and management patiently wait for profitability in spite of years of losses. As the Dot-Com bubble illustrated, profits and returns become front and center to investors' minds. In that spirit, one must evaluate the future of Fintech in EU savings and investors by our project's 2 business lines:

Consumer Banking: Neobanks, Challenger Banks, and iBanks

The consumer banking sector appears to be robust and promising for Fintech. Regulatory support for added competition has grown after the 2007-2009 Global Financial Crisis, the Euro Debt Crisis, and the CMA report findings. As a result, over 18 banking licenses in the UK have been issued recently including one for Atom bank (with £45 million), the first fully digital mobile-only bank. Digital banking in Spain has grown significantly – not surprising considering that the wave of consolidation after the Euro Debt Crisis in Spain merged several *cajas* into one bank that became BBVA. In 2016, EVO Bank of Spain increased account balances by 21%. ING Direct increased account balances by 11% to €27.5 billion – staggering, considering Deutsche Bank only had under €8 billion in online deposits in 2016. Challenger bank (BankInter), meanwhile, held €5 billion in Spain. Bancopopular-e, increased accounts balances by 578% from last year's €305 million to over €2 billion in 2016 (Banks.EU, 2016).

An objective measure of a bank's profitability and managing assets would be return on assets (ROA). Standard financial metrics are difficult to find since Neobanks and Challenger banks not associated public companies are private entities and thus not required to provide financial disclosures. Challenger banks, however, often disclose that they reach relatively high profitability in terms of returns on equity (ROE). One should use caution using this metric since figures may be distorted with higher leverage.

A PwC paper on Challenger Banks (Lyons, 2017) identified ROE as a proxy for profitability and grouped mid-size full-service Challenger Banks as having an ROE of -15% to 5%; Specialist Fintech firms are estimated to be more highly profitable in the space with an ROE of 10-40%; Non-bank brands such as Tesco and Virgin are assessed as having moderate profitability with an ROE of 0-15%. Digital-only Neobanks and Challenger Banks have an indeterminate level of profitability but are in a business growth stage.

Fintech savings institutions face different prospects depending on the type of bank they are. Challenger banks aligned with retailers and mobile phone operators like Virgin Mobile with Virgin Money may be the more successful in the short and intermediate-term given their advantages with an initial brick-and-mortar presence and opportunities integrating banking with rewards programs. Further, the opportunity of having a physical branch nearby will make this a more appealing option for non-Millennials. Ultimately, the question of size versus profitability will emerge. Few companies can navigate successful long-term positions by "serving many masters." In the case of mobile and retail companies, being a customer-centric company is not sufficient as a mission statement or as a long-term strategy to effectively compete at the same level of legacy banks. In the long-term, digital-only Challenger Banks and Neobanks will make the greatest gains in terms of higher profitability, if not necessarily size. Millennials will help spearhead this growth. While an outright change in banks isn't the norm, EU customers are prone to add new accounts based on desired features and products. When the customer-centric features can be integrated with multiple accounts from different institutions in a one-stop-shop portal, it would not be surprising that adoption will lead to commoditization of savings accounts. The question is how legacy banks will respond to erosion of their margins by having to play catch-up with more profitable competitors while lacking of engagement and opportunity for cross-selling of higher-margin services. Ultimately, the fate of savings Fintech firms will hinge on regulation, speed of growth, and their attractiveness.

The looming PSD2 will determine restrictions Fintech firms will have on account data: how they can use it and the kinds of liability they may face. To those firms able to navigate the coming regulatory hurdles, they will be pressed to accelerate customer growth and demonstrate their ability to challenge legacy banks. Ultimately, it is reasonable to expect the most successful Fintech Challenger and Neobanks will be M&A targets by legacy banks.

Wealth Management

The wealth management Robo-Advisory service has potential to grow rapidly and prosper as independent businesses without being acquired by legacy wealth management firms or diversified high-street banks. With respected investors such as Warren Buffett and Jack Vogel speaking out about the high costs of active asset managers and mutual funds charging high annual fees for underperformance, it is expected that people will seek to exercise added control of their investments. Millennials who are more inclined to seek out information for themselves and make their own investment decisions seek the lowest possible costs to achieve the performance of market averages. The low cost of investment assets like ETFs and the low initial investment requirements by Robo-Advisor Wealth Management firms provide a service to investors and the financial services industry that is long-overdue. Were it not for human judgement, the prospects of the Robo-Advisement sector of Fintech would be more certain.

Ultimately, as investors like Benjamin Graham, Warren Buffett, and Charlie Munger long argued, investment by individuals is determined largely by temperament. Investor risk tolerances are easy to determine when market volatility is low and gains appear certain. During times of market volatility, however, risk tolerances seem to shift dramatically lower prompting second-guessing and poor investment decisions. Investors have a tendency to share the sentiment Larry Bird once opined, "I hate to lose more than I love to win." Such behavior explains why retail investors during market vicissitudes tend to sell low and buy higher. Successful Robo-advisor firms will be ones where investors are better educated, where human assistance is readily available to "talk people off the ledge" during downturns, and where investors tend to transact the least. Their business may be devoid of many features of legacy wealth management firms that overload clients with market data that induce greater volume of transactions. Prosperous Robo-Advisor firms will see growth in assets under management due to both increased asset valuations from more successful investment strategies, a lack of redemptions in spite of market volatility, positive word of mouth, and favorable marketing.

The only other question regards the adoption of fully-automated Robo-Advisement versus the hybrid model. The dominance of full-automation appears far away, but continued positive performance by global indices may begin shifting sentiment towards a fully automated Robo-Advisement model.

Fintech Companies and Valuing Their Business

The Fintech industry's success depends in part on investment into these institutions, rounds of venture capital funding, and even M&A activity. The viability of the savings and wealth management sectors requires rapidly-growing business fundamentals including revenues and costs. Mainly, there is the costs for operations and new client acquisitions. With client churn figures still indefinite due to the youth of the industry, it is reasonable to assume that many Fintech products (primarily in savings which is a sticky product) may have high retention rates and acquisition costs will be amortized over a long period of time. In banking, such a cost-per-new client is relatively straight-forward to calculate.

Robo-Advisor firms are easier companies to value. Their earnings via assets under management, low cost, low capital expenditures, and high ROE apply higher valuations than previous M&A transactions in the asset management space (such as the Aberdeen-Standard Life merger in March 6, 2017).

Summary

As ING's 2016 annual report stated, "the bank of the future will not look like banking as we know it today… in particular in the development of digital ecosystems that connect people and businesses. They include features like shopping and paying for goods and services to become the one stop for all of their users' needs. This trend makes it important [to] develop an open platform where customers can go not only for their financial needs but also for relevant offers and services from third parties, and even for an overview of their accounts with other financial services providers… [T]his will provide them with an enhanced customer experience and us with opportunities to generate fee and other revenue from services beyond traditional banking… [D]igital systems must be capable of integrating seamlessly into other ecosystems [and to be] present where… customers spend their time online – available anytime and anywhere."

Size and scale remains to be the main advantage of large established financial firms. As PayPal demonstrated through acquisitions of firms such as Venmo, its scale can be leveraged to buy smaller Fintech competitors before they grow too big too fast. The ability of these firms to influence regulation through lobbying will work against the same banks that seek to develop the same capabilities. The main advantages of Fintech firms aren't their clean-slate approach to banking functions like savings, lending, or wealth management. The ease and accessibility itself is not the main advantage Fintech firms have. The main advantage of Fintech firms is the customer-centric nature of these new businesses. The truly successful Fintech firms – like any truly successful startups – are the firms that focus on the customer and her or his needs. Features such as access and capabilities are merely conduits that allow the EU customer to regain control that disappeared with two devastating financial crises in a span of 5 years. The Fintech firms that are less enamored with their technology and more obsessed with their market will be the firms that mount the biggest challenge to legacy financial institutions. Fintech firms also lack the baggage and robust infrastructure that is often hard to modify.

Fintech provides many new opportunities for banking and finance. Through specialization and customization, cost reductions may be feasible which facilitates the unbanked population to join the formal banking system (at least at higher than current levels). The costs seen in the two sectors of Fintech discussed earlier will benefit existing consumers and provide an avenue to introduce yet more people to the formal financial system. For the ability to provide services like wealth management that once were unattainable by less-affluent savers and helping savers capture more of the gains from their hard-earned money, the Fintech industry already succeeds where decades of legacy banking failed.

REFERENCES

Akanji, O. O. (2002). Micro Finance as A Strategy for Poverty Reduction. *CBN and Financial Review*, *39*(4), 1–30.

Arnold, M., & Brunsden, J. (2017, January 31). Fintech sector fears dilution of EU 'open banking' legislation. *Financial Times*. Retrieved from https://www.ft.com/content/65342eb2-e7ab-11e6-893c-082c54a7f539

Banks, E. (2016). *What is a neobank and why is it our future?* Retrieved from http://eng.banks.eu/news/info/2289/

Banks.eu Staff. (2016, January 12). *What is a Neobank and Why Is It Our Future*. Retrieved from http://eng.banks.eu/news/info/2289/

Barba, R. (2017, February 3). Valuing a digital bank isn't so Simple. *American Banker*. Retrieved from https://www.americanbanker.com/news/valuing-a-digital-bank-isnt-so-simple

Barber, B. M., Lee, Y., Liu, Y., & Odean, T. (2008). Just How Much Do Individual Investors Lose by Trading. *The Society of Financial Studies*, *22*(2), 610–632.

BBVA Compass Staff. (2014, February 20). BBVA acquires Simple to accelerate digital banking expansion. *BBVA Compass PR Newswire*. Retrieved from http://newsroom.bbvacompass.com/2014-02-20-BBVA-acquires-Simple-to-accelerate-digital-banking-expansion

BetterFinance. (2016). Long Term & Pension Savings: The Total Return, 2016 Edition. *Better Finance*. Retrieved from http://betterfinance.eu/fileadmin/user_upload/documents/Research_Reports/en/Pension_Savings_-_The_Real_Return_-_2016_Better_Finance_Presentation.pdf

Binham, C. (2015, December 4). European watchdogs to scrutinise robo-advice industry. *Financial Times*. Retrieved from https://www.ft.com/content/36d17f2c-9a7a-11e5-a5c1-ca5db4add713

Buck, G. (2016, December 13). MasterCard: Millions of Europeans still unbanked. *GTNews*. Retrieved from https://www.gtnews.com/2016/12/13/mastercard-millions-of-europeans-still-unbanked/

Calomiris, C. W. (2010). *Banking Crises Yesterday and Today. Financial History Review, 1(1)*, 3–12. Retrieved from https://jeffludwig.wordpress.com/2012/01/02/how-to-cite-white-papers/

Delbecque, B., & Carrol, A. (2017, May). *Asset Management in Europe: 9th Edition Facts and Figures*. European Fund and Asset Management Association. Retrieved from http://www.efama.org/publications/statistics/asset%20management%20report/efama%20asset%20management%20report%202017.pdf

Demirguc-Kunt, A., Klapper, L., Singer, D., & Van Oudheusden, P. (2015). *The Global Findex Database 2014 Bringing the 2 Billion Unbanked into the Financial System*. Retrieved from http://pubdocs.worldbank.org/en/113791483565360488/N2UnbankedV5.pdf

Dietz, M., Khanna, S., Olanrewaju, T., & Rajgopal, K. (2015). *Cutting Through the FinTech Noise: Markers of Success, Imperatives for Banks*. McKinsey & Co. Retrieved from https://www.mckinsey.com/~/media/mckinsey/industries/financial%20services/our%20insights/cutting%20through%20the%20noise%20around%20financial%20technology/cutting-through-the-fintech-noise-full-report.ashx

ELINET. (2015). *Literacy in Europe: Facts and Figures*. ELINET and the European Commission. Retrieved from http://www.eli-net.eu/fileadmin/ELINET/Redaktion/Factsheet-Literacy_in_Europe-A4.pdf

EU GDPR. (2017). GDPR Key Changes. *EU GDPR*. Retrieved from http://www.eugdpr.org/key-changes.html

European Banking Authority (EBA), The European Securities and Markets Authority (ESMA), & The European Insurance and Occupational Pensions Authority (EIOPA) Joint Committee. (2015, December 4). *Joint Committee Discussion Paper On Automation In Financial Advice*. JC 2015 080. European Banking Authority (EBA), the European Securities and Markets Authority (ESMA) and the European Insurance and Occupational Pensions Authority (EIOPA). Retrieved from http://www.eba.europa.eu/documents/10180/1299866/JC+2015+080+Discussion+Paper+on+automation+in+financial+advice.pdf

European Banking Authority (EBA), The European Securities and Markets Authority (ESMA), & The European Insurance and Occupational Pensions Authority (EIOPA) Joint Committee. (2015, December 4). *Joint Committee Discussion Paper On Automation In Financial Advice*. JC 2015 080. European Banking Authority (EBA), the European Securities and Markets Authority (ESMA) and the European Insurance and Occupational Pensions Authority (EIOPA). Retrieved from http://www.eba.europa.eu/documents/10180/1299866/JC+2015+080+Discussion+Paper+on+automation+in+financial+advice.pdf

European Central Bank. (2014). Structural Indicators for the EU Banking Sector. *ECB*. Retrieved from https://www.ecb.europa.eu/stats/pdf/140711_ssi_table.pdf?a8644f6780aee14061e973e596141fac

European Central Bank. (2017). *Single Supervisory Mechanism*. Retrieved from https://www.bankingsupervision.europa.eu/about/thessm/html/index.en.html

European Fund and Asset Management Association. (2016). *The Pan-European Personal Pension Product*. EFAMA. Retrieved from https://www.efama.org/Publications/PEPP_flyer.pdf

Eurostat. (2017, March). *Asylum statistics*. Retrieved from http://ec.europa.eu/eurostat/statistics-explained/index.php/Asylum_statistics

Eurostat. (2017, March). *Migration and migrant population statistics*. Retrieved from http://ec.europa.eu/eurostat/statistics-explained/index.php/Migration_and_migrant_population_statistics#Migrant_population

ING Group (2016). *Annual Report*. Retrieved from file:///C:/Users/mcaesar/Downloads/2016%20Annual%20Report%20ING%20Groep%20N.V..pdf

GSMA. (2015). *The Mobile Economy: Europe 2015*. London: GSM Association. Retrieved from https://www.gsma.com/mobileeconomy/archive/GSMA_ME_Europe_2015.pdf

Gullapalli, D., Anand, S., & Maxey, D. (2008, Sept. 17). Money Fund, Hurt by Debt Tied to Lehman, Breaks the Buck. *Wall Street Journal*. Retrieved from https://www.wsj.com/articles/SB122160102128644897

Hagstrom, R. (2013). *What is the difference between investing and speculation?* Retrieved from https://blogs.cfainstitute.org/investor/2013/02/27/what-is-the-difference-between-investing-and-speculation-2/

History.com staff. (2010). *Automated Teller Machines*. Retrieved from http://www.history.com/topics/inventions/automated-teller-machines

Intelligence Staff, B. I. (2017). *French Retail Giant Launches Online Bank Accounts*. Retrieved from http://www.businessinsider.com/french-retail-giant-launches-online-bank-accounts-2017-3

Kumok, Z. (2016). *7 Top Non-U.S. Robo-Advisors*. Retrieved from http://www.investopedia.com/articles/financial-advisors/032216/7-top-nonus-roboadvisors.asp (Investopedia)

Lyons, J. (2017). *Who Are You Calling a Challenger?* Retrieved from http://pwc.blogs.com/files/challenger-bank-2017-.pdf

Malkiel, B. G. (2017, June 5). Index Funds Still Beat 'Active Portfolio Management'. *Wall Street Journal*. Retrieved from https://www.wsj.com/articles/index-funds-still-beat-active-portfolio-management-1496701157

McDermott, T., & Roxburgh, C. (2016, March 1). *Financial Advice Market Review: Final Report*. London, UK: Financial Conduct Authority. Retrieved from https://www.fca.org.uk/publication/corporate/famr-final-report.pdf

McFarlan, F. W., & Tempest, N. (1999, September). *Charles Schwab Corporation (A). Harvard Business School Case 300-024*. Harvard Business School. Retrieved from http://www.hbs.edu/faculty/Pages/item.aspx?num=26491

Mellinghoff, M. (2017). *Map of Robo Advisors in Europe and Germany*. TechFluence. Retrieved from http://www.techfluence.eu/investtech.html

Nowodazkij, E. (2014, July 11). Data shows slight decline in number of bank branches in most EU countries. *ECB*. Retrieved from https://www.ecb.europa.eu/press/pr/date/2014/html/pr140711.en.html

Pisani, B. (2010). *Man Vs. Machine: How Stock Trading Got So Complex*. Retrieved from http://www.cnbc.com/id/38978686

Reichheld, F. F. (1993). *Loyalty-Based Management. Harvard Business Review. March–April 1993 Issue*. Cambridge, MA: Harvard Business School.

Rubini, A. (2017). *Fintech in a Flash: Financial Technology Made Easy*. London, UK: Banking Innovations.

Runde, D. (2015) *M-Pesa And The Rise Of The Global Mobile Money Market*. Retrieved from https://www.forbes.com/sites/danielrunde/2015/08/12/m-pesa-and-the-rise-of-the-global-mobile-money-market/#62172bcd5aec

Rysdal, K. (2017). *PayPal's Schulman says the future may not be cashless, but it's definitely mobile*. Retrieved from https://www.marketplace.org/2017/05/24/business/corner-office/paypals-schulman-says-future-may-not-be-cashless-its-definitely-mobile

Ryssdal, K. (2017, March 24). PayPal's Schulman says the future may not be cashless, but it's definitely mobile. *Marketplace*. Retrieved from https://www.marketplace.org/2017/05/24/business/corner-office/paypals-schulman-says-future-may-not-be-cashless-its-definitely-mobile

Shoster, S. (2016, June 24). The U.K.'s Old Decided for the Young in the Brexit Vote. *Time Magazine*. Retrieved from http://time.com/4381878/brexit-generation-gap-older-younger-voters/

Sullivan, B., Garvey, J., Alcocer, J., & Eldridge, A. (2014). Retail Banking 2020: Evolution or Revolution. *PwC*. Retrieved from https://www.pwc.com/gx/en/banking-capital-markets/banking-2020/assets/pwc-retail-banking-2020-evolution-or-revolution.pdf

The Pew Research Center. (2015, April 2). *The Future of World Religions: Population Growth Projections, 2010-2050*. The Pew Research Center. Retrieved from http://www.pewforum.org/2015/04/02/europe/

The Telegraph Staff. (2011, March 31). Ireland's Banking Crisis: Timeline. *The Telegraph*. Retrieved from http://www.telegraph.co.uk/finance/financialcrisis/8419616/Irelands-banking-crisis-timeline.html

The World Bank Group. (2017). *Global Financial Inclusion Database*. Retrieved from http://databank.worldbank.org/data/reports.aspx?source=1228#

Thorpe, J., & Keal, A. (2017, July 10). *Britons are Europe's most prolific digital banking users according to new study*. Innofact AG and commissioned by Mastercard. Retrieved from https://newsroom.mastercard.com/eu/press-releases/digital-banking-study-show-britons-are-europes-most-prolific-digital-banking-users/

Tomtom Staff. (2017). *Tomtom Traffic Index*. Retrieved from https://www.tomtom.com/en_gb/trafficindex/list?citySize=LARGE&continent=EU&country=ALL (Tomtom)

Tsosie, C. (2017). *The History of the Credit Card*. Retrieved from https://www.nerdwallet.com/blog/credit-cards/history-credit-card/

Tumpel-Gugerell, G. (2009). *Business models in banking: Is there a best practice?* Retrieved from https://www.ecb.europa.eu/press/key/date/2009/html/sp090921.en.html

Wall, St. Journal Staff. (2007, October 18). Why a Run on Northern Rock but Not Countrywide. *The Wall St. Journal*. Retrieved from https://blogs.wsj.com/economics/2007/10/18/why-a-run-on-northern-rock-but-not-countrywide/

Wang, S. (2017, February 1). SoFi Buys Mobile-Banking Startup Zenbanx for About $100 Million. *Bloomberg*. Retrieved from https://www.bloomberg.com/news/articles/2017-02-01/sofi-buys-mobile-banking-startup-zenbanx-for-about-100-million

Wingfield, N. (2004, January 28). Amazon Reports Annual Net Profit For the First Time. *Wall Street Journal*. Retrieved from https://www.wsj.com/articles/SB107523879334513159?mg=prod/accounts-wsj

Wolff, J. (2017, July 10). *New European Digital Banking Study by Mastercard highlights convenience as the greatest advantage of digitized banking solutions*. London: Mastercard. Retrieved from https://newsroom.mastercard.com/eu/press-releases/new-european-digital-banking-study-by-mastercard-highlights-convenience-as-the-greatest-advantage-of-digitized-banking-solutions/?utm_source=Triggermail&utm_medium=email&utm_campaign=Post%20Blast%20%28bii-fintech%29:%20Crunch%20time%20for%20UK%20neobanks%20%E2%80%94%20IBM%20runs%20another%20blockchain%20POC%20%E2%80%9-4%20Santander%20partners%20Tradeshift&utm_term=BII%20List%20Fintech%20ALL

World Bank. (2017). *Literacy rate, adult total (% of people ages 15 and above)*. United Nations Educational, Scientific, and Cultural Organization (UNESCO) Institute for Statistics. World Bank Database. World Bank. http://data.worldbank.org/indicator/SE.ADT.LITR.FE.ZS?locations=EU

ENDNOTES

1 BI Intelligence Staff. (2017). *French Retail Giant Launches Online Bank Accounts*. Retrieved from http://www.businessinsider.com/french-retail-giant-launches-online-bank-accounts-2017-3 (BI)

2 Buck, G. (2016, December 13). MasterCard: Millions of Europeans still unbanked. GT News. Retrieved from https://www.gtnews.com/2016/12/13/mastercard-millions-of-europeans-still-unbanked/

3 The World Bank Group. (2017). Global Financial Inclusion Database. Retrieved from http://databank.worldbank.org/data/reports.aspx?source=1228#

4 Lyons, J. (2017). "Who are you calling a 'challenger bank': How competition is improving customer choice and driving innovation in UK retail banking". PwC UK. London. Retrieved from https://www.pwc.co.uk/industries/banking-capital-markets/insights/challenger-banks.html

Chapter 5
Inclusive Disruption:
The Role of Financial Technologies in Filling Financial Inclusion Gaps in Russia

Oksana Smirnova
Moscow School of Management SKOLKOVO, Russia

Vladimir Korovkin
Moscow School of Management SKOLKOVO, Russia

Evgeny Plaksenkov
Moscow School of Management SKOLKOVO, Russia

ABSTRACT

This chapter discusses the important socioeconomic role of financial technologies in the emerging market which is Russia today. While the issues of financial inclusion are of recognized importance for the developing markets, until recently they were seen largely as areas of affirmative regulatory action, not of competitive play by private market actors. However, the advent of fintech companies changes the paradigm. Many fintech companies in Russia view the gaps in financial inclusion as attractive market niches and formulate relevant consumer offers. This chapter reviews their strategic approaches based on the study of five business cases, and introduces an analytical matrix mapping the approaches to existing inclusivity gaps. The model strengthens the existing policy aimed at developing financial inclusion as it allows a targeted cost-benefit analysis of market players' actions. As Russia demonstrates many of the financial inclusivity challenges seen in other countries, the findings of this chapter have certain applicability in the context of both emerging and advanced economies.

DOI: 10.4018/978-1-5225-4148-6.ch005

INTRODUCTION

Financial inclusion arguably lies at the foundation of economic inclusion, where finance is recognized as a backbone of any modern economic system. The access to modern financial instruments such as accounts, transfers, deposits and loans may not only bring immediate economic benefits to consumers, but also largely define their capabilities to pursue effective economic activity either through employment or through entrepreneurship. In this case financial services work as infrastructure systems like roads or communications networks, which are obvious public goods. What makes modern financial services unique is that their suppliers can legally discriminate between customers, refusing to provide services, either implicitly (through price barriers) or explicitly. Explicit discrimination is inherent in the current procedures of bank lending, where the supplier judges the ability of a customer to make certain future action (repayment of loan) through a proprietary process, which is largely non-transparent to the customers and the results of which are commonly irreversible for the affected customer. As was said, such discrimination is of broad consequence, as it not only precludes some customers from gaining immediate economic benefits, but also limits their ability to operate as actors within the modern market economy. While the system of discrimination is absolutely necessary within the current approaches to risk management in financial systems, it has unwanted social side effects, which are especially manifest in lower income economies.

Within this context the issues of financial inclusion are largely viewed as being a key to the overall agenda of inclusive growth. According to World Bank, financial inclusion is defined as the state when *all interested individuals and businesses have access to useful and affordable financial products and services that meet their needs delivered in a responsible and sustainable way* (World Bank, 2013). Financial inclusion can be considered as the key enabler of reducing poverty, as has been shown by empirical research, in that it improves macroeconomic indicators including economic development and stability (Beck et al., 2007). However, as seen from the definition, financial inclusion (or its absence) is a complex phenomenon, which may include various combinations of barriers: physical, social, economic, legal, etc.

Practically in every country of the world the traditional banks are not reaching the whole of financially active population, creating groups of "financial exclusion". The size of these groups relative to the population and the reasons for exclusion differ widely, mostly correlating with the overall state of economic development. For instance, in the low income developing countries up to 80% of the population can be excluded mostly due to poverty and lack of financial infrastructure. In the advanced countries the excluded groups would comprise some 10 to 15% of the population, and the reasons for exclusion would be more complex (Demirgüç-Kunt et al., 2015).

Whatever the reason, solving the problem of exclusion was usually viewed as requiring affirmative action stimulated by the policies of the national financial regulators. However, the advancement of the modern financial technologies companies, the so-called fintech, gave a new perspective to the issue, allowing the forces of a free market to fill many of the niches which had not been covered by traditional banks. Such examples include, for instance, the Kenyan M-Pesa, AliPay in China, Digibank in India. Here is the arena of disruption in the financial services markets: providing products which cannot be profitable if offered by the traditional banks to the customers against whom the traditional banks tend to discriminate. One can see obvious social benefits coming from such disruption (on top of the economic gains for the customers and suppliers), however, like in any disruption there are important risks. The worst-case scenario can be the degree of disruption of the overall financial system, which will make it unprofitable for most of the incumbent players to operate, following which the system will lose stability and the trust of all participants. This scenario is unlikely to develop in full, however it is generally kept in mind by the relevant regulatory bodies, like national Central Banks and their international associations.

Thus, from a scholarly prospective the key question is: will the private actors of financial technology ("fintech") working in the generally free market environment create enough of the social effects of financial inclusion without generating unacceptable social and economic risks? Finding the answers will advance our knowledge of how regulated markets work. It will also be of high practical importance for national and international decision making on the approaches to the issue of regulation of the new disruptive business models. Answering this question requires, among other things, an understanding of the possible motives and strategies of the entrepreneurs who provide fintech products to the market. One way to obtain an understanding is by case studies, verified and enriched by offsite analysis of statistical data.

Russia is, in a sense, a model case for the development of inclusive disruptive financial technologies as it combines certain aspects of financial inclusion of both developing and advanced economies. Russia is an upper-middle income country, with a mostly urban population, almost compulsory primary literacy and a high share of skilled employment (IMF, 2016b). The penetration of modern communication technologies such as the Internet and mobile communications is above average, though the bandwidth is often lower than in the most digitally advanced countries. Russia commands a vast pool of talent in digital technologies, being one of the world's leading producers of students of mathematics, programming and engineering.

At the same time, the country's geography creates certain pockets of population that are difficult or unprofitable to access with a traditional banking infrastructure, and the tightening of bank risk policies during the recent economic recession has led to a growing withdrawal of banks from operations involving the lower income strata. Also, due to the historic legacy of the Soviet era, when the consumer finance services in the country were rudimentary, the population has a relatively low level of experience and understanding of modern finance, especially in the middle- and older age strata.

This peculiar situation requires a country-specific model of financial non-inclusivity, which serves as a basis for analysis. The model outlines segments of non-included population based on different barriers to inclusivity, such as physical (remote areas, disabled persons), social (lower income strata, self-employed, migrant workers, etc.), competences and skills (elder ages, undereducated), and SMEs (small and medium enterprises) financial services (especially for businesses in the early stages of development). Various issues of inclusion emerge in relation to different types of financial services: payments and transfers, loans, deposits, insurance, etc.

The resulting segments differ in the number of participants, size of opportunity and cost of action to overcome the barriers. Modeling by segment allows for targeted cost-benefit analysis of possible measures aimed at increasing financial inclusion. While in some segments the existing market actors are more or less effective with their current operations, other segments require strong affirmative actions. So, based on this overview, analysis of some specific business cases suggests a way to target the pockets of financial exclusion in Russia by fintech.

It is important that fintech companies should see filling the "inclusivity gaps" as one of the factors that shape their business strategies, actually as guidance to the available market niches. From the perspective of fintech companies, there are four primary domains of action, corresponding to the financial "mechanics" underlying the business model: (1) make payments more accessible for customer-to-business (C2B); (2) facilitate payments in the business-to-business (B2B) segment, including small businesses; (3) make credit more accessible; and (4) provide mass-scale financial advice, including personal budgeting.

The existing inclusivity gaps and the view of fintech companies on the core domains of action form sort of a matrix mapping the fintech effects on financial inclusion in Russia; this matrix can be adapted to other markets, both emerging and advanced.

Thus, the objectives of this Chapter are the following:

- Study the background for the role of technologies in the financial inclusion phenomenon based on international experience.
- Review the financial inclusion gaps and controversies in Russia.

- Study five financial technologies business cases aimed at covering particular exclusivity pockets.
- Build a specific business model based on a dual-dimensional "inclusivity-gaps–business strategy" matrix.
- Discuss business strategy and policy implications based on the model developed.

FINANCIAL TECHNOLOGIES AS A DRIVER FOR FINANCIAL INCLUSION: BACKGROUND OVERVIEW

By the end of 20th century there came a common understanding that economic growth, as manifested in the increase of GDP per capita, cannot by itself achieve all the social goals of humanity. As the important gaps among the nations and even within the nations in terms of quality of life were expanding, a new concept of "inclusive" growth was developed and adopted leading to a series of economic, social and political reforms. Involvement of the population in financial services arose as an important element of the new policy as it became clear that it might be a crucial element for the successful growth of other spheres: the development of the financial sector is associated not only with overall economic growth, but also with more equal distribution of its benefits in the society (Delis et al., 2013).

Following this idea, the "financial inclusion" phenomenon emerged as a political issue in 1997 in the United Kingdom and was discussed worldwide in 2010, when the Global Partnership for Financial Inclusion (GPFI) was created (Financial Inclusion Commission, 2017; GPFI, 2017). The latter has become an integrating platform for coordinating the efforts of all participating countries to improve financial accessibility as a factor of socioeconomic development. The joint work of this group resulted in the development of a comprehensive definition of financial inclusion, which goes beyond having an account in a bank, as well as its complex classification which includes access, quality, usage and welfare dimensions.

Additionally, a number of policy papers and action plans in financial inclusion all over the world were introduced aimed at solving the problems in each of the dimensions of the above classification (Arun & Kamath, 2015). Although practical actions towards financial inclusion do not have a long history yet, empirical research has revealed a list of positive microeconomic[1] and macroeconomic[2] effects in favor of the hypothesis that the growth of inclusive financial systems is a significant component of general development progress.

Among other drivers, such as proper regulation and policies, technology stands out as a clear enabler of financial inclusion because it can solve the associated issues for all types of economies including low-income, advanced and developing.

Traditionally, the technological development in the financial sector was led by the "licensed banks" as national regulators generally discouraged financial intermediation on a large scale by other types of institutions. However, in recent years the global phenomenon of fintech – innovative technological solutions in finance sphere – has become prominent.

Actually, fintech in its early forms originated in 19th century, with the invention of the telegraph in 1838 which became the fundament for financial globalization. Then, first credit cards (1950) and ATMs (1967) were introduced. This step is described by Arner, Barberis and Buckley (2015) as the "analogue-to-digital" stage. From 1967 to 2007 "development of traditional digital financial services" took place, with the establishment of Clearing House Interbank Payments System (1970), the Society for Worldwide Interbank Financial Telecommunications (Swift) (1973), online brokerage (1982) and the ubiquitous penetration of Internet banking in major USA banks (1998). After almost a decade during which there were no major financial innovations, the era of "democratizing digital financial services" began in 2008, and brought fintech in its contemporary form: electronic money, mobile banking, blockchain, etc. (Arner, Barberis & Buckley, 2015).

So, in recent years the financial technology industry has developed rapidly. A number of successful cases of an increased access to safe and affordable financial services through fintech solutions expanded the diversity of products as well as lowered the costs associated with the usage of financial services. As EY's Fintech adoption index shows, in 2017 the average percentage of digitally active consumers regularly using fintech services reached 33% across the surveyed 20 markets, compared to 16% in 2016 (EY, 2017).

This, in turn, leads to the disruption in the traditional banking model. As shown by the PricewaterhouseCoopers survey (2016), 80% of respondents consider consumer banking is likely to be disrupted by fintech by 2020. In addition, some 60% and 40% respectively of respondents also considered fund transfers/payments and investment/wealth management as being highly likely to be disrupted.

In order to examine the role of digital channels and technologies in inclusivity in more detail, several cases worldwide have been considered, and analyzed by the level of economy development and income levels.

High-Income Economies

over#all the high-income economies serve as benchmarks of financial inclusion for other nations. Still, while financial inclusion in advanced economies has increased significantly over the past years and almost reached 100%, innovations in the financial industry help to build on these positive trends and allow consumers who already have bank accounts to access innovative new products that can improve

Figure 1. Comparison of fintech adoption in 2015 vs 2017 and progress of fintech adoption in 2017, measured as share of population regularly using fintech
Source: (EY, 2017)

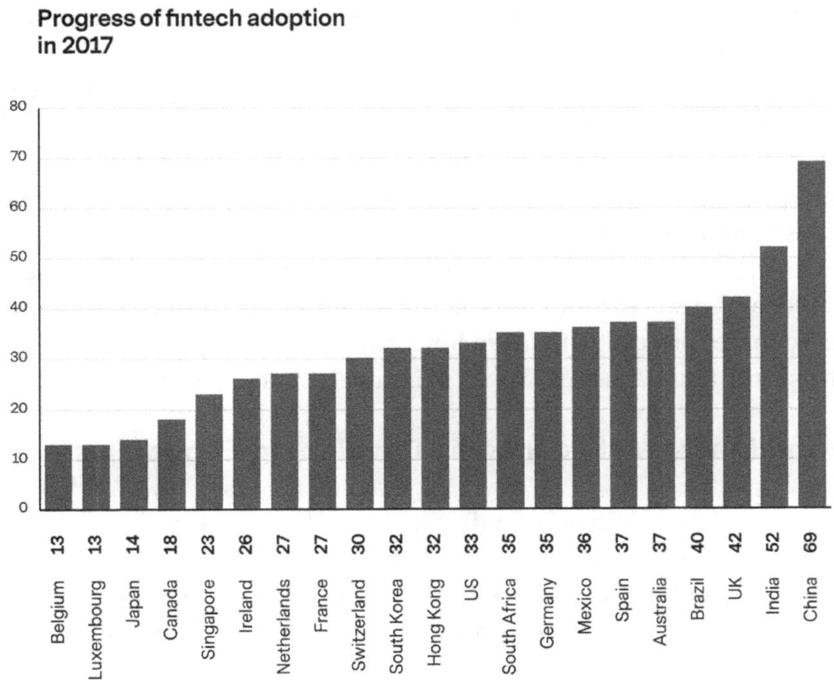

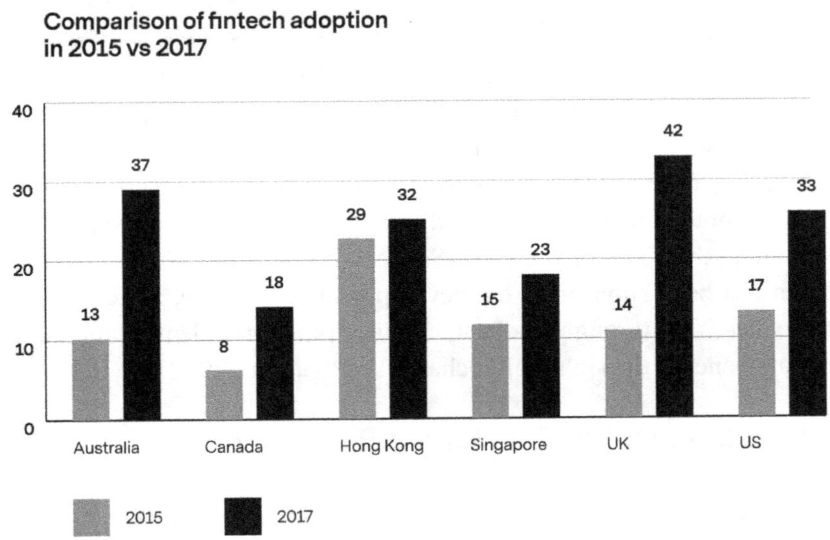

their overall financial health. For instance, the amount of fintech startups in the U.S. reached 2,000 in 2016, with a sufficient number of platforms working towards involving excluded segments of consumers (i.e. Autism Expressed, a learning platform teaching marketable skills to empower individuals with disabilities such as autism); helping to reduce the cost of serving low- and moderate-income users (like Benefit Kitchen, a benefit screening and financial literacy tool that provides eligibility information to low-income families; PYT Funds, a debt reduction model for student loans); and expand access to safe and affordable products (for example, eCredable, a platform for consumers without credit connecting them to affordable financial services). Moreover, there is a list of fintechs from advanced economies aimed at improving financial inclusion in low- and lower-middle income countries, such as money transfer services SimbaPay from the United Kingdom working with African countries or Lenddo based in Singapore, ascertaining financial stability of customers from Latin America, South Asia and South East Asia.

Low- and Lower-Middle Income Developing Economies

In these countries financial and general economic exclusion is prevailing in the society for a number of reasons. There is a lack of physical infrastructure, on the one hand, and a large proportion of population cannot afford the costs of traditional banking products, on the other. This stimulates the development of financial technologies that work across weak infrastructure and provide products at a very low cost - the so-called "microfinance" (Korovkin, 2014). Digital channels like mobile banking (m-banking) are likely to provide better coverage and more cost-effective services to the unbanked population of low- and lower-middle income economies. Conventional banking might not be cost-effective for low-ticket-size transactions; hence, financial inclusion might not be feasible. The most well-known example of an m-banking solution is M-Pesa, a mobile payment service in Kenya, Tanzania and some other markets which is operated through a private telecommunications provider with a nationwide coverage independent of traditional banks. M-Pesa provides financial services to more than 75% of Kenyans (Sangeetha & Koushik, 2015). In India, the government has been supporting the development of a viable, large-scale digital ecosystem since 2009 through Aadhaar, a unique biometric identification aimed at pushing state benefits through digital channels (Parussini, 2017).

Upper-Medium Income Developing Economies

Similar to the low- and lower-middle income economies, in the upper-middle income developing economies (more often referred to as the "emerging markets[3]"), the field for the application of financial technologies as a solution to financial inclusion is wide. Due to the large number of problems of both supply and demand, fintech has huge opportunities to improve the situation. All sectors of financial services are covered by inclusive fintech solutions, including Payments (for instance, e-money in Peru, cashless governmental benefits through South African Social Security Agency in South Africa), Merchant services, Insurance, Lending (for example, an Argentinian bitcoin and digital payments startup BitPagos which has a consumer credit product, enabling those with limited banking history to get credit through the use of its consumer wallet service), Personal finance/savings (like personal finance application GuiaBolso from Brazil), Money transfers, Tools for Financial Institutions (Association of Banks, 2016). The range of technologies underlying the solutions are also wide, including mobile and internet, contactless and NFC payments, electronic money, cloud systems, bitcoin, etc. (Bourreau & Valletti, 2015). Still, the problem that developing markets face is that the market players have strong incentives to focus on the relatively broad base of an affluent population, and are thus less motivated to fill the pockets of financial exclusion, compared to the low-income markets. The combination of these opposite issues – a sufficient number of underserved groups of population and relatively high financial risks of working with them compared to serving the affluent population – makes this group of countries particularly interesting for a more detailed analysis.

Summing up all the cases of inclusive fintech worldwide, the four main advantages of fintech in promoting financial inclusion are of note:

1. As the review of international cases shows, fintech is penetrating all types of traditional banking operations, using a number of technologies and innovations which give financial technologies a certain advantage in covering the gaps in financial inclusion that are inaccessible to banks, by developing access and usage of financial services, and improving their quality, suitability and welfare effect.
2. Interoperability and open application programming interfaces (APIs) of most solutions make inclusivity an intrinsic feature of fintech. Due to open APIs, it is possible to utilize the same technology for a range of inclusion problems in many countries because it allows small innovators to develop, test and refine services for consumers at a very low cost and develop good solutions relevant for the particular inclusivity issue based on a shared technology or platform.
3. Compared to traditional financial services, fintech solutions provision is not limited to only ‘licensed’ operators as there may be parallel or alternative financial institutions among fintech players including fintech start-ups and digital companies. The latter drive innovation and boosts financial inclusion in places where traditional financial institutions have reached their potential, which is often the case in the advanced economies (Breloff & Parker,

2011), while in low-income economies, mobile operators take on the role of financial institutions. At the same time, while large banks still remain key players influencing financial inclusion development in emerging economies, technologies often disrupt their traditional business models compelling them to place digitalization at the center of their strategy to stay profitable (Cheston et al., 2017).

4. Fintech easily adapts to the existing infrastructure, as it utilizes a wide range of technologies, from mobile technologies, which are already developed even in most low-income countries, to more sophisticated ones involving blockchain. At the same time, it can drive improvements in the financial infrastructure through development and growth.

DIGITAL FINANCIAL INCLUSION BACKGROUND IN RUSSIA

Economic inclusivity in Russia is one of the important issues of socioeconomic development. The report by the World Economic Forum on the overall economic inclusivity released in 2017 puts the Russian Federation among the more inclusive of the emerging economies (number 13 out of 78), yet labels it as "slowly advancing" (Samans et al., 2017). The report also notes that Russia's inclusivity position is lower than its GDP ranking, suggesting that there is significant room for improvement. As was discussed above, financial inclusivity is an essential part of the overall inclusivity of socioeconomic system and is currently of high priority in Russia.

As was said, there are basically two groups of barriers to financial inclusion, physical and socioeconomic. The former depends largely on the size of the country and the density of its population, as it is definitely more effective economically to create infrastructure for compact populations. The latter barriers are more manifest in the lower income countries. Thus, to understand the relative level of development of Russian financial services, the country – which is vast geographically and belongs to upper-middle income economies – should be compared to two sets of peers. First, the large-size countries with a relatively low population density, like the USA, Argentina, Canada, Australia and Mongolia – for the comparison of the development of banking infrastructure such as the number of branches, ATMs or POS terminals for payment cards. Secondly, countries that are close in GDP per-capita, like Israel, Portugal, Malaysia, Kazakhstan, Chile - for the comparison of the penetration of banking services, such as the share of population with bank accounts and debit cards.

In general, the 2015 survey results show that between 26%[4] and 66%[5] of the population is not included in formal financial services at all (National Agency for Financial Studies & BDO, *2016;* Central Bank of Russia, 2016). Nevertheless, the analysis of statistics on access points for banks branches, ATMs, payment terminals, client service standards and the price of financial products and services shows that there are no general problems with accessibility of the financial infrastructure. For example, in 2016, Russia had about 17,000 financial organizations including 834 banks with more than 44,000 branches with the average density of 38 bank branches per 100,000 people. While in Canada, Australia, and Argentina, countries with large territories and uneven population density and social demographic structure, this indicator is 23.6, 28.7 and 13.2 branches per 100,000 people, respectively (see Table 1). It follows that there is no apparent correlation – either direct or inverse – between the size of the country and its density of population and the degree of development of its financial infrastructure.

Analyzing countries comparable with Russia by GDP per capita, one can see that based on the penetration of bank accounts, Russia is closer to its less advanced comparables, with 67% penetration level compared to 54% and 63% in Kazakhstan and Chile, respectively. In other countries listed in Table 2 more than 80% of population have bank accounts. As for the penetration of debit cards, the tendency is for moderate to low levels, except for Portugal, which stands out with 67% compared to 32-54% in other countries.

Despite the fact that the general level of financial inclusion is moderate and even relatively high compared to emerging countries, at least 26% of population are still unbanked. This figure is further broken down into small groups of excluded, the so-called "exclusivity pockets" or "inclusivity gaps", due to a number of barriers that might arise due to potential lack of knowledge of financial services, or readiness to use them, distrust in the formal banking system or failure of banks to serve the needs of particular groups of population.

Table 1. Data on access to financial infrastructure in 2016;population density comparable countries

Country	Area, km^2	Population Density, by km^2	Branches, per 100,000 Adults	ATMs, per 100,000 Adults	POS Terminals, per 100,000 Adults
Russia	17,098,242	8.33	38	187	1,117
USA	9,826,675	32.45	32.9	–	2,156*
Argentina	2,780,400	15.47	13.2	61	–
Canada	9,984,670	3.49	23.6	221	2,202*
Australia	7,741,220	2.91	28.7	165	3,939*
Mongolia	1,564,116	1.89	70.4	72.75	448*

*Latest available data is for 2009

Source: (The World Factbook, 2016; National Agency for Financial Studies & BDO, 2016; Central Bank of Russia, 2016; IMF Financial Access Survey, 2016a)

Table 2. Data on penetration of financial services in 2014; GDP per capita comparable countries

	GDP per Capita, in USD Dollars	% of Population With Bank Account	% of Population With Debit Card
Russia	26,100	67%	44%
Israel	34,800	90%	32%
Portugal	28,500	87%	67%
Malaysia	27,200	80%	41%
Kazakhstan	25,700	54%	32%
Chile	24,000	63%	54%

Source: (Factbook, 2016; Demirgüç-Kunt et al., 2015)

Physical Barriers

Although, on average, Russia outperforms most of the comparable countries in terms of bank branches accessibility, the statistics on distribution of credit organizations by size of the settlement shows that while there are more than 412 organizations per city with more than 1m inhabitants, their numbers are decreasing to 8.5 in cities with the population below 100,000 people, and many small towns and villages – even with several thousands of inhabitants – have no branches at all. This is all due to some special features of demographics in Russia where 68.3% of Russians live in the European part of Russia, which accounts for only 20.66% of the territory. The density of the population in the European part of Russia is 27 h/km^2, and that in Asia is 3 h/km^2. Urban population is 74.27% (Rosstat, 2016, 2017).

Moreover, there are infrastructural barriers in the Internet and mobile penetration which are important for the availability of digital financial services. In 2016, the Russian Ministry of Telecommunications reported 27m subscribers to fixed broadband Internet and over 100m subscribers to the mobile Internet (Ministry of Telecommunications of Russia, 2017). As confirmed by consumer research, only 70% of the total population in Russia can go online (FOM, 2017). These distinctive features of Russia's territorial and urban structure hamper the provision of financial services by traditional financial institutions, often making it impossible for them to set up contact with potential users.

Social Barriers

There is a range of socially unsecured groups of population due to their income or education level, age or disabilities. The population structure in Russia shows a huge disparity in access to social benefits and income inequality. However, the situation is getting better over time, the wealth of the Russian population is growing and so is the demand for appropriate financial services. For example, in Russia, there are 20m people (over 14% of population) close to the poverty line, and for most of them traditional financial services are too expensive or inaccessible, i.e. they do not have an appropriate credit history to be eligible for credit from a bank. This group faces a price discrimination barrier, being unable to bear the costs of traditional financial services.

Retired people form another large excluded group, having both a lower income – less than Rub 15,000 (approx. $ 250) a month on average – and a lack information about financial products, compounded by gaps in financial competences and skills. Additionally, there are about 10 million working migrants in Russia, who remit their salaries to their families. The volume of remittances to neighboring countries (e.g. Azerbaijan, Kyrgyzstan, Tajikistan) from Russia reaches $ 20 billion, while the transaction costs charged by traditional providers are very high and include the fee, accessibility and safety of remittances. Social barriers also prevent such groups as students (due to financial instability) and long-term ill or disabled people (because of low accessibility to financial services) from inclusion to financial services. This second type of social barrier, is discrimination by banks based on high operational risks expectations: in most banks, operational strategy often deems services for low-income social classes as both commercially unfeasible (i.e. the costs of maintenance of deposit accounts with lower than average amounts are higher than any potential return on the deposited amounts) and risky (for instance, because of the high possibility of loan default).

Competences and Skills

Although, the level of education is high in Russia, with a 99.7% general literacy rate (measured as the percentage of the population aged 15 and older who can, with understanding, both read and write a short simple statement on everyday life) (UNESCO Institute for Statistics, 2016); the 2015 financial literacy survey showed that while 57% of respondents considered themselves financially literate, only about half of the questions were answered correctly (measured as the proportion of correct answers in the test for knowledge of basic financial concepts like inflation, interest rate, etc.) (National Agency for Financial Studies & BDO, 2016). Additionally, about half of the population may be considered financially dependent – based on the same survey, 43% of population get a bank card for their salaries automatically from their employers and due to low financial literacy they might not understand the nature of the card, for example, that it is linked to a bank account. As a result, those people mostly use the card to withdraw cash and are not able to make independent decisions regarding financial services and products (National Agency for Financial Studies & BDO, 2016).

Moreover, by virtue of the law and regulation in Russia, children over 6 are allowed to make small transactions, and adolescents aged 14 to 18 can independently manage their income (earnings, scholarships, etc.). So, this group of more than 22 million school children needs proper financial services in order to transact with money and at the same time accumulate knowledge and experience to be prepared for the modern financial market when they come of age.

Small and Medium Enterprises (SMEs) Financial Services

Last, but not least, there are also exclusivity issues connected not only to personal finance, but also to access to financial services for *small and medium sized businesses*. For example, there are 5,523,765 entrepreneurs (Resource Center of Small Business, 2016) running small and medium businesses in Russia who do not easily fulfill the requirements of the formal financial sector, mainly when it comes to credit, and especially in the early stages of development.

Summing up Table 3 estimations and taking into account the overlapping of inclusivity gaps, approximately up to 43% of the population (62m adults) in Russia are limited in their access to modern financial services for one or more reasons. At the same time, the "pockets of exclusion" happen for a number of reasons and are relatively isolated, thus, no single straightforward measure can be offered to deal with the problem, and the possible impact of affirmative action policy by authorities is limited. On the other hand, private market players may find it an attractive business idea to fill in the existing market gaps. Further analysis will focus on the examples of such cases.

FINANCIAL INCLUSION CONTROVERSIES IN RUSSIA

Though usually classified as an "emerging market" the Russian Federation faces inclusivity challenges that are more associated with advanced economies. The population is mostly urban, with almost 100% basic literacy, one of the highest tertiary education enrollment ratios in the world, with a high share of skilled employment and relatively low unemployment. At the same time, due to the historic legacy of the Soviet era, when the consumer finance services in the country were rudimentary, the population has a relatively low level of experience and understanding of modern finance, especially in the middle- and older age strata.

This raises the question on how wide the financial services access should be which is given to a financially illiterate and low-income population in order to

Table 3. Inclusivity gaps in Russia;summary table

<table>
<tr><th rowspan="2">Barrier Type</th><th rowspan="2">Exclusivity Pocket Type</th><th rowspan="2">Causes</th><th colspan="2">Estimated Size of Pocket</th></tr>
<tr><th>as a % of Total Population*</th><th>in Million People</th></tr>
<tr><td rowspan="2">Physical</td><td>Demographic features</td><td>Uneven distribution of population. Large disparity between development of urban and village territories</td><td>26%</td><td>38</td></tr>
<tr><td>Infrastructural problems</td><td>Uneven penetration of mobile and internet network among territories</td><td>30%</td><td>44</td></tr>
<tr><td rowspan="5">Social</td><td>Price discrimination</td><td>Disparity in access to social benefits and income inequality</td><td>14%**</td><td>20</td></tr>
<tr><td rowspan="4">Discrimination by banks based on high operational risks expectations</td><td>Migrant workers using remittance services with high transaction costs</td><td>–</td><td>10</td></tr>
<tr><td>Students lacking proper services and experience</td><td>5%</td><td>7</td></tr>
<tr><td>Elderly lacking proper services and experience</td><td>29%</td><td>42</td></tr>
<tr><td>Long-term ill or disabled people lacking proper services and experience</td><td>8%</td><td>12</td></tr>
<tr><td rowspan="2">Competences and skills</td><td>Financial independency</td><td>Large share of those who have got an account at a bank are through salary card from their employers</td><td>43%</td><td>63</td></tr>
<tr><td>Financial Literacy</td><td>Disparity of general level of education versus financial literacy, young people aged 6-18 lacking proper services and experience</td><td>43%</td><td>63</td></tr>
<tr><td>SME Financial Services</td><td>Small and medium businesses</td><td>Lack of financial services for small and medium businesses, especially in the early stages of development</td><td>4%</td><td>6</td></tr>
</table>

* Total population is 146,804,372 people

**Measured as people having less than the living wage monthly income which was Rub 9,828 ($160) as of 2016 ($160)

Source: (Rosstat, 2015, 2016, 2017; Ministry of Education, 2014; FOM, 2017; National Agency for Financial Studies & BDO, 2016; Central Bank of Russia, 2016; IMF Financial Access Survey, 2016a)

overcome social risks. The main problem lies in the expanding availability of credit as the key strategy for increasing banks' revenues, which in the situation of financial illiteracy can lead to terrible consequences. The subprime crisis in 2008 in the USA as well as India's 2010 Andhra Pradesh microfinance crisis can be seen as crises of financial inclusion, when uncontrolled growth in access to financial services can contribute to financial disturbances and social discontent (Hannig & Jansen, 2010; Raghuram, 2010). Russia is now facing a similar problem, with the annual growth of 'bad' debts of 32% in 2016 (United Credit Bureau, 2016).

Moreover, there is an additional financial inclusion controversy in Russia in that survey results are ambiguous and indicate a low level of financial services penetration: - for example, two surveys conducted in Russia late in 2015 show different results on "bank account" possession – 44% versus 74% of respondents (National Agency for Financial Studies & BDO, 2016; Central Bank of Russia, 2016). At the same time, the number of cards per capita, at almost 2 per person, is much higher, demonstrating that people do not recognize the relationship between possession of a card and a bank account.

Overall financial non-inclusion in Russia has traits of both advanced economies (relatively rare and separated cases of non-inclusion in specific social groups) and emerging economies (strong disparities between regions in term of development). It is important that in terms of the modern infrastructure like telecommunications Russia definitely belongs to the advanced economies. This peculiar situation requires development of a country-specific model of financial non-inclusivity based on a financial technologies solution which serves as the basis for analysis and strategic planning.

FINTECH MARKET PLAYERS FILLING THE INCLUSIVITY GAPS IN RUSSIA

Online semi-formalized questionnaire interviews of 37 fintech companies shows that 82% of fintech start-ups surveyed self-attributed their services as having a positive effect on financial inclusion, with 54% claiming this effect to be high which shows that the fintech market players are, on the one hand, aware of the existing gaps in availability of financial services (SKOLKOVO School of Management, 2016). At the same time, in their business strategy planning these companies take a different approach, usually starting from the possible technological solution and then developing a financial service, i.e. an offer to the customers. In many cases these offers fill the existing gaps in financial inclusivity. This duality leads to Figure 2.

Here are five short case-studies of fintech solutions from Russia trying to fill in the niches of financial inclusivity.

Figure 2. Some ways in which the fintech companies fill the inclusivity gaps
Source: Authors' analysis

Inclusivity gaps ▾ / Type of financial service ▸	C2B payments accessibility	B2B payments accessibility	Accessible credit	Financial advice
Physical	Payments through mobile phones, payment cash-in		Peer-to-Peer lending	
SME financial services	Payments through mobile phones, payment cash-in	Payments at lower transaction costs	Credit products with appropriate terms and conditions for SME and early-stage business	Targeted services for financial consultations
Competences and skills	Micro-payments with low commission, trans-border payments		Credit scoring solution for better loan portfolios	Targeted services for financial education
Social	Easy-to-execute payments		Credit products with transparent terms and conditions	Learning-by-doing financial services and products

Financial Technologies for Cash to Non-Cash Transfers for Remote Areas Underserved by Bank Branches: QIWI Terminals and Wallets[6]

Inclusivity Pocket: Physical Barriers

History of the Problem

As of 2000 the traditional banking system in Russia was still in its infancy, it had no resources to provide the public with proper financial services to cover their needs for payment operations. For example, with a widespread pre-paid mobile phone tariff plans, people looked for an easy and cheap way to top up their phone balance. Although several solutions emerged, including the so-called 'scratch' cards distributed by retail chains, they were unable to provide complete coverage, particularly in the remote areas due to territorial barriers. The solution was in dealing with small individual shops as agents to build a wide network; however, telecoms were reluctant due to concerns over receiving payment.

Technical Solution and Business Model

While searching for a solution to the problem, in 2004 QIWI, a company founded through the merger of a 'scratch' card producer and an instant payments provider, introduced the network of cash-in machines (QIWI, 2017). Any small shop could install these machines, or terminals, operated by QIWI acting as an intermediary in money-collection for telecoms and other partners, including utilities and communal services etc. Each payment is subject to a commission from 0% to 5% depending on the sum. Although not a pioneer in the business of cash-in terminals, the main advantage of QIWI over its competitors was a convenient user interface and the fact that company undertook a risk in expanding its network to work in distant geographic locations and succeeded. In 2017, the company has more than 160,000 cash-in terminals with 56m users monthly, making QIWI terminals network comparable with the ATMs network.

In 2008, as a response to the need of even more convenient payment service, QIWI launched e-wallets with extended functionality compared to the terminals, such as easy access from any device connected to the Internet, payments to any merchant or P2P transactions, and lower fees. QIWI commission ranges from 0% to 0.75% based on the amount and the bank that issued the card that is linked to the wallet. After acquiring a full license, the company launched its QIWI Visa Plastic, enabling the usage of e-money offline. The card's price includes a commission for the issue of 2.5% (but not less than 25 rubles or at least $1 when buying the card in US dollars). Payments using the card are commission-free. In 2017, there have been about 17m users of QIWI Wallet service.

Benefits and Results

The network of cash-in terminals which are installed even in the remotest and smallest settlements still remains the important solution to the social and physical inclusivity gaps in Russia. At the same time, QIWI's e-money solutions may be integrated by small online retailers' payments, who usually cannot afford expensive and sophisticated traditional services provided by banks allowing them to expand their customer base. Moreover, both terminals and e-wallets are easy to use and either do not require special competences and skills or are characterized by a learning-by-doing feature.

Alternative Credit Scoring for Boosting the Quality of Microfinance Portfolios: Axicredit.ru

Inclusivity Pocket: Barriers for SME Financial Services (Microcredit Organizations), Price Discrimination Social Barriers

History of the Problem

Microcredit plays a dramatically important role in providing underprivileged social groups and remote geographies of Russia with credit. In Q12017, there were about 900,000 microcredits issued (half of the amount of loans in banks issued at the same period), and the demand shows positive dynamics compared to the same period last year – the growth is 8% (United Credit Bureau, 2017). However, recently microcredit has been facing regulatory issues with the introduction of two laws: one limiting the interest accrual and the other the so-called "anti-collection law", limiting hard methods of debt collection. This raised the question of profitability, as MFOs need a moderate quality loan portfolio together with a less formal credit management, compared to banks, to maintain inclusivity of credit – to remain profitable. In order to be sustainable, the model should not be based on higher risk tolerance, so credit scoring is often managed manually, which is flexible, but expensive, compared to computer-based scoring systems at banks.

Technical Solution and Business Model

Axiomatica, a Russian start-up company launched in 2012, introduced a "cloud technology" approach as a solution to the dilemma. Microfinancial companies get access to the platform called AxiCredit (www.axicredit.ru) providing all the processes of credit management and, in particular, scoring. The FICO Origination Manager Decision Module is a foundation of Axicredit's credit risk evaluation model. At the centre of the company's solution is an architectural innovation which introduced a separation of the Scoring Strategy module. This allowed it to make the module's decision rules transparent and easy to manage directly by clients. By 2017, AxiCredit client list included no less than 10 MFOs and banks, getting revenues from a fee from application processing (about $1 per application).

Benefits and Results

Despite the fact that technology is fairly novel to the market, the company has already managed to form a base of highly enthusiastic clients both from the microcredit sector and small banks. The benefits are numerous. First, the system can be easily installed and gives high-quality decisions in a short time due to its cloud architecture. Secondly, the platform uses transparent rules, allowing client companies to change their credit acceptance and interest rate criteria in response to a market situation. This allows it to maintain the equilibrium between the liquidity supply, cost and market demand. So, the service is beneficial to small financial companies as well as its potential clients.

Financial Technologies to Manage the Risks of Cash Handled by Children: "Ladoshki" project

Inclusivity Pocket: Competences and Skills – School Children

History of the Problem

In Russia, like elsewhere, parents should either regularly (weekly or monthly) pay or pass cash to their children for nutrition at school in advance or provide the child with cash for every-day purchase of food of the child's choice at the school cafeteria as well as pocket money for small everyday transactions. Additionally, school children often use public transport to reach the school, so they have to pay for transport in cash as well. However, this obviously has risks for children, often inexperienced in financial services: loss, theft/extortion with associated psychological consequences, and incorrect cash management.

In 2014, the "School Card", a pre-paid card or bracelet with NFC technology used for payments or pass, identification information and photo of the student, connected to the parents' internet bank account and SMS alerts was piloted in some regions. It bundles the permit/crossing pass, cashless nutrition and other school payments, cashless transport payments as well as cashless, out of school "pocket money" transactions. Although it showed some benefits and was used by more than 200,000 schoolchildren in 2017, there is a sufficient disadvantage as parents are claiming that children either lose or damage their cards, so that parents should pay for them to be reissued. So, a much better instrument for child's identification for financial transactions was needed.

Technical Solution and Business Model

In 2015, the project "Ladoshki" based on the biometric technology of payment with the palm of the hand was introduced through a collaboration of pioneers in innovative finance in Russia – the government, a big state bank, an international payments system in partnership with Yandex.Money (the company's case is discussed further in this Chapter) – with a goal to help schools and parents overcome the risk of cash handled by school children.

To pay, a child simply needs to enter the amount and just place his/her palm on the sensor. It analyzes the palm "print" ie the structure of the capillaries in the child's palm, and a special optical system integrated in the sensor identifies the payer, after which the amount is automatically deducted from the parent's account. So, the solution is based on sharing the parent's bank account with a child. The account, in turn, may be replenished without commission in a variety of ways: by linking the bank card with the child's hand, and the account will be replenished automatically, via the Internet bank or the bank's mobile application as well as at ATMs.

This instrument is also free of fraud as biometric data is not stored: palm scans are converted into a digital code, which is transmitted to the data center via secure communication channels.

Benefits and Results

So, by 2017, the "Ladoshki" project has been successfully introduced in at least 150 schools, including schools in cities with a population under 500,000.

Parents point at the benefits of this project: better control over attendance, control over the expenses of the child, convenience in replenishing the card and the possibility of transferring money to the child, and even more, ability to pay for their child's nutrition in credit. School children are able, in turn, to use modern payment systems and manage their personal finance, which increases the level of their financial culture and literacy as well as improves security due to the decrease in the use of cash. Bank service providers also claim that parents whose children are involved in these projects also use financial products and services more often and with better understanding. Local authorities have the opportunity to control the funds allocated from the budget for school children. Nutrition organizations and educational institutions, in turn, reduce the turnover of cash, reduce the cost of collection and fraud risk, and have an instrument which helps to plan nutrition menus.

Financial Technologies for Fundraising/Crowdfunding of "Micro-Causes" in P2P and P2B: Yandex.Money

Inclusivity Pocket: Social, Physical, Competences and Skills Barriers

History of the Problem

The ways of peer-to-peer (P2P) money remittances offered by traditional providers are often pricy and/or difficult to access. E-wallets overcome these two major disadvantages allowing money to be sent and received faster and at lower cost. Moreover, the digital revolution in P2P products had a number of significant social effects, for example, crowdfunding: collection of funds from a large number of people for a business project or a cause. Although money contributions from the public using various channels have always been the center of charity organizations' mode of operations, the process used to be connected with excessive management costs. Due to these costs, small organizations and private civil volunteers often gave up fundraising campaigns. This is in contrast to international practice in crowdfunding, where, for example, more than 125,000 projects collected more than $3bn on the biggest crowdfunding platform Kickstarter.com (a popular American funding platform). The Russian public is not eager to contribute much to early-stage, 'would-be' projects. So, there is a need for a service, which allows to funds to be collected in stages and to show the project's progress in order to raise money for its development.

Technical Solution and Business Model

Yandex.Money, a big financial innovations market player in Russia since 2002, has introduced a number of products for crowdfunding management by both individuals and professional organizations, based on e-money and other digital instruments. The fundamental idea is that fundraising is, of its nature, a social activity, putting communication management on the same level of importance as the transfer of money. The services are integrated into popular social network platforms like Facebook and VKontakte (a popular analogue of Facebook in Russia and the CIS) with posts promoting the cause and eye-catching instruments like instantaneous reporting of the sum collected or progress achieved.

In 2014, the company launched a platform for personal money collection vmeste.yandex.ru ("vmeste" means "together" in Russian), which in 2016 was relaunched as yasobe.ru (an original way of saying "I will raise"). The service allows funds to be raised for a business project, for an art or musical project, for extracurricular activities in a school class or even for a civil urbanist initiative. About 200,000 people visit the yasobe.ru web site monthly.

Moreover, the core Yandex.Money activities, its B2B solution Yandex. Kassa (translated as Cashbox) and Yandex.Wallet also play an important role in crowdfunding. The main goal of Yandex.Kassa is to integrate all possible means of payment in one stream providing single-source accounting statements. By 2017, it gathered more than 86,000 retailers, both major online market players and the small and micro- businesses including charity organizations. So that one can easily contribute to charity organizations using Yandex services, Yandex.Wallet service enables the remittances of e-money to any private person or business, connected to Yandex.Kassa.

Benefits and Results

A digital crowdfunding service adapted to Russian realities and introduced by Yandex.Money allows getting group financing for small, but important, activities, like microbusinesses, art projects and so-called "microcauses" – private initiatives aimed at contributing for the society. The analysis of the Yandex.Money service shows that, in general, Russians have invested in crowdfunding projects 70% more funds in 2016 compared to 2015. At the same time, the number of those who transfer money to projects through the service doubled - 290 million rubles were collected through Yandex.Kassa and yasobe.ru.

Financial Technologies in Personal Finance Budgeting: Easyfinance.ru

Inclusivity Pocket: Competences and Skills

History of the Problem

As mentioned earlier, Russia has an issue with financial literacy leading to problems with personal budgeting, low skills of usage of financial services and even such severe problems as overborrowing. Based on expert opinion, only 10 to 20% of population want to become financially educated and skilled, and 20% are getting involved, including those who want to possess personal budgeting skills.

Since correlation between financial literacy and usage of financial services is positive and significant, there is a possibility of increasing financial inclusion by providing appropriate solutions to the demand.

Technical Solution and Business Model

Following this idea, the service easyfinance.ru was launched in 2009. Easyfinance.ru combines a personal finance budgeting interface with a financial consultation service based on the expenses and gains listed by the user. In contrast to other personal budgeting solutions, this project is developed as a platform, not just an application. It allows other applications, such as internet-bank applications or other financial services to connect, it has an open APIs, and promotes integration with the ecosystem.

At the beginning, the project was adapted to the Russian reality of the cash economy, providing instruments to keep records easily not only for cashless transactions by cards, for which there is an SMS or Internet-bank reporting, but also for money flows in cash, in contrast to existing solutions in other countries where cashless transactions were more developed. Currently, Easyfinance.ru combines the control of both cash and cashless flows and aggregates information from more than 200 banks for 300,000 clients. There are 4 types of accounts: Free (with no fee), Light Pro, Medium Pro, Full Pro, which differ in price (from $1 to $2.5) and the number of available services, such as expanded budget, advanced reports, the ability to synchronize with a mobile application, or online backup. About 4% of users had paid accounts at the end of 2015.

The service also covers the B2B-segment, providing the banks with financial management solutions, meeting the demand of banks to give value to their clients and deepen the usage of their services by clients, compared to the current situation when loans and deposits are the most popular products.

Benefits and Results

Analysis of service users shows that 55% noticeably improved their financial condition, 25% got rid of debts, and 22% of users achieved their goals within a year of use. Moreover, the clients of Easyfinance.ru showed a tendency to increase the share of cashless transactions to 80% compared to cash (20%). Ability to collect such statistics allows financial services providers to understand better the behavior of users.

Summing up the presentation of successful case studies of digital financial products, a specific model of business can be developed. Such a model is two-dimensional:

- Dimension of "Inclusivity gaps" outlines the segments of non-included population based on different barriers to inclusivity, like physical, social, competences and skills, or institutional as discussed above.

- Type of financial service dimension is the basis on which the fintech solution is built.

Case-studies also show that different market actors are working in each segment to achieve financial inclusion through technological solutions: smaller and regional banks, government-owned banks with an explicit mandate for inclusivity, MFOs and credit cooperatives, telecom operators, fintech companies.

POLICY AND STRATEGIC IMPLICATIONS

In general, despite its significant role, fintech is only part of the financial sector development. There is also a sufficient change in traditional banks' strategies, with fintech's disruptive role as one of the reasons. For banks that incorporate financial inclusion into operations, digital payments become an important gateway to new customers, so that they get involved in technological advancement. As a result, in Russia, the main consumers of fintech are banks competing among themselves for the level of service. Moreover, big new players are emerging as Internet companies and telecoms are becoming banks to a certain extent, issuing bankcards and providing financial services.

Figure 3. Cases of fintech solutions filling the inclusivity gaps
Source: Authors' analysis

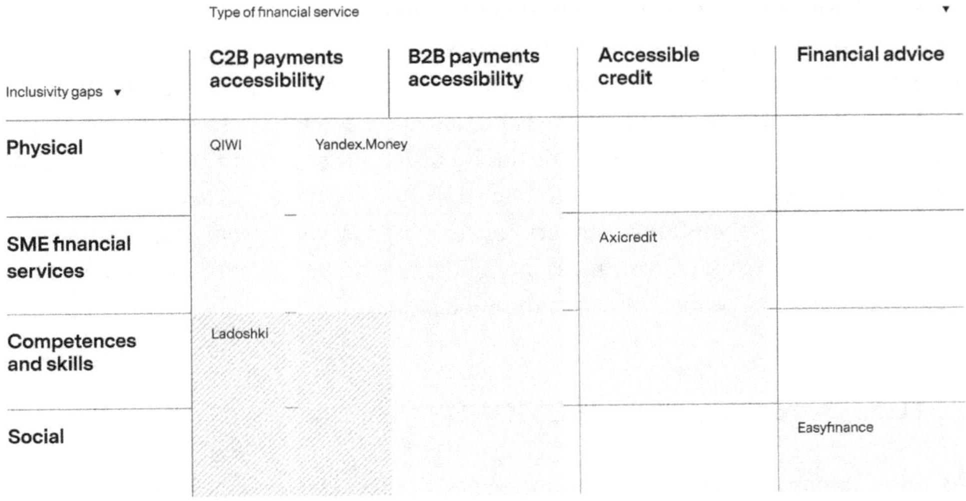

Thus, with all these developing technologies and innovations together with the growing number of agents involved in financial markets, there is a need for regulatory approaches that will stimulate the offering of market products that close niches in inclusiveness: physical, social, competences and skills as well as SME financial services barriers, without creating unacceptable risks. Regulation is needed to protect customer funds held inside the digital transactional platforms, support privacy and security of users' financial data, ensure transparency of financial products and services, guarantee proper customer rights as well as provide protection for financial market players against operational risks (CGAP, 2014). One effective solution might be the development of risk-proportional regulatory initiatives in order to manage the disruptive consequences or potential effects on the financial market. Additionally, governments can support the penetration of fintech solutions by introducing them in G2C (government-to-consumer) and G2B (government-to-business) payments (like Aadhaar program in India).

Moreover, currently, there exists an under-estimated feature that technology requires the development of distribution channels. To access digital financial services, access to a mobile connection is important, but it is equally important to be able to convert cash to digital money and, at least for now, back into cash again. So mobile phones have been important in places like Kenya, but the real game changer has been the emergence of large and well-functioning agent networks. So, there is an increasing role for government policies and other market players actions in promoting financial technologies through providing proper infrastructure.

Additionally, following the Russian experience, fintech entrepreneurs from emerging markets, although having brilliant ideas on how to cover the particular financial inclusion gap, often lack proper competences and skills. It could be IT skills, a barrier to product prototype realization, experience in the legal field, necessary for the product to comply with laws and regulations, or business modeling, essential for economical success. Moreover, there is no one approach to access fintech projects' success and value as these projects are totally different from traditional projects and companies. As a result, it is hard for investors in emerging countries to choose projects with potential. It is therefore necessary for such big market players as governments, banks and other financial institutes, large business, experienced practitioners and even academia, to support fintech entrepreneurs.

FUTURE RESEARCH DIRECTIONS

As this Chapter is one of the first attempts in the academic sphere to present an overview of the fintech and its (disruptive) role in financial markets and, in particular, in financial inclusion, there is still room for future research on the topic.

First, the analysis may be deepened by introducing additional case studies of fintech projects aimed at inclusivity gaps mentioned above. This will allow for generalizations and representativeness, which are essential for the development of well-substantiated theories on the issue. Alternatively, in order to get more quantitative data on the role of fintech projects in financial inclusion and financial market as a whole, a survey might be conducted, which opens up a possibility for statistical and econometric analysis. Secondly, as some comparable countries were mentioned in current research, the analysis could go wider and study the situation in other countries and regions, with a focus on their peculiarities and similarities in terms of fintech disruptive and inclusive roles.

An additional direction of further research could be to go deeper in studies of consumer behavior and analyze how consumers combine services of different types of financial organizations or choose between them. For example, to get more qualitative data, financial diaries as a survey instrument may be applied in each particular country or region. Financial diaries have already been introduced in South Africa, India and the U.S. (Collins, 2008; Kamath, Mukherji & Ramanathan, 2010; Hannagan & Morduch, 2015). As a result of such research, both market players and national and international regulators get information on the ways of effective introduction of financial products and services in response to the existing challenges of financial inclusion at acceptable levels of risks and costs.

CONCLUSION

As was shown, financial inclusion is increasingly seen in the world as an important driver of socioeconomic development. In the advanced economies meeting the challenges of inclusion is viewed as an important contribution to the creation of a just society of equal opportunities. In the context of developing markets, financial inclusion is also a powerful lever of economic growth that allows many of the existing informal financial practices to be institutionalized and thus made more effective.

Until recently it was widely believed that financial inclusion challenges should be solved primarily through the affirmative actions of governments, including the policies of the national financial regulators: establishing the rules and incentives for licensed banks and other traditional financial players to work with the "excluded" population. However, modern digital financial technologies make working with the "bottom of the pyramid" consumers effective for private market players. This advancement disrupts the established paradigm and offers an important new class of solutions to the problem.

As this Chapter demonstrates, the private market players are capable of filling any of the existing gaps in financial inclusivity in markets like Russia, with a very complex model of "exclusion". Though this solution is not universal – the gaps are not fully filled – it deserves the attention of both researchers and policy makers. The proposed analytical approach allows a prioritized set of targeted actions to be outlined for the possible measures, increasing financial inclusion based on cost-benefit analysis. This analysis provides for an effective combination of actions by market players and strong affirmative regulatory actions. Such actions should aim at promoting means of non-direct market participation - regulatory environment and targeted operational stimuli to the private and public market players.

REFERENCES

Arner, D. W., Barberis, J., & Buckley, R. P. (2015). The evolution of Fintech: A new post-crisis paradigm. *Geo. J. Int'l L.*, *47*, 1271.

Arun, T., & Kamath, R. (2015). Financial inclusion: Policies and practices. *IIMB Management Review*, *27*(4), 267–287. doi:10.1016/j.iimb.2015.09.004

Association of Banks. (2016). *With the launch of electronic money starts a new era in financial inclusion.* Retrieved March 10, 2017, from http://www.asbanc.com.pe/Paginas/Noticias/DetalleNoticia.aspx?ItemID=209&lang=en

Beck, T., Demirgüç-Kunt, A., & Levine, R. (2007). Finance, inequality and the poor. *Journal of Economic Growth*, *12*(1), 27–49. doi:10.100710887-007-9010-6

Beck, T., Demirgüç-Kunt, A., & Peria, M. S. M. (2007). Reaching out: Access to and use of banking services across countries. *Journal of Financial Economics*, *85*(1), 234–266. doi:10.1016/j.jfineco.2006.07.002

Bourreau, M., & Valletti, T. (2015). Enabling Digital Financial Inclusion through Improvements in Competition and Interoperability: What Works and What Doesn't? *CGD Policy Paper*, *65*, 1–30.

Breloff, P., & Parker, S. R. (2011). Financial Inclusion in the U.S.: Spending Time In Our Own Backyard. *CGAP Blog*. Retrieved February 11, 2017 from http://www.cgap.org/blog/financial-inclusion-usspending-time-our-own-backyard

Buera, F. J., Kaboski, J. P., & Shin, Y. (2012). *The macroeconomics of microfinance (No. w17905)*. National Bureau of Economic Research. doi:10.3386/w17905

Bull, G. (2017). Four Drivers of Change for Financial Inclusion in 2017. *CGAP Blog*. Retrieved February 11, 2017 from http://www.cgap.org/blog/four-drivers-change-financial-inclusion-2017

Central Bank of Russia. (2016). *Review of the status of financial inclusion in the Russian Federation in 2015*. Retrieved January 30, 2017 from https://www.cbr.ru/finmarkets/files/affordability/rev_fin_20161110.pdf

CGAP. (2014). *Digital Financial Inclusion: Implications for Customers, Regulators, Supervisors, and Standard-Setting Bodies*. Retrieved January 9, 2017, from https://www.cgap.org/sites/default/files/Brief-Digital-Financial-Inclusion-Feb-2015.pdf

Cheston, S., Conde, T., Bykere, A., & Rhyne, E. (2017). *The Business of Financial Inclusion: Insights from Banks in Emerging Markets*. Retrieved February 11, 2017 from Http://Www.Centerforfinancialinclusion.Org/Storage/Documents/Iif_Cfi_Report_Final.Pdf

CIA Factbook. (2016). *The world factbook*. Retrieved June 17, 2017 from: https://www. cia.gov/library/publications/the-world-factbook

Collins, D. (2008). Debt and household finance: Evidence from the Financial Diaries. *Development Southern Africa*, *25*(4), 469–479. doi:10.1080/03768350802318605

Dabla-Norris, M. E., Ji, Y., Townsend, R., & Unsal, D. F. (2015). Identifying constraints to financial inclusion and their impact on GDP and inequality: A structural framework for policy (No. 15-22). International Monetary Fund.

Delis, M., Hasan, I., & Kazakis, P. (2013). Bank Regulations and Income Inequality: Empirical Evidence. *Review of Finance*.

Demirgüç-Kunt, A., Klapper, L. F., Singer, D., & Van Oudheusden, P. (2015). *The global findex database 2014: Measuring financial inclusion around the world*. Academic Press.

Duflo, E., Banerjee, A., Glennerster, R., & Kinnan, C. G. (2013). *The miracle of microfinance? Evidence from a randomized evaluation (No. w18950)*. National Bureau of Economic Research.

Dupas, P., & Robinson, J. (2013a). Savings constraints and microenterprise development: Evidence from a field experiment in Kenya. *American Economic Journal. Applied Economics*, *5*(1), 163–192. doi:10.1257/app.5.1.163

Dupas, P., & Robinson, J. (2013b). *Daily needs, income targets and labor supply: evidence from Kenya (No. w19264)*. National Bureau of Economic Research. doi:10.3386/w19264

EY LLP. (2017). *EY FinTech Adoption Index 2017: The Rapid Emergence of FinTech.* Retrieved July 11, 2017 from www.ey.com/Publication/vwLUAssets/ ey-fintech-adoption-index-2017/$FILE/ ey-fintech-adoption-index-2017.pdf

Financial Inclusion Commission in the U.K. website. (n.d.). *History*. Retrieved March 21, 2017, from http://www.financialinclusioncommission.org.uk/history

FOM. (2017). *Dynamics of Internet Penetration in Russia. Survey Results*. Retrieved May 1, 2017 from http://fom.ru/SMI-i-internet/13585

Global Partnership for Financial Inclusion website. (n.d.). *About*. Retrieved March 21, 2017, from https://www.gpfi.org/about-gpfi

Goetzmann, W. N., & Jorion, P. (1999). Re-emerging markets. *Journal of Financial and Quantitative Analysis*, *34*(1), 1–32. doi:10.2307/2676244

Hannagan, A., & Morduch, J. (2015). *Income gains and month-to-month income volatility: Household evidence from the US Financial Diaries*. Academic Press.

Hannig, A., & Jansen, S. (2010). *Financial inclusion and financial stability: Current policy issues*. Academic Press.

International Monetary Fund. (2016a). *World economic outlook database*. Retrieved April 1, 2017, from https://www.imf.org/external/pubs/ft/weo/2017/01/weodata/index.aspx

International Monetary Fund. (2016b). *Financial Access Survey 2015*. Retrieved March 2, 2017 from http://data.imf.org/?sk=E5DCAB7E-A5CA-4892-A6EA-598B5463A34C

Kabakova, O., Korovkin, V., & Plaksenkov, E. (2015). *Digital platforms and the ecosystems of financial inclusion. The Russian experience' report*. Retrieved April 3, 2017 from https://iems.skolkovo.ru/downloads/documents/SKOLKOVO_IEMS/Research_Reports/SKOLKOVO_IEMS_Research_2015-11-11_en.pdf

Kalinin, A., Shapenko, A., Korovkin, V., Zaitseva, N., Remyga, O., Novak, K., & Michael, B. (2016). *Emerging Markets Decoded: The Four Domains of Development for Growth*. Academic Press.

Kamath, R., Mukherji, A., & Ramanathan, S. (2010). Ramanagaram financial diaries: Cash patterns and repayments of microfinance borrowers. *Enterprise Development & Microfinance*, *21*(2), 101–117. doi:10.3362/1755-1986.2010.010

Korovkin, V. (2014). *Microcredit in Russia. On the eve of a boom or a crisis?* Retrieved May 24, 2017, from https:// iems.skolkovo.ru/downloads/documents/SKOLKOVO_IEMS/Research_Reports/SKOLKOVO_IEMS_Research_ 2014-12-06_en.pdf

Mehrotra, A. N., & Yetman, J. (2015). *Financial inclusion-issues for central banks. BIS Quarterly Review*. Basel: Bank for International Settlements.

Ministry of Telecommunications in Russia. (2017). *Report for the IV quarter of 2016*. Retrieved January 30, 2017 from http://minsvyaz.ru/ru/pages/statistika-otrasli/#section-403

Moscow School of Management SKOLKOVO. (2016). *FinTech in Russia: Analysis of Survey Results*. Author.

National Agency for Financial Studies & BDO. (2016). *The level of availability of financial services in the Russian Federation*. Report on the results of a comprehensive study among consumers and financial service providers. Retrieved May 25, 2017 from http://nafi.ru/projects/finansy/uroven-dostupnosti-finansovykh-uslug-v-rossiyskoy-federatsii/

Parussini, G. (2017). India's Massive Aadhaar Biometric Identification Program. *Wall Street Journal*. Retrieved April 29, 2017, from https://blogs.wsj.com/briefly/2017/01/13/indias-massive-aadhaar-biometric-identification-program-the-numbers/

PricewaterhouseCoopers (PwC). (2016). *Global FinTech Report 2016*. Retrieved May 4, 2017 from https://www.pwc.ru/en/banking/publications/fintech-global-report-eng.pdf

QIWI. (2017). *Corporate history*. Retrieved January 1, 2017 from https://corp.qiwi.com/en/company/history.action

Rajan, R., & Lines, F. (2010). *How hidden fractures still threaten the world economy*. Princeton, NJ: Princeton University Press.

Resource Center of Small Business. (2016). *Statistics on Small and Medium Enterprises*. Retrieved May 10, 2017 from http://rcsme.ru/ru/statistics

Rosstat. (2015). *Statistics on Demographics 2015*. Retrieved May 15, 2017 from http://www.gks.ru/free_doc/new_site/population/demo/Popul2015.xls

Rosstat. (2016). *Statistics on Demographics 2016*. Retrieved May 15, 2017 from http://www.gks.ru/free_doc/new_site/population/demo/Popul2016.xls

Rosstat. (2017). *Statistics on Demographics 2017.* Retrieved May 15, 2017 from http://www.gks.ru/free_doc/new_site/population/demo/Popul2017.xls

Sahay, R., Čihák, M., N'Diaye, P., & Barajas, A. (2015). Rethinking financial deepening: Stability and growth in emerging markets. *Revista de Economia Institucional, 17*(33), 73–107.

Samans, R., Blanke, J., Hanouz, M. D., & Corrigan, G. (2017). The inclusive growth and development report 2017. *In Geneva: World Economic Forum.* Retrieved February 11, 2017 from http://www3.weforum.org/docs/WEF_Forum_IncGrwth_2017.pdf

Sangeetha, C., & Koushik, K. S. (2015). *Catalyst for financial inclusion. Bank the unbanked with digital channels. White paper.* Retrieved March 30, 2017, from https://www.infosys.com/industries/financial-services/white-papers/Documents/catalyst-financial-inclusion.pdf

Sanyal, P. (2014). *Credit to Capabilities: A Sociological Study of Microcredit Groups in India.* Cambridge, UK: Cambridge University Press. doi:10.1017/CBO9781139939461

UNESCO Institute for Statistics. (2016). *ICF Macro Demographic and Health Surveys and UNICEF's Multiple Indicator Cluster Surveys.* Retrieved January 15, 2017 from http://data.uis.unesco.org

United Credit Bureau. (2016). *Report on credit structure and trends in Russia.* Retrieved May 14, 2017 from http://www.bki-okb.ru

World Bank Group. (2013). *Global Financial Development Report 2014: Financial Inclusion.* World Bank Publications.

KEY TERMS AND DEFINITIONS

Crowdfunding: Collection of money from a large group for a business or a cause.

Electronic Money: Electronic store of monetary value on a technical device that may be widely used for making payments to entities other than the e-money issuer.

Financial Inclusion: The situation where everyone is provided with a full suite of high-quality financial services and is able to use them, while there exists appropriate infrastructure and ecosystem of financial market, which together enable these services to improve personal and social welfare.

Financial Literacy: The knowledge and understanding of financial concepts (i.e. interest rates, credit and insurance, discounting), ability to use financial services and products, competences in private financial decisions.

Financial Technologies: An innovative technology in the financial industry that changes, breaks, substitutes, supports or evolves parts of/the whole value chain of the traditional players through offering easier and cost-efficient solutions to businesses or consumers.

Inclusive Growth: A phenomenon that advances fair opportunities for economic participants during economic growth with benefits sustained in each sphere of society.

Inclusivity Gap/Exclusivity Pocket: The sector of population not covered by financial services due to particular reasons.

Mobile Banking: An access to financial services through a mobile phone allowing transactions like payments, remittances, etc. to be executed.

Online Banking: An access to financial services through any device connected to the Internet allowing transactions like payments, remittances, etc. to be executed.

ENDNOTES

1 Microeconomically, access to finance influences both individuals and firms. Firstly, the lack of access to financial services may lead to poverty traps and inequality (Beck, Demirg-Kunt & Levine, 2007). Moreover, a growing literature focuses on the positive consequences of access to financial services which are an increase in savings, productive investment growth (Dupas & Robinson, 2013a), consumption stimulation (Dupas & Robinson, 2013b) and female empowerment (Sanyal, 2014). As for the firms, it was shown by empirical research that small businesses gain advantages from access to credit products (Banerjee et al., 2013).

2 Some research also shows the positive effects of financial inclusion on macroeconomic indicators (Sahay et al., 2015): economic stability, measured by aggregate consumption volatility (Mehrotra & Yetman, 2015), growth (Dabla-Norris et al., 2015), consumption and output (Buera, Kaboski, & Shin, 2012).

3 Such economies are often called the "emerging markets", which is a complex phenomenon. The latter is more than just an amalgam of well-established metrics including GDP per capita or growth rate, it includes the combination of opportunities and risks for trade and investment in those markets. The concept arose in the late 1980s, reflecting the unprecedented scale of business opportunities in the countries around the globe that were opening to international trade (Goetzmann & Jorion, 1999). The main characteristics of such countries were industrialization, with higher level of human capital, substantial technology potential and the ambition to be competitive (Kalinin et al., 2016).

4 The share of adult respondents who answered positively to the question of using at least one open account in credit organization, survey of 5,000 users.

5 The share of adult respondents who answered positively to the question of using at least one open account in credit organization, survey of 1,589 respondents.

6 This and the latter, Axicredit and Yandex.Money cases are based on the results of a case-study conducted by the authors of this Chapter in 2016 as part of Digital Platforms and the Ecosystems of Financial Inclusion. The Russian experience report (Kabakova, Korovkin & Plaksenkov, 2015).

Chapter 6
Cleantech and Water Treatment as a Case of Disruptive Innovation

Vincent Sabourin
University of Quebec, Canada

ABSTRACT

Shifts in the disruptive innovations and the growing impact of consumer cleantech are best understood by looking at megatrends in water management. Trends are global long-term forces of development that impact society, economy, environment, culture, and personal lives. Trends are continuously changing the dynamic landscape in which companies operate. This chapter examined commercial opportunities for the internet of things in the market of cleantech with respect to water treatment. The chapter theoretically looked at the four clusters of water management as relates to the application of internet of things: metering, piping, reclamation, and agricultural clusters. An analysis of market strategies and competitive strategies is conducted leading to a brief development of innovation frontiers in water management.

DOI: 10.4018/978-1-5225-4148-6.ch006

INTRODUCTION

Disruptive innovations such cleantech are swiftly capturing the interest of scholars and industry practitioners due to its immense economic and social contributions. The emergences of disruptive waves have significantly revolutionized the cleantech sector especially in water treatment (Grady, 2016). According to Markets and Markets (2017), extensive market research and analysis of this fast paced technology as well as its trends, forecasts and the profiling of major actors show how vital, the growth of disruptive innovations in cleantech has become. Huston (2015) had argued that the effectiveness and efficiency of internet of things has led to great discoveries in the niche of communication and business management. Internet of things through the lens of cleantech market has contributed to a great environmental impact in the areas of water treatment and this has decreased the amount of harmful pollutants that cause erosion of the environmental systems thus creating imbalances on the normal functioning of the ecosystem (Lakmaaker & Goehring, 2017; Markets & Markets, 2017). Lakmaaker and Goehring (2017) further asserts that internet of things technologies play an increasingly vital role in different markets and industries of the world economy.

For instance, Huansheng, Hong, Jianhua, Laurence, Yueliang, Xiaozhen and Runhe (2016) argue that internet of things has been adopted in almost all sectors of the business and social functioning. Swan (2012) had established that internet has been enabled on devices with an estimate more than human beings population. According to Swan (2012), this number is expected to grow to 50 billion by 2020.

It now emerges that internet of things ecosystem is slowing coming into being to offer support processes that connect tangible features such as roads, human bodies, buildings and human appliances, to the internet. This application is enabled through microprocessor chips and sensors which record and produce useful data such as temperature variations, objects movements as well as sound waves (Culumbus, 2015; Markets & Markets, 2017). According to Finley (2015), the current surge in the explosion of internet-linked sensors and microchips implies that new breeds of technical application and capability are being developed. For instance, Swan (2012) identifies that new data analytical and simulation behaviors such as anomaly detection, high frequency data processing and correlation assessment are now emerging as human population begins to adjust to the various natures of data flows necessitated by the internet of things. This is the new boom for internet of things especially its four functional areas of data development, data generation, decision making and action implementation (Swan, 2012, Huansheng et al., 2015). However, for the purpose of this research, we endeavor to limit ourselves to the role of the internet of things in cleantech markets with a specific focus on the water treatment innovations. The review of literature has shown that very few researches have been worked on to categorize and compare different technologies that are disruptive and its adoption in water treatment. Thus the main objective of this chapter is to study and categorize the role of the internet of things in cleantech markets with a specific focus on the water treatment with a survey of examples from economic press. As defined by Cohen & Lefebvre (2005), categorization is the process in which ideas and objects are recognized, differentiated, and understood categorization implies that objects are grouped into categories, usually for some specific purpose. A research effort of categorization is fundamental in language, prediction, inference, decision making and in all kinds of environmental interaction (Frey & Saake 2011).

BACKGROUND

According to Lakmaaker and Goehring (2016), cleantech refers to a company, or a section of an industry that is specifically engaged in the identification, development, and commercialization and selling of smart technologies, innovations and/or processes that lower or eliminate the adverse effects of the environment in the air, land or water in a more disruptive front. Disruptive innovations in cleantech market ranges from smart grids, solar panels, bioplastics, biofuels, lithium batteries, energy management systems, smart water treatment and carbon capture technologies. For instance, Lakmaaker and Goehring (2016) note that in Canada, massive investments in disruptive innovations are being initiated to adopt cleantech solutions and products into Canadian extractive industries and companies to increase efficiency in operations and also in Eco business models in water treatment. On the global scenes, Lakmaaker and Goehring (2016) stills find that new cleantech investments are driven by the need for sustainable and disruptive innovations such as, smart technologies such as smart water treatment, smart cities' and the Internet of Things.

MAIN FOCUS OF THE CHAPTER

Literature Review

Disruptive Innovation and Cleantech Markets

It is prevalent that cleantech is going to be an embodiment to water treatment in the future. This is due to the current decline in the costs of smart innovations, growing functioning space in data analytics and unrivaled disruptive ecosystems remain to be driving force in the growth of cleantech markets (Cabrera & de Sousa, 2016). All these patterns put together, implies that the world is entering into a new era of cleantech market where massive venture capital will be directed towards businesses that finds ways to link, optimize data and offer high impact services to consumers and businesses (Condry & Nelson, 2016).

Perhaps one question to be asked is: what does this shift in cleantech industry mean for water treatment segment? We note from Woo, Piralta, and Matthews (2015) that the ability of network innovations will lead to possibility of automating decision making process in a wide spectrum to lead efficiency in key applied areas such as, water management and agricultural management. This in different from analogue laden water management process, where key decisions have been arrived at without adequate backup information. Research finds that all business models of cleantech industry in these sectors, especially in water treatment, will enhance efficiency and offer advantage of lowering technology risks (Waleed & Munam, 2016; Grady, 2016; *Greguras, 2016*).

One of the reasons for the proliferation of cleantech innovations into natural markets such as water industry is due to surging population growth in the urban centers. According to OECD report (2016), about two thirds of global population will be living in cities, and this high population is likely to cause a 70% raise in the greenhouse gas emissions besides increased resource consumption. Therefore, in order to tackle these challenges countries will require new technologies and solutions. Cleantech innovations are the only available option that is likely to help transform urban environments into green cities, with abilities to use disruptive characters to tackle these challenges, improve quality of water as well as achieve sustainable green targets in urban areas (Grady, 2016; Markets & Markets, 2017). For instance Weber (2016) argues that cleantech innovations can facilitate better coordination and management of water, its transportation besides tracking its utilization. Cleantech based software and sensor networks have been found to help reduce environmental, social and economic costs by increasing data laden management, accountability and transparency (Nonnecke et al., 2016). In this view, the use and installation of innovative cleantech applications in homes and buildings can help to manage the costs of water treatment (Sonderlund, Smith, Hutton, & Kapelan, 2014). Cleantech applications have with immediate effect increased water utilization efficiency besides reducing water wastage effects (Hartman & LeBlanc, 2015). Equally important, Shaw (2014) notes that the presence of cleantech laden sensors in water channels are able to provide real-time information on water levels, its availability and quality, reducing cases of waste and overconsumption. Cleantech industry is swiftly capturing the interest of scholars and industry practitioners due to its immense economic and social contributions, particularly in water treatment. Thus emergence of disruptive technologies such cleantech as and its swift commercialization has significantly revolutionized naturally inclined segments such as water treatment and other dependent sectors (Grady, 2016). According to Markets and Markets (2017), extensive market research and analysis of cleantech industry as well as its trends, forecasts and the profiling of major actors, show how vital the growth of the sector has become. Huston (2015) for instance argues that cleantech through the lenses of sensors and real time analytics has contributed to a great environmental impact in areas of water treatment and this has decreased the amount of harmful pollutants that cause erosion of the environmental systems thus creating imbalances on the normal functioning of the ecosystem (Lakmaaker & Goehring, 2017; Markets & Markets, 2017).

Cleantech and Water Treatment

Research has established that most countries lose unaccounted amount of water through leakage and wasteful use (Adler, 2015). This massive water loss is caused by lack of properly designed and technologically managed water and sewerage system and infrastructures in most urban centers (Forer & Staub, 2013). Therefore, the need for adoption of cleantech innovation and improvement in water collection, conservation and distribution is especially vital owing to the fact that the depletion of freshwater resources in numerous areas in many parts of the world and the shifts in water tables due to climate change, caused by global warming. Research has established that some parts of the world will experience persistent drought and frequent depletion of water resources due to encroachment (Culumbus, 2015). Other regions will experience heavy rains and therefore cleantech technology will have new opportunities to capture and recycle rain water so as manage to freshwater resources. In this regard, cleantech innovation can provide better opportunities to help cities and urban centers in better use, and conserve water more efficiently and effectively through smart water management applications. According to Hollands (2008), cleantech technology by is an important factor in the creation of smart water management, particularly water treatment and waste water management ecosystems and platforms because the use of information communication technology (ICT) helps to change the nature of water management decisions and use within home setting. Therefore, this review finds that a well-working water management infrastructure is a necessary pillar but not as essential as current smart technological applications. Key applied cleantech technological applications include: smart metering, smart piping, smart land reclamation and smart agricultural practices (Nonnecke, Bruch, & Crittenden, 2016).

Therefore, from this literature it is quite clear that cleantech and its disruptive applications remain to be taken with greater interest as being the prerequisite to developing and implementing a smart water management initiative (Gilpin, 2015). However, it is quite clear that the real implementation is not possible without actual sacrifice by public utility entities and the willingness of the different stakeholders to cooperate and collaborate to achieve these visions, a rejoinder which can be well realized under public-private partnerships (PPP) (Komninos, 2015).

Most research on the activities of smart water management highlight the need of technological infrastructure as well their enabling technologies in smart water infrastructure development. The focus by scholars on utilization of technological infrastructure places great emphasis on the level of accessibility as well as availability of support systems offered by cleantech technology (Giffinger & Gudrun, 2010; IBM, 2010). Although research has not yet documented whether emerging disruptive innovations such as cleantech through smart meters are more effective in lowering industrial and consumer water usage than other alternative expensive interventions, early studies indicate that connecting smart meters with an option to relay real-time information can help reduce consumption to a larger extent (Sonderlund et al., 2014). In one study, it was established that customers who had a smart-meter-paired visual displays installed in their houses alarm sensors to alert when a set level of water usage was exceeded, reduced water consumption and usage in the home by an average of 25% (Sonderlund et al., 2014).

The key take-away point in our review is that cleantech technology will be the bedrock of future water management efforts. In this regard, smart water applications, do seek to offer real time information on the current water events and advanced analysis to help service providers and consumers make right decisions from the collected intelligence regarding new alternatives and courses of action to be adopted to optimize water consumption and while at the same time minimizing on the cost implications (Washburn et al., 2010). One the unique ability of cleantech innovations and technologies is to use data analysis and simulation to improve wastewater reclamation and sewage treatment. This has been considered to be useful particularly in hotspot areas that face frequent threats of storms and or droughts. For instance according to Lux Research, a new technology, Zero-liquid discharge (ZLD), an approach to wastewater treatment that prevents any liquid waste from flowing out of a power plant or factors, will grow at a 12% annual rate into a $2.7 billion market in 2030.

METHODOLOGY

This research design framework is based on action-response research design (Chen et al. 1991; Chen et al., 1992; 1992). This Chapter's analysis focuses on the adoption of disruptive innovations in cleantech with a focus on water treatment. This research relied on a similar research design that other scholars in previous analysis on relationships of cleantech and water treatment (Chen et al., 1992; Smith et al., 1991; Tang & Shapira 2011) had adopted in their researches.

Data Collection and Sample

To collect information on this analysis' cases the author followed five steps:

Step 1: The author collected information and surveyed 9 innovations for their disruptive character from a technological and economic perspective. The following Figure 1 illustrates in a preliminary manner of these 9 disruptive innovations. The data collection was completed with a survey of reports and articles on the topic of disruptive innovation with the following lists and sources: Accenture; EAI Catalyst Cleantech Sustainability, (2015); CNBC Disruptors 50; Forbes, Interaction Design Foundation; McKinsey Global Institute, Fortune; Intelligent HQ; Observer (2005); Mashable.com; PC World (2008); Solis (2015) and MIT Technology Review.

Step 2: All these reports were supported with economic data (generally quantitative) and technological data (generally descriptive) describing their disruptive nature. The McKinsey Global Institute has produced an integrative and very complete report and also a full set of detailed reports surveying and assessing specific disruptive innovation for specific innovations. These reports were completed from other sources to reach a list of 9. Since the objective of the research was not to provide an assessment the technological and economic impact, the author listed these disruptive innovations by alphabetic order.

Step 3: The author also provided a categorization using the dominant drivers according to the description of the technological and economic impact of the literature.

Step 4: The disruptive character of the innovation was assessed from a technological perspective and from the economic perspective. From a technological perspective, the disruptive character was assessed by the radical nature of the substitution provided by the innovation. When the innovation could be accommodated and easily integrated by the incumbent it was not considered as being disruptive. When it has to be assimilated and not easily integrated (measure by the ease of substitution) and therefore required a transformation of the cleantech technology in water treatment, it was considered as being "disruptive".

Step 5: The selection was completed using a technological and an economic perspective. From a technological perspective the author assessed the disruptive character of these innovations in terms of efficiency, water purification and waste water treatment. From an economic perspective, this scholarly analysis relied on economic impact assessment. For instance, the McKinsey Global Institute has provided a full set of economic impact for a large number of innovations into its report and the author relied on these economic evaluations to assess the disruptive aspect of the innovations.

Data Selection

The author provided a categorization using the dominant drivers according to the description of the economic and the technological impact provided by the economic press. The selection was completed using both a technological and an economic perspective. From a technological perspective the author assessed the disruptive character of these innovations in terms of functionalities (products), distribution and competencies based on the sources from engineering, computer science and software and life science. For an economic perspective, the author relied on economic impact assessment. For instance, the McKinsey Global Institute has provided us with a full set of economic impact for a large number of innovations into its report and the author relied on these economic evaluations to assess the disruptive aspect of the innovations. For the final selection, the author only kept innovations that combined these three dimensions.

Firstly, the author rejected some of disruptive innovations that were considered for the study, which had a high technological impact but no significant economic impact, and were not really disruptive from a technological point of view such as OLED/LED technology and advanced photonics.

Secondly, for instance, the author included artificial intelligence but several innovations such as the next generation nuclear, flexible displays or Uber type disruptive were not considered for final evaluation. The author corroborated selection with this research's information sources.

Thirdly, some innovations had a large economic impact but were not disruptive from technological perspective and were thus not considered. This was the case of new feature of smart phone or the IPad or the smart televisions that has economic impact but was not really disruptive from a technological perspective.

Data Analysis

The researcher analyzed each component of the conceptual framework drivers for each innovation using the economic literature on the subjects. For each disruptive innovation the author identified a predominant driver among the cleantech segment with impact on water treatment. Finally, the author developed short background case studies to build on the systematic review of literature.

Table 1. Survey of disruptive innovations by alphabetic order

Disruptive Innovations
1. Advanced Materials
2. Advanced Robotics
3. Artificiel Intelligence
4. Big data advanced Analytics
5. Cloud technology
6. Internet of Things
7. Mobile Internet
8. Mobile Payment
9. Peer-to-Peer market (P2P)

Source: compiled from Accenture; EAI Catalyst Cleantech Sustainability, (2015); CNBC Disruptors 50; Forbes, Interaction Design Foundation; McKinsey Global Institute, Fortune; Intelligent HQ; Observer, 2005; Mashable.com; PC World (2008); Solis (2015); MIT Technology Review

Categorizing Established Issues

Growth Rate of Cleantech Markets

According to Markets and Markets (2017), cleantech market size is estimated to grow from $157.05 Billion in 2016 to $661.74 Billion by 2021, at a Compound Annual Growth Rate (CAGR) of 33.3% from 2016 to 2021. All cleantech large companies are experiencing a huge transformation in a bid to move toward efficient, affordable, accessible, and quality water products and services to their customers. New applications and use cases, a result of competitive edge technology and innovations, are being designed and developed to address the changing industry dynamics and economic needs in cleantech markets. According to Wang et al. (2016), IoT, in combination with data analytics, cloud computing and big data simulations, is creating lucrative opportunities for cleantech markets especially as relates to waste and water treatment. Researches indicate that data management applications are gaining high market momentum with the help of cloud and predictive analytics to manage data generated from IoT software (Roth, 2016). This data is evaluated through data management software to extract useful information that helps organizations venturing in water treatment, transport management and wind powers in decision-making (Condry & Nelson, 2016; Roth, 2016). Markets and Markets (2017) projects that the global IoT commercialization in Energy market, especially wind power is expected to grow from $ 923.59 billion in 2017 to $ 1.22 trillion by 2025, at a

Compound Annual Growth Rate (CAGR) of 24.1% within the forecasted timeline. In the current projection, North America is considered to become the largest single market on the basis of market size and technological advancements in the application of IoT in cleantech markets especially in waste and water treatment, smart transport networks and efficient use of energy (Grady, 2016). The rise in security challenges against cyber-attacks and demand for increased operational efficiency is expected to propel the internet of things in energy market in the coming decade (Woo et al., 2015; Edwards, 2016). As a result, solutions and software developers and providers are shifting focus towards IoT security solutions to offer maximum security to the devices connected to the smart energy grids (Edwards, 2016). Internet of things-based solutions are further enabling companies in cleantech markets to scale up their operational efficiency in waste and water treatment systems, not forgetting transport systems (Advancement in next generation sensors is another major driving factor of the IoT in Energy market. However, it has been established that limitation in the supply of skilled personnel and increasing issues regarding data security and privacy remains to be an impediment to the growth of the internet of things in the cleantech market (Hartig & Kahn, 2016). According to reports from GE Digital wing (2017), the current heavy investment in Internet of Things is likely to grow to about $2.12 trillion by 2030. This will hold over 50 billion in assets especially in water equipment, turbines, machines and rolling stock. Industries and different companies report that a growth prospect of $2.12 trillion is such a huge figure, however looking at the coverage area in terms of potential application and use of industrial internet of things in different sectors such as transport, energy, water management, aviation, manufacturing, healthcare, environment, and how much these sectors contribute to world economies, the number is still within reach by 2030 (Balaji & Roy, 2016).

Figure 1. Projected growth rate of cleantech in water treatment

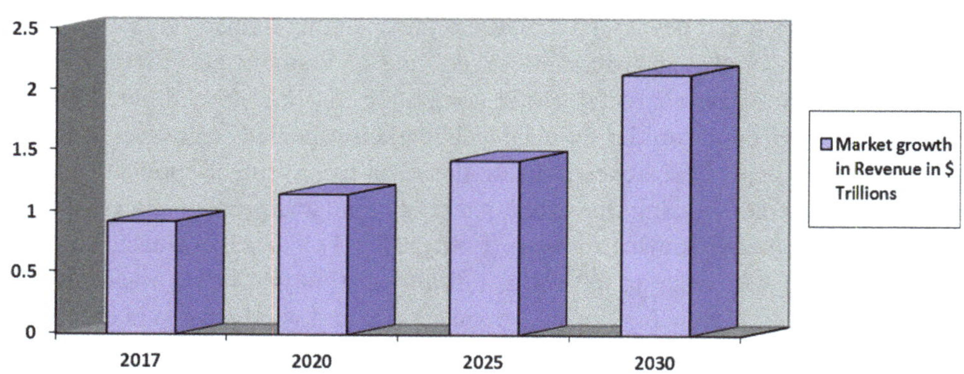

Market Trends in Cleantech Markets

This research has noted that as more and more devices economies invest in cleantech, the potential for industrial cleantech to drive efficiencies, increase productivity and promote economic growth is becoming a new reality (Markets & Markets, 2017). On a more particular note, scholars and practitioners have established that cleantech is fueling innovation and creating opportunities in the development of smart technology in water management, and helping redefine the sustainability of water supply management (Roth, 2016). Current research continues to indicate that there are emerging trends in the markets of cleantech as relates to application and use of clean technologies in water management (Adler, 2015). Qui et al., (2014) argue that in all markets with technological vocation, these heavy sound trends are important driving elements to explain the growth of the market of the internet of things. Research and industrial evolution indicate three key trends in the cleantech markets in the years ahead (Altmann, Butzin, Balla, Golatowski & Timmermann, 2015). The single largest one is the connectivity of information through the abilities of emerging innovations. Cleantech technology is currently stimulating massive demand of sensors and real time trackers in water treatment process. Noting from Frost and Sullivan (2014) there has been a significant trend over the past few years in cleantech investment portfolios especially in water management. The industry has experienced a dynamic shift away from highly priced capital intensive renewable energy which has been supported through debt to a less capital intensive energy efficient innovations and technologies. According to Bures, Plasil, Kit, Tuma and Hoch (2016), this new evolution is now referred to as the cleanweb. One of the reasons for this evolution has been to employ efficiency in energy sector, as it plays a significant role in software driven innovations and technologies leading to reduction in water resource consumption. Players in this sector have witnessed a massive reduction in technological risks as well as low scaling costs. Software that makes resource demand more flexible and reduces waste is the new black (Grady, 2016). Another trend that is driving the growth of cleantech technologies is the surging population growth in the developing economies. According to Ritola et al., (2015), the developing world is growing at faster rate, with close to 90% of the population growth expected to take place in Asia and Africa. This implies that there is a need for industries to better understand how to meet the increasing, diverse and unique needs of their customers', while at the same time scaling down on their use of natural resources such as water to lower the current population pressure.

Global environmental challenges affect all nations and urban areas also face their own local challenges that are linked to the ever-growing population density and thus increasing emission rates. Deteriorating quality of air, water pollution, and noise pollutions are typical pains of urbanization (UN, 2014). According to an OECD report (2014), the traffic congestion in almost all U.S. cities and urban centers created \$78 billion annual drain on the U.S. economy in the form of 4.2 billion lost hours and 2.9 billion gallons of wasted fuel. Despite their smaller size, European urban centers and cities are experiencing similar challenges in terms of costs surge and CO2 emission (Grady, 2016).

We therefore note that these challenges deteriorate the quality of life of billions around the world and exert heavy pressure on the management of huge infrastructure systems, such as water management. According to Ritola et al., (2015), a growing demand for new types of smart and clean solutions to manage these challenges are the driving force in the creation of smart technologies and innovations in water companies across the world. The global superpowers with rapidly expanding populations and huge sustainability impediments recognize the need for clean solutions in form of technologies and innovations.

Strategies of Adoption of Internet of Things in Waste and Water Treatment

Research indicates that smart waste and water treatment is essential, whether there is no rain or there is rain calling for a need to foster conservation efforts to the foreseeable future (Grady, 2016; Roth, 2016). Cleantech innovation can ensure that the water supply from the rains patterns is used more efficiently and with less waste. Regarding the adoption, different actors both from public and private sectors are in the process of establishing commercial units to focus on the specific end markets for cleantech products, who can be new buyers of these products (Po-Chiang Lin, 2015). Their strategic teams are focused on mapping cleantech into their respective operations to ensure that they remain aligned with their customers' changing needs, and anticipate market trends smart water management different from the traditional customers (Lakmaaker & Goehring, 2017). According to Huansheng et al., (2015), one of the challenges that have been affecting the effective implementation of smart technologies in waste and water management is lack of properly maintained water pipes, meters and drainage systems. Governments and public utility companies have now moved with speed to engage smart cleantech developers through public private partnerships to ensure that there are minimal effects of water wastage.

For instance, Adler (2015) argue that currently, an estimated 2.1 trillion gallons of clean, treated water is lost each year to leaks in the infrastructure of America's 52,000 water utilities. The need for innovation in water collection, use and conservation is very critical given the depletion of freshwater resources in numerous areas of the world and the shifts in water availability due to climate change caused by dropping sea levels as well as global warming (Forer & Staub, 2013).

Cleantech Application Areas in Water Treatment

According to *Greguras (2017) of the water online,* the cleantech sector is currently transitioning from a niche market to a market that has the potential to impact other sectors of the economy such as water supply sector. Recent advances in cleantech innovations offer exciting opportunities to develop next-generation water supply systems. The current water treatment, distribution, and discharge practices, which heavily rely on conveyance and centralized systems, have been found to be unsustainable. The highly efficient, modular, and multifunctional processes enabled by cleantech are envisaged to provide high performance, affordable water and wastewater treatment solutions that rely less on large infrastructures (Qu et al., 2013). In terms of commercialization and product segmentation, cleantech technology finds applications in water treatment under the following categories: smart metering, smart piping, smart water reclamation and smart agriculture.

Smart Meters

According Tilley (2015b), smart water meters involve fitting water drainage pathways and homes with meters that offer real time data on the level of usage and wastage. Through smart cleantech innovations, it's now possible to track any possible leakages whether in homes or commercial setting. According to Wang (2015a), while traditional meters are manually read either on a monthly or yearly basis, smart meters capture consumer use in real time, and can relay this information to both the utility providers and the consumer, offering opportunities for improved waste and water management. For utilities providers, smart meters allow system managers to detect any water leaks along the chains. For residential settings, detecting water leakage is straightforward. In a commercial setting that use water all the time, utilities providers use algorithms generated from the software vendors to identify indications of water leakage. In nutshell, Finley (2015) reports that real-time data on usage offered by smart meters allows utilities to monitor water-wasting or any unwarranted behavior, such as outdoor watering or non-essential usage (Finley, 2015).

Smart Piping

Research on disruptive innovations indicate that cleantech technologies enable water supply companies to make their water pipe systems "smart" by installing remote sensors to, track and monitor water flow and prevent any possibility for leakage (Nehls, 2015). According to Adler (2015), leading companies in this segment such as AquamatiX and i2O have moved with speed to develop embedded sensors that can be installed in water pipes throughout the water chain system and connected to central pumping control systems. The sensors regulate water pumping by monitoring water flow and minimizing the amount of water in the pipes, reducing both the amount of water lost through leakage and the quantity of energy needed for pumping.

Water Reclamation

According to Jassby (2016), since freshwater resources continue to increasingly diminish due to climate change and at the same time face demand, smart water reclamation will be considered as an important option in sustaining the water needs in smart cleantech market. For instance, Opti, a leading developer is using cleantech technology to combine information from weather predictions and storage spaces to explore more space in the water system to properly process and capture excessive water during heavy runoff caused by storms. One of the challenges that come with reclaiming waste water is the presence of complex pathogens in the water distribution system. According to Adler (2015), cleantech could improve wastewater reclamation and sewage treatment, which will prove especially critical in areas with threatened freshwater resources.

Agriculture

Most commercial agriculture relies on irrigation water and this poses a great challenge of competition for the limited resource. Equally important, as reported by WHO (2012) there is room for abuse when farmers use water to irrigate their farms. However, with the development of internet of things with diverse market applications in agriculture, farmers and governments can be able to plan on how to use water for farming purposes in a more efficient way while at the same time limiting wastage and costs.

Finley (2015) argues that agriculture consumes about 40 percent of the freshwater available in any country with a large amount being wasted by leaky irrigation systems, inefficient field application methods and the planting of water intensive crops in the wrong growing location. As a key segment with strong market potential, cleantech has great potential to make water use smarter for the agricultural industry, more specifically in ensuring that there is irrigation efficiency (Markets & Markets, 2017).

Innovation Frontiers in the Water Sector

The ultimate aims of the business of water sector are to offer customers with safe and reliable water supplies and to safely dispose of wastewater, in compliance with water standards and quality regulations. Several hypotheses have historically directed water managers and administrators in meeting these targets. First, water managers believed that demand for clean water would increase with population growth and that the only way to have a balance between demand and supply was to find new sources of supply so as to ensure that there is sustainability. The focus of this paper therefore was to enhance supply rather than managing demand. Water managers, additionally, turned to large-scale, centralized infrastructure projects to increase their supply, on assumption that large-scale projects would generate large economies of scale and provide greater operational efficiency (Ajami et al., 2012; Hering et al., 2013). As a result, enhancement of supply meant the creation of large dams, Market reservoirs, aqueducts, and water and wastewater treatment units. Water managers are rethinking all these assumptions, and the new waterscape has opened up opportunities for a variety of new disruptive technologies. Rather than seeking to enhance supply, water managers are now placing increasing emphasis on demand management by leveraging on the usefulness of new disruptive technologies such as cleantech in water treatment. Studies have demonstrated that new technologies, supported with incentives and education, can significantly lower water use. As a result, there is increasing interest in disruptive technologies that are more water efficient and in technologies that can help encourage greater conservation among consumers. It can be observed from the Table 2 that the scope of water treatment based on supply side is focused on key areas, some of which include but not limited to: efficient use of water, water purification and waste water management.

- **Efficient Use of Water:** This research has established that water efficiency is reducing water wastage by measuring the amount of water required for a particular purpose and the amount of water used or delivered. Water efficiency differs from water conservation in that it focuses on reducing waste, not limiting water use.
- **Water Purification:** This research has established that water purification as a supply element in water treatment seeks to remove undesirable substances, biological contaminants suspended particles and poisonous gases from water. The aim of this process is to produce water that fit for a specific purpose. Practitioners in this sector have argued that most water is disinfected for human

drinking, but purification may as well be designed to meet other objectives including fulfilling the requirements of chemical, pharmacological, medical and industrial applications.

- **Waste Water Treatment**: Waste water treatment as a process is used to convert waste water, which is no longer required or suitable for use into an effluent form that can be either reused with minimal environmental effects or be returned to the water cycle. When applying cleantech innovations, during the treatment process, pollutants are removed or broken down.

A summary of this process is highlighted in Table 2.

DISCUSSION

This chapter's discussion and thoughts offer unique insights in terms of impact of cleantech on water management. The chapter sought to categorize and present the impact of cleantech innovations on key sectors of an economy in terms of their influence on the tactics and commercialization potential from a disruptive innovations perspective. In a general manner, this chapter has indicated that disruptive innovation plays a key role on the functionality of water management sector. The author also noted incumbent water-oriented companies and emerging companies in cleantech sector need to keep up with their industry's technologies which are disruptive, changing dynamics in skill sets, and competencies that present challenges to their existing business model.

Table 2. Innovation frontiers in water treatment

Cleantech	Internet of things	Big Data and Advanced Analytics	Advanced Robotics	Smart Advanced Materials	Mobile Applications	Cloud Management	Uber-type Applications	Peer2Peer market
Water								
Efficient use of water	•	•	•	•	•	•	•	•
Purification	•	•	•	•	•	•		
Waste water treatment	•	•	•	•	•	•		

From the literature review, the author had found that new cleantech investments are driven by the need for sustainable and disruptive innovations such as, smart technologies such as smart water treatment, smart cities' and the Internet of Things. According to Lux research, new startups and dominant incumbents in water technology enable a growing market for technologies that can eliminate all liquid waste from power plants and other facilities. For instance according to Lux Research, a new technology, Zero-liquid discharge (ZLD), an approach to wastewater treatment that prevents any liquid waste from flowing out of a power plant or factors, will grow at a 12% annual rate into a $2.7 billion market in 2030. The cleantech market will be boosted by technology innovations, rising water cost, and regulations due to growing concerns over surface water contamination. Leading Zero-liquid discharge companies like GE, Veolia, and Aquatech International have a large chunk of the market, but startups with emerging technologies are addressing cost and energy issues to enable even wider adoption of cleantech in water treatment. Emerging trends indicate that a wave of startup companies is rising up to address the growing need for wastewater treatment and monitoring solutions, part of a $600 billion industry that has traditionally been dominated by large companies such as GE, Veolia, Siemens and Suez. An even interesting force is that the industry as a whole is generating an average operating profits of 12.9% as reported by Lux Research (2017). Given the $6.4 trillion market opportunity for cleantech products, this research has established that countries that effectively support local clean-technology industries are poised to capture important economic and environmental benefits, while increasing agricultural output and improving access to clean water. In many cases, it turns out; investing the right things for the environment is also investing the smart things for the economy.

The take away point in our review is that IoT will be the bedrock of future water management efforts. In this regard, smart water applications, do seek to offer real time information on the current water events and advanced analysis to help service providers and consumers make right decisions from the collected intelligence regarding new alternatives and courses of action to be adopted to optimize business operations and while at the same time minimizing on the cost implications. One of the unique ability of internet of things is to use data analysis and simulation to improve wastewater reclamation and sewage treatment. This has been considered to be useful particularly in hotspot areas that face frequent threats of storms and or droughts.

RESEARCH IMPLICATIONS

Strategic Policy

From the strategic analysis, this paper puts forward new policy recommendations to scale up innovation and deployment of new disruptive technologies such as internet of things in the water sector. In this context, it is vital to discuss the emergence of strategic issues and policy implications for companies and economies affected by an emerging sector of cleantech. To this far, it is quite evident that cleantech is next innovative investment that is set destabilize the traditional markets. This paper has offered key insights on the growth trajectories and profitability of the sector which indicates that in order to achieve sustainable competitiveness, global economies should invest infrastructural and technological funds in cleantech markets. Cleantech technologies and innovations such as Internet of things are everywhere around us: from communications and health, to lighting and transport and to everyday use like in water and irrigation. As a strategic ramification, cleantech technology particularly, internet of things which is among the fastest-growing business sector globally, should be given the highest considerations by all stakeholders who seek to have clean environment.

This strategic analysis has also established that cleantech sector is a very dynamic and vibrant business sector that holds the potential for huge market growth in the environment sector, with a specific focus on water treatment segment. This implication is as a result of increased demand for greener technology and carbon neutral energy generation, for ever greater data bandwidths, for better and cheaper food solutions for an aging population or for better and higher quality water resources provide huge opportunities for rapid market growth, in particular for broadening the applications space.

This paper presents that the key challenges for economies is to make an economic assessment of where cleantech can be used to create regional growth and jobs in water treatment sector. For instance, a European Commission report of 2013 this will allow economies to identify where are the economic niches and competitive advantages in cleantech development and deployment activities. Regions should be aware of key guiding principles, such as: the industrial and residential needs with regard to cleantech technologies, main players in their regions, the potential customers of these technologies, their level of cooperation along the value proposition chains and the promising business studies that can be developed in terms of market development and exploitation in water treatment.

On a more specific implication, this paper calls for:

- First, the analysis calls for a review in pricing of water products to better level the economic and marginal cost of furnishing water and as well to strive for an increase private-sector innovation. Lack of proper water pricing mechanisms denies both an opportunity for incentives for water-conserving innovations by water stakeholders and the financial and monetary stability required to finance both the implementation and adoption of new water innovation and technologies by the water providers.
- Second, this paper urges for policy and regulatory changes at the country-specific level to create a more robust innovative environment. As part of this policy implication the paper suggests that some economies could benefit from the establishment of new water innovation centers to engage, coordinate and support planned innovation policies and regulation. The author argues that a majority of current regulations frequently impede the adoption and use of cost-effective innovations and technologies.
- Thirdly, this paper and its analysis calls for a public utility charge on water to generate more public funding for water innovations and technology adoption. Summed together, these major policy implications could massively improve technological innovations in water segment and lead to more-efficient yields in water sector globally.

Practical Implications

It has been noted from this research that the premise behind smart services such as cleantech is that companies can provide innovative and differentiated offerings as part of a service or product offering by connecting to their products in the field to a centrally managed server and extracting insight remotely. In other words, companies can provide better service in a more cost-effective manner when they have visibility into how their products are being used in real time or near real time. In lieu to this argument, this paper proposes that following practical applications

Sustained Service Performance

It has been noted from the research that there has been a dramatic shift towards service performance across virtually all businesses areas in recent years, Cleantech product companies included. Whether service is provided directly by the water vendors or through the channel, Cleantech companies stand to improve their financial performance through improved service performance of their products towards their applied markets of water treatment. Rather than relying on time-based routine service calls and reactive emergency service calls, the water management organization can proactively determine service priorities, remotely diagnose and trouble shoot problems, and optimize the dispatch of their service resources through adoption of cleantech.

Sustained Product Performance

Product manufacturers, especially water supply companies know the most about how to get maximum performance from their products—the challenge however is that they rarely know how their products are being operated once in the field. By gaining access to real-world product performance information through cleantech technologies, water manufacturers can identify opportunities and challenges to improve their product performance, both by changing operational practices, improvement activities, and by renovating better products for the future.

Improved Customer Value

Improving the value delivered to customers through product benefits ought to be a primary business goal of any products-based business. Therefore smart disruptive innovations open up an entirely new array of possibilities. This could be as simple as knowing precisely when to change the oil and oil filters on a product, regardless of variations in production, and without necessarily having to be onsite. In this regard, customers may gain through reduced downtime, longer equipment life, and/or improved operating efficiency. Equally important product manufacturers may gain by capturing the aftermarket business for consumables that were previously provided by third party pirates. And hence the possibilities of delivering improved customer value are extensive and boundless.

Competitive Advantage

Cleantech product manufacturers stand to build competitive advantages by implementing smart disruptive and innovative services. By combining their product with a unique smart service value proposition, manufacturers of smart cleantech are able to differentiate their offering. One of the clearest examples of this is found in water treatment industry, where monitoring expensive, water assets has long been a critical element of commercial water management. Today these types of capabilities are now trickling down into the consumer space. A practical example can be found in the case study on Siemens Water Technologies (SWT), the world's leading water and wastewater treatment company providing comprehensive water treatment systems and services for industrial, institutional and municipal markets.

FUTURE RESEARCH DIRECTIONS

This chapter analysis has provided reliable insights and opened a perspective for additional research to better categorize water management innovations according the benefits derived from the use of cleantech technologies. However, additional work will be provided to structure this research perspective.

Based on the concept of cleantech and its application in water management, this chapter concludes the a diverse research on the concepts discussed might facilitate better the categorization of disruptive innovations such as internet of things and leads to better study this complex economic and social phenomenon using a market/industry perspective.

CONCLUSION

This research has established that cleantech represents the convergence of digitization, networking, and the cleantech revolution in water treatment and wastewater management. Although cleantech is still in its infancy, and significant challenges have to be overcome in building enormously complex ecosystem linking together hardware, software, databases, applications and new business models, the author of this chapter believes that 'smart' will become an increasingly fundamental part of key manufacturing and service processes in energy production, manufacturing, building operations, water, and agriculture. The core promise of cleantech in this case is to improve the productive use of resources while reducing negative ecological and environmental impacts- creating smart real-time systems in key industrial activities will be integral to delivering that promise. Therefore, as cleantech technology evolves into its own distinct industry category, smart services will prove an invaluable means to efficiently track and control the first generations of widely dispersed physical assets needed to provide and sustain relevant solutions to water management sector. In conclusion, research on cleantech as disruptive innovations in water treatment has begun to change the concept showing its usefulness from a conceptual perspective to a more commercial perspective and there are few empirical works to categorize and compare different innovations that are disruptive on the various sectors of the economy such as in water treatment. Thus the objective of this chapter was to identify, present and discuss the concept of disruptive internet of things and cleantech technologies to help drive solutions to challenges that potential users of water sector are facing.

REFERENCES

Alexandratos, N., & Bruinsma, J. (2012). *World Agriculture Towards 2030/2050.* Retrieved from www.fao.org/docrep/016/ap106e/ap106e.pdf

Alliance for Water Efficiency. (2017). *Smart Water Application Technology (SWAT) Initiative*. Retrieved from:http://www.allianceforwaterefficiency.org/Smart_Water_Application_Technology_(SWAT)_Initiative.aspx

Altmann, V., Butzin, B., Balla, R., Golatowski, F., & Timmermann, D. (2015). A BACnet gateway for embedded Web services. *Emerging Technologies & Factory Automation (ETFA) 2015 IEEE 20th Conference on*, 1-6. 10.1109/ETFA.2015.7301563

Balaji, M. S., & Roy, S. K. (2016). Value co-creation with Internet of things technology in the retail industry. *Journal of Marketing Management*, 1.

Bradley, J., Barbier, J., & Handler, D. (2013). *Embracing the Internet of Everything To Capture Your Share of $14.4 Trillion - More Relevant, Valuable Connections Will Improve Innovation, Productivity, Efficiency & Customer Experience*. Cisco. Retrieved from: http://www.cisco.com/web/about/ac79/docs/innov/IoE_Economy.pdf

Bures, T., Plasil, F., Kit, M., Tuma, P., & Hoch, N. (2016). Software Abstractions for Component Interaction in the Internet of Things. *Computer*, *49*, 50-59.

Cabrera, F. L., & de Sousa, F. R. (2016). Achieving Optimal Efficiency in Energy Transfer to a CMOS Fully Integrated Wireless Power Receiver. *Microwave Theory and Techniques IEEE Transactions on*, *64*, 3703-3713.

Cirani, S., & Picone, M. (2015). Wearable Computing for the Internet of Things. *IT Professional*, *17*, 35-41.

Condry, M. W., & Nelson, C. B. (2016). Using smart edge IOT devices for safer rapid response with industry IOT control operations. *Proceedings of the IEEE*, *104*, 938-946.

Culumbus, L. (2015). Roundup Of Internet Of Things Forecasts And Market Estimates. *Forbes*. Retrieved from: https://www.forbes.com/sites/louiscolumbus/2016/11/27/roundup-of-internet-of-things-forecasts-and-market-estimates-2016/#29064e7e292d

Edwards, L. (2016). *Privacy, security and data protection in smart cities: A critical EU law perspective*. Retrieved from http://papers.ssrn.com/sol3/papers.cfm?abstract_id=2711290

Finley, K. (2015). Smart meters shame you for how much water you're wasting. *Slate.com.* Retrieved from http://www.slate.com/blogs/future tense/2015/06/15/in California smart water meters help spot cases of extreme waste.html

Forer, G., & Staub, C. (2013). *The US water sector on the verge of transformation.* Ernst & Young, Global Cleantech Center White Paper. Retrieved from http://www.ey.com/Publication/vwLUAssets/Cleantech_Water_Whitepaper/$FILE/Cleantech-Water-Whitepaper.pdf

Frost & Sullivan. (2014). *Mega Trends Driving Change in Business Environments.* Retrieved from http://www.investinbsr.com/ipaforum/wp-content/uploads/Iain-Jawad-IPA-Forum-2014-Presentation.pdf

Giffinger, R., & Gudrun, H. (2010). Smart cities ranking: An effective instrument for the positioning of cities? *ACE: Architecture, City and Environment, 4*(12), 7-25. Available at http://upcommons.upc.edu/revistes/bitstream/2099/8550/7/ACE_12_SA_10.pdf

Gilpin, L. (2015). *How cleantech-as-a-service will drive renewable energy adoption.* Retrieved from: http://www.techrepublic.com/article/how-cleantech-as-a-service-will-drive-renewable-energy-adoption/

Goldenstein, T. (2015, May 5). Smart water meters help users, agencies gauge usage. *Los Angeles Times.* Retrieved from http://www.latimes.com/local/california/la-me-smart-meter-explainer-20150505-story.html

Goldenstein, T. (2015, May 5). Smart water meters help users, agencies gauge usage. *Los Angeles Times.* Retrieved from http://www.latimes.com/local/california/la-me-smart-meter-explainer-20150505-story.html

Grady, B. (2016). *Driving battery storage, IoT, electric transportation mainstream.* GreenBiz Group. Retrieved from: www.techrepublic.com/article/how-cleantech-as-a-service-will-drive-renewable-energy-adoption/

Greguras, F., (2016). Water And the Internet of Things: 2016. *Wateronline.* Adopted from: https://www.wateronline.com/doc/water-and-the-internet-of-things-0001

Hartig, T., & Kahn, P. Jr. (2016). Living in cities, naturally. *Science, 352*(6288), 938–940. doi:10.1126cience.aaf3759 PMID:27199417

Hartman, B., & LeBlanc, B. (2015). In pursuit of the perfect portal: Smart meters, big data, and customer engagement. *E Source.* Retrieved from https://www.esource.com/ES-WP-19/Smart-Meter-Portals https://www.tekes.fi/globalassets/julkaisut/consumer_cleantech_report.pdf

Huansheng, N., Hong, L., Jianhua, M., Laurence, T. Y., Yueliang, W., Xiaozhen, Y., & Runhe, H. (2015). From Internet to Smart World. *Access IEEE*, *3*, 1994-1999.

Huston, G. (2015, April 30). *The internet of stupid things*. APNIC Labs. Retrieved from http://blog.apnic.net/2015/04/30/the-internet-of-stupid-things/

Kemp, S. (2015). *Digital, Social & Mobile Worldwide in 2015*. Retrieved from http://wearesocial.net/blog/2015/01/digital-social-mobile-worldwide-2015/

Lakmaaker, M., & Goehring, M. (2017). *Taking Canada's Clean Tech Industry to the Next Level. An Opportunity for Canada's Clean Technology Sector: The Pan-Canadian Framework on Clean Growth and Climate Change.* Adopted from: http://www.national.ca/Bold-Thinking/Taking-Canadas-Clean-Tech-Industry-to-the-Next-Level.aspx

Lakmaaker, M., & Goehring, M. (2017). *Taking Canada's Clean Tech Industry to the Next Level. An Opportunity for Canada's Clean Technology Sector: The Pan-Canadian Framework on Clean Growth and Climate Change.* Adopted from: http://www.national.ca/Bold-Thinking/Taking-Canadas-Clean-Tech-Industry-to-the-Next-Level.aspx

Lin, P.-C. (2015). Optimal smart gateway deployment for the Internet of Things in smart home environments. *Consumer Electronics (GCCE) 2015 IEEE 4th Global Conference on*, 273-274. 10.1109/GCCE.2015.7398715

Markets & Markets. (2017). *Internet of Things (IoT) in Energy Market worth 22.34 Billion USD by 2020.* Adopted from: http://www.marketsandmarkets.com/PressReleases/iot-energy.asp

Nonnecke, B., Bruch, M., & Crittenden, C. (2016). *IoT & Sustainability: Practice, Policy and Promise*. Center for Information Technology Research in the Interest of Society & the Banatao Institute University of California.

OECD. (2014*). Trends in urbanizations and urban policies in OECD countries: What lessons for China?* Retrieved from http://www.oecd.org/urban/roundtable/45159707.pdf

Qu, X.L., Brame, J., Li, Q., & Alvarez, J.J.P. (2013). Nanotechnology for a safe and sustainable water supply: enabling integrated water treatment and reuse. *Accounts of Chemical Research, 46*(3), 834-843.

Ritola, M., Annala, M., Hulkkonen, S., Lahtinen, V., Latti, R., Noponen, E., ... Hietaniemi, J. (2015). Cleantech takes over consumer markets. *Demos (Mexico City, Mexico).*

Roth, B. (2016). Four Drivers Propelling Clean Tech's Growth. Retrieved from http://www.earth2017.com/clean-tech-ai-iot-growth/

Shaw, A. (2014). Accelerating sustainability: Demonstrating the benefits of transportation technology. *The Intelligent Transportation Society of America.* Retrieved from http://digitalenergysolutions.org/dotAsset/933052fc-0c81-43cf-a061-6f76a44459d6.pdf

Sonderlund, A. L., Smith, J. R., Hutton, C., & Kapelan, Z. (2014). Using smart meters for household water consumption feedback: Knowns and unknowns. *Procedia Engineering*, *89*, 990–997. doi:10.1016/j.proeng.2014.11.216

Waleed, R., & Munam, A. S. (2016). Performance evaluation of IoT network infrastructure. *Automation and Computing (ICAC) 2016 22nd International Conference on*, 348-353.

Wang, F., Hu, L., Hu, J., Zhou, J., & Zhao, K. (2016). Recent Advances in the Internet of Things: Multiple Perspectives. *IETE Technical Review*, 1.

Weber, M. (2016). Our water system: what a waste. *The New York Times.* Retrieved from http://www.nytimes.com/2016/03/22/opinion/our-water-systemwhat-a-waste.html

Woo, D., Piralta, K., & Matthews, J. (2015). Towards sustainable water supply: Schematic development of big data collection using Internet of Things (IoT). *Procedia Engineering*, *118*, 489–497. doi:10.1016/j.proeng.2015.08.465

KEY TERMS AND DEFINITIONS

Cleantech: Cleantech refers to a company, or a section of an industry that is specifically engaged in the identification, development, and commercialization and selling of smart technologies, innovations and/or processes that lower or eliminate the adverse effects of the environment in the air, land or water in a more disruptive front.

Disruptive Innovation: Disruptive innovation is a term that was developed by Christensen to explain the process through which a product/service goes through right from development to commercialization in the market place, to a period of displacing competitors.

Internet of Things: This is an ever-growing technology involving physical computer structures, machines and digital objects that are assigned unique identifications to help transmit data over an identified network or ecosystem with human interaction.

Chapter 7
3D Printing Technology Diffusion:
A Revolution or an Illusion?

Kemal Yayla
Ege University, Turkey

Basak Ozdemir
Independent Researcher, Turkey

Serhat Burmaoglu
Izmir Katip Celebi University, Turkey

Haydar Yalcin
Izmir Katip Celebi University, Turkey

ABSTRACT

3D printing technology has been considered one of the most potentially groundbreaking technologies for the future, as the customer expectations, market requirements, and the competition grows in a global scale. In order to understand the potential effect of 3D printing technology and if it is a disruptive innovation that will change the traditional manufacturing paradigm, it is essential to examine the diffusion of knowledge in this area. In this study, 3D printing technology has been reviewed and patent analysis regarding 3D printing technology has been conducted in order to understand the diffusion of 3D printing technology. The results of the patent analysis indicate that the diffusion of 3D printing technology which is represented by the patents of four key methods not expected to fit with Bass diffusion model. According to the findings, it can be concluded that 3D printing technologies are in a situation where a state of maturity has not been reached, yet the growth still continues.

DOI: 10.4018/978-1-5225-4148-6.ch007

INTRODUCTION

In advanced economies, the range of economic variables, such as firm performance and technological specialization, play important role in explaining economic growth. The essence of this argument is that innovation, which involves resembles aspects of evolutionary process of accumulating and consolidating existing technologies into new technologies, is a natural process. However, the diffusion of knowledge plays an active role in shaping the direction of innovation, technological and economic evolution. It is only through the dissemination of information that the knowledge that a product or service can be created in the medium in which it could generate an additional opportunity for the economic units in every society (Hur, 2017). The diffusion of new ideas to different environments through recombination of existing technologies depends on the flow of technology in between different sectors (Fleming, 2001; Fleming & Sorenson, 2001). At this point, digital transformation creates opportunities and advantages for the global dissemination of any kind of information that has been produced (Kiesling, Günther, Stummer, & Wakolbinger, 2012; Rifkin, 2012; Samuelson, 2000).

Digitalization is rapidly transforming all sectors that it touches. Although it is not yet clear how digitalization finds a precise response in the society, it seems to affect the certain layers of society more rapidly. Production technology is at the forefront of this change. Cloud Computing, Internet of Things and Robotics are regarded as major technologies, which have emerged as the benchmarks of digitization (Manyika et al., 2013). All of these technologies are considered as technical innovations which generally create changes in business processes, product functions or customer-oriented services, rather than creating value directly.

Among the varied emerging technologies, 3D printing technology stands out as a phenomenon within the concept of technological innovation. 3D printing applications are used in a wide range of areas, from direct production of digital goods to hobby products or biomedical applications (Forrest & Cao, 2013; Gibson, Rosen, & Stucker, 2015; Pierrakakis, Gkritzali, Kandias, & Gritzalis, 2015). 3D printing promises a new wave of innovation that will enable the production of wide variety of objects for a wide user range, from amateurs, to new business ventures, and large-scale companies (Desai & Magliocca, 2013).

3D printing technology is predicted the one of driving forces of technological revolution, which will have a substantially profound economic impact on the manufacturing sector. In addition, 3D printing will spark a new industrial revolution and, technology will have transformational effects in the coming years. 3D printing is generally the nomenclature of mass media, marketers of the technology, and decision makers. The similarity between Additive Manufacturing and 2D printing has given rise to the alternate common name of 3D printing. However, AM (Additive

Manufacturing) is broader term that embraces all the technologies within the name of a single one (Bourell, 2016). There has been an accelerating amount of interest in AM technologies since the early 1990s. At the same time, scholars in the field of technology of management have been increasingly attracted the business model and technological development perspectives.

In particular, there is clearly articulated perception by both scholars and practitioners that AM technologies have the potential to change the traditional manufacturing paradigm. However, there are only a small number of early global adopters in the new business model of AM systems, which have seen predominantly in sectors that produce small batches of products and/or require customization (Jiang, Kleer, & Piller, 2017; Rogers, Baricz, & Pawar, 2016; Steenhuis & Pretorius, 2016; Weller, Kleer, & Piller, 2015). It appears to be significantly unrealistic to expect AM technologies will disrupt the existing manufacturing and innovation practices (Rayna & Striukova, 2016). To be specific, this paper aims to answer one research question as "What is the diffusion characteristic of AM technologies?"

In order to answer the research question, the outline of this chapter is as follows: a brief literature on disruptive technology and concept of disruptive innovation is discussed in the first part of background. Moreover, to demonstrate how disruptive technologies penetrate in real life problems, AM technologies are investigated. Patent analysis, diffusion models and growth curves are reviewed in the methodology section. Analysis is performed and findings are demonstrated. The chapter concludes with findings and possible future research directions regarding to AM technologies.

BACKGROUND

Technology refers to the theoretical and practical knowledge and skills that can be used to develop products or services for production and distribution systems. In this sense, technological change has the power to transform the way we live. Many technologies have potential to bring about broad economic value to be affected, and have high-tech change potential for disruptive economic impact. Technological change can affect lifestyles, the way people work, and it can also provide new business opportunities for economic growth (Abernathy & Clark, 1985; Bower & Christensen, 1996; Cheng, Huang, Ramlogan, & Li, 2016; Danneels, 2004; Kostoff, Boylan, & Simons, 2004; Manyika et al., 2013; Walsh, 2004; Walsh & Linton, 2000).

The concept of disruptive technology, first introduced by Bower and Christensen (1996) has become a new feature in continuous or accelerated change rates for customer demands in the mainstream market, providing new technological features in terms of price / performance ratio, and rate of change for the products in the product portfolios (Cheng et al., 2016; Manyika et al., 2013). Regarding disruptive technologies Christensen argues that “disruptive technologies are typically simpler, cheaper, and more reliable and convenient than established technologies’’ (Danneels, 2004). In other words, disruptive technologies can be regarded as scientific discoveries that transcend ordinary product / technology capabilities, and form the basis of a new competitive paradigm (Kostoff et al., 2004). Disruptive technologies attract the attention of niche market customers because they have penetrated the market due to their new features. When disruptive technology enters mainstream markets, the performance of features that disruptive technology meets the expectations of the main market (Vecchiato, 2016)

Although the terminology used to define the concept is varied such as hurricanes, typhoons or emerging technologies have been made the notion of disruptive technology has been accepted as general term (Walsh & Linton, 2000). Disruptive technologies create new significant growth opportunities in industries where those with different skill sets and firms penetrate existing sectors by adding value through stepping, or by creating innovations from the value they provide (Kostoff et al., 2004). Research on successful disruptive technology cases indicate that they create new values for customers who find existing production paradigm less satisfactory.

Disruptive technology first coined by Christensen, described the technologies that were rather simple or cheap when introduced, but have a potential to replace the existing competitors, however the term itself was variously defined, and this situation caused debate among the academics, as it is became unclear what constitutes a disruptive technology (Danneels, 2004; Schmidt & Druehl, 2008; Weeks, 2015; Yu & Hang, 2010). The term “Disruptive Innovation” took place of “disruptive technologies”, as the concept grew and the cases described exceeds the mere example of technological changes, and started to includes other forms of disruption cases (Yu & Hang, 2010). In this sense, as Yu and Hang suggests, disruptive innovation implies a process (Yu & Hang, 2010).

Disruptive innovation frequently results from changing social practices, reframing the problems from their very roots. In order to create a common understanding, Schmidt and Druehl proposed a framework that would clarify the patterns and degrees of diffusion of disruptive innovations, as well as managerial guidelines, how to assess whether an innovation is sustaining or disruptive (Schmidt & Druehl, 2008). The framework indicates that each of the 4 types of innovation can be equivalent to one of 5 types of “encroachment” (Schmidt & Druehl, 2008) which have been used to define a cannibalization of an older product or service by according to whether diffusion starts the low-end market or the high-end market (Schmidt & Druehl, 2008). Danneels suggested that the term ‘disruptive technology’ is somewhat vaguely defined and therefore is not able to help managers to predict which technologies will be disruptive earlier on (Danneels, 2004). As disruptive innovation has been described as a process (Yu & Hang, 2010), identifying potentially disruptive technologies is essential for companies to be able to survive and thrive in their respective areas. After reviewing the disruptive technology and innovation concept, which is the focus of the chapter and the most appropriated AM case in literature, is reviewed from various perspectives from historical roots to methods and application areas.

MAIN FOCUS OF THE CHAPTER

Historical Roots of Additive Manufacturing Technologies

AM has come to the fore as a popular field, as well as scientific research conducted in many different areas and its history can be traced to three technological developments dating back around 90–150 years: photo sculpture, topography, and material deposition. In view of the historical development of additive manufacturing, the roots of technologies originated with photo sculpture (1860) and topography (1890) in contrast to the general belief that the methods emerged only with the widespread use of modern production techniques (Gao et al., 2015; Küchler, 2014). One of first example of additive manufacturing is a method of producing three-dimensional landscape images from wax plates patented by J.E Blanther in 1862 (Bertling & Rommel, 2016; Bourell, Beaman, Leu, & Rosen, 2009). The method consisted of smoothing the topographic contour lines across a series of wax plates, cutting the wax along the contour lines, and then smoothing out the wax sections. It was possible to create a three-dimensional map of the shape in accordance with the negative and positive gradients of the surface. The method is regarded as having a similar approach to creating objects with current three-dimensional printing technologies (Bourell, 2016).

Although methods for 3D printing were developed from the early 19th century, the commercialization of methods has increased considerably since late 1950s. The first patent was filed by Munz (1956), which possessed current features of stereolithography (SLA) technique (Bourell et al., 2009). The method proposed by Munz, was designed to produce three-dimensional objects with photochemical substance that displays the image on the squeeze cylinder in the transparent photo emulsion of the layered structure of the cross-section of the scanned object. Another method proposed by Swainson (1977), suggests creating a three-dimensional figure by means of three-dimensional polymerization at the intersection of a laser-sensitive polymer and two laser beams, had similar properties. Laser, electron beam or plasma particles were used to create a continuous layer in a laboratory environment, but not for a commercialized production apparatus (Bourell, 2016; Bourell et al., 2009).

The commercialization of 3D printing methods started in mid 1980s. It started with the first well documented additive manufacturing patent, stereolithography (Hull, 1986), recognized as the most influential patent in additive manufacturing technologies. Further patents included laminated object manufacturing (Feygin, 1988), selective sintering (Deckard, 1991), fused deposition modeling (Crump, 1992), three-dimensional printing (Sachs, Haggerty, Cima, & Williams, 1993) and Laser engineered net shaping (Jeantette, Keicher, Romero, & Schanwald, 2000). More recently, the AM systems has been slowly gaining attraction, specifically within design, however, expiration of patented methods have amplified potential growth and there have become accessible to a range of small businesses and supportive communities also to individual designers.

Modern Additive Manufacturing Technologies

ASTM F42 Technical Committee defines additive manufacturing (AM) as "layered manufacturing process for forming objects with three-dimensional (3D) model data" (Simon Ford & Despeisse, 2016; Gross, Erkal, Lockwood, Chen, & Spence, 2014; Guo & Leu, 2013). Unlike mechanical machining processes such as grinding or milling, additive manufacturing is a group of emerging technologies that make up objects by adding layers of material in cross-sections (Guo & Leu, 2013; Pierrakakis, Kandias, Gritzali, & Gritzalis, 2014).

Different additive manufacturing technologies have been applied for decades in a wide variety of sectors such as aviation, biomedicine, automotive and digital arts. Additive manufacturing technology entered rapid period of change at the end of the 1980s, with the widespread use of selective laser sintering printers. However, the definition of additive manufacturing has evolved since then as consequences of developing new methods and business models. So far, there are a range of AM techniques ranging from well-established methods, employed in industrial settings to those still under development in research laboratories designed for more specific applications. In the literature, alternative terms used for AM techniques include rapid prototyping (Bak, 2003), free form fabrication (Hutmacher, Sittinger, & Risbud, 2004), rapid manufacturing (Levy, Schindel, & Kruth, 2003), additive fabrication (Giannatsis & Dedoussis, 2009) and direct digital manufacturing (Chen et al., 2015). Although AM is described as a synonym for such terms, it can be considered as an umbrella term for the concept of layered production (Coon, Pretzel, Lomax, & Strlič, 2016; Jin, Ji, Li, & Yu, 2017). AM processes are divided into three main categories, liquid, powder and solid states, which are related with the initial state of materials, landmark AM methods include Stereolitograhpy (SLA), selective laser sintering (SLS) fused deposition modeling (FDM) and laminated object manufacturing (LOM). In order to understand diffusion characteristics and technological development of

AM, these major methods are investigated in detail (Coon et al., 2016; Han & Sohn, 2015; Huang, Liu, Mokasdar, & Hou, 2013; Levy, 2010).

SLA is the oldest liquid-based manufacturing process and uses a light-sensitive monomer resin and a UV laser to form layers in the production of the base part (Bourell, 2016). Supports are needed to connect the component to the building platform. On each layer, the laser beam follows the cross section of the particle on the surface of the liquid resin to solidify the pattern. Subsequently, the building platform is lowered to completely cover the part then a blade is used to wipe off the resin and resulting in a level that leaves a layer of resin on top of the part. The part is then lowered with a single layer and left until a flat surface is obtained before the next layer is formed. When the part is complete, support can be removed manually. SLA is particularly appropriate for the manufacturing industry because it reduces the time required to produce a prototype part and achieves good surface coverage. The SLA-generated part was constructed from cross-sectional data obtained from the stereolithography file (.STL), which is the standard format for all AM technologies (Coon et al., 2016; Gross et al., 2014; Huang et al., 2013). SLA has advantages and disadvantages in terms of application procedure. The advantage in terms of dimensionally correct parts production is the limited risk of shrinkage dimensional change after thermal degradation of the thermo set photopolymers used in the production process, and the main limitation is that the product size is relatively small, generally less than a 2-meter cube.

Another approach is powder-based methods, which are based on the principle that the material is selectively formed by a local heat source to form each layer of powdered material. In first place, powdered materials is used to spread in a layer and then selectively scanned by a laser called as Selective Laser Sintering (SLS) method. Three dimensional parts can be created with SLS having relatively wide powder form materials, including wax, polymers, polymer / glass composites, polymer / metal powders, metals and ceramics (Coon et al., 2016; Levy, 2010). SLS creates a three-dimensional CAD model by cross-sectioning on a powder bed and layering with the laser. The laser is used to heat the powder particles to the melting temperature rather than to the complete melting state, as this should cause deformation. The powder material bed used during production reduces thermal degradation and melting temperature to ensure coalescence between the layers (Gibson et al., 2015). SLS does not require support holder during the production of parts, since the manufactured part is surrounded by undiluted powder (Gao et al., 2015). A number of factors may inhibit the production of quality prototypes, such as the different thermal properties of different materials, the presence of powdered polymers with appropriate particle size and morphology, large thermal processing windows, and constraints in laser sintering systems (Simon Ford & Despeisse, 2016). The density and porosity of the sintered part will depend on the size of the particles and how often they are packed in the structural part. Other parameters, such as laser wavelength, laser energy and temperature distribution in the cartridge, can also affect the part's mechanical properties, such as porosity or anisotropy (Caulfield, McHugh, & Lohfeld, 2007).

Materials extrusion (Fused Deposition Modeling (FDM)) and sheet lamination (Laminated Object Manufacturing (LOM)) are two others methods used for solid state materials. An AM process, trademarked as Fused Deposition Modeling(FDM), creates layers mechanically on the principle of phase change of a thermoplastic material, this requires high operating temperatures, and the finished prints typically exhibit high porosities (Coon et al., 2016; Gao et al., 2015). A thermoplastic filament (mostly an ABS filament) is heated to a semi-molten state, and extruded by a nozzle moving around on the x–y plane, depositing traces on the working platform. Since the material is extruded in a semi-molten state, the newly deposited material fuses with adjacent material that has already been deposited (Ahn, Montero, Odell, Roundy, & Wright, 2002; Gao et al., 2015). Lowering the platform in the z direction to a specified distance allows for deposition of subsequent layers. A second nozzle extrudes support material where required. The molten material fuses with the layer underneath. This layering in different orientations gives FDM parts anisotropic properties (Coon et al., 2016). LOM is another AM process designed to build a part sequentially from layers of paper, plastic, metal, ceramics or composite sheets, all coated with a thermally activated adhesive (Mueller & Kochan, 1999; Park, Tari, & Hahn, 2000). Generally, the LOM technique was developed for processing paper. The process begins with the lamination of 20 layers of paper onto the retractable platform. The slices of material sheets are cut in the required shape from a roll of material using a laser beam. The cutting material sheet is laid on machine platform and bonded to the previous sheet using a hot roller, which activates a heat sensitive adhesive. After addition of all sheets, the solid part of the material is removed from the platform. Surrounding material and material in regions of the part that are hollow must be removed in a post processing step (Gao et al., 2015). Finally, it is apparent that these methods were developed for solving practical real-life problems. To understand the AM concept from the output side, the AM applications in different sectors are reviewed and briefly explained in the following section.

Applications of Additive Manufacturing Technologies

In the late 1980s and early 1990s a plethora of AM methods appeared. In the emerging stage of AM technology, the research community generally applied the processes and process variations in novel ways in order to address to a wide variety of research problems in diverse technical areas, whether the application is technically possible or economically feasible largely depends on its production volume, part size, complexity, and material cost (Berman, 2012; Rayna & Striukova, 2016). The impacts of AM technologies continue to grow, in terms of both commercial and scholarly activities. Double digit annual growth is expected in the global additive manufacturing market for both rapid prototypes and functional end-use parts in 2016 (Gordon, 2015). In addition, giant leading manufacturers all over the world have applied AM technologies in many fields, including the aerospace & defense, automotive and medical industries.

The aerospace and defense (A&D) industry was an early adopter of AM technology. There are two main reasons why the aerospace industry is a pioneer of adaptation of AM technologies. First, the industry requires parts that are lightweight, strong, and geometrically complex. Second, the lot sizes in aviation are typically produced in small quantities (Uriondo, Esperon-Miguez, & Perinpanayagam, 2015). Many of these parts are costly to produce in high quantities and AM can minimize the amount of materials used in stringent operating conditions. In addition, AM has significant impact for the parts in complex geometries relative to other conventional manufacturing processes (Lyons, 2014; Paritala, Manchikatla, & Yarlagadda, 2017). The advantages of AM in A&D industry include reducing the expenses of physical warehousing, the capability design to complex tools, the low cost of parts, waste reduction and part simplification. Furthermore, AM machines produce less scrap than traditional machines and increase the availability of parts at the point of use.

Aerospace firms are increasingly adapting to the AM technologies to reduce the costs of developing models and of creating components. Several aerospace companies have started to adjust supply chain operations to accommodate AM parts (Cotteleer & Joyce, 2014; Sharon Ford, 2014; Kellner, 2014; Lyons, 2014). Current AM applications in the A&D industry have ranged from arm rests to more complex parts such as engine components. For instance, GE has been using additively produced parts inside commercial jet engines since approval of FAA in 2015 (Steenhuis & Pretorius, 2017). Airbus has also updated certain cabin brackets for the A380, making this the company's first commercial plane to use AM components (Bogue, 2013). In addition, NASA has already built its prototypes of tools and instrument parts for current missions using AM methods (Kietzmann, Pitt, & Berthon, 2015).

The modern automotive industry was another early adopter, and has used AM technology to make tool prototypes and small custom parts with short production runs for decades. Since production volumes in the automotive industry are often high, companies in this industry use AM technology to offers an increasing variety of different car models that are customizable due to mass customization, vast amounts of components create considerable overheads during production. AM could provide a novel approach to creating new products and reach the market faster than traditional subtracting manufacturing (Gibson et al., 2015; Jiang et al., 2017). Despite many advantages, however, most manufacturers have not incorporated to

AM parts on their mass-produced car models. In fact, AM methods can be cost effective for specialty cars or low-volume manufacturers. The American company KOR Ecologic has applied AM technologies to mainly produced their car Urbee for only 50 components (Bechthold et al., 2015). There also have been some niche applications of AM technologies, some racing teams started to use AM parts for their cars. The racing organization Formula 1 has been using AM for rapid prototyping since the early to mid-2000s (Bechthold et al., 2015; Bogue, 2013; Gibson et al., 2015). Additionally, AM could make spare part inventories become obsolete, it means manufacturers need only to keep the CAD files of the parts, and then manufacture the spare parts demand.

Another application area is medical sector. AM methods have been used for medical applications when this technology was first commercialized. In recent years, AM technologies have gained momentum in the healthcare sector by providing new solutions in the areas of surgical guides, medical manufacturing and bioengineering (Giannatsis & Dedoussis, 2009; Gibson et al., 2015). The use of AM technologies for diagnostic purposes was presumably the first medical application. AM has provided opportunities to meet the demand for individualized health care products based on the data obtained from various well-developed techniques since the digitalization of patient data. For instance, AM methods are effective in transforming medical imaging data into bio models' due to the organic nature of the complex forms (Bechthold et al., 2015). AM methods have found widespread use as a tool to bioengineered tissue, varying in composition from bone and teeth to vascular and organ scaffolding. It currently uses a variety of biomaterials with different physical and biochemical properties to construct organic scaffolds for the tissue repair process (Hutmacher et al., 2004). Thus, customized AM manufactured scaffold allows for individualized treatment for tissue regeneration while avoiding the need to sample from the patient's own tissue for scaffold formation (Gross et al., 2014). It also allows for the design of customized prosthetic implants suited to individuals need, including size, shape and mechanical properties. AM methods also eliminate the need for time consuming steps of prosthetic fabrication (Jardini et al., 2014).

It is clear that AM has disruptive effect on different sectors from these application areas. It can be seen that AM has diffused into many sectors with different motivations and potential advantages. For instance, AM technologies are expected to present substantial positive effects on the structure of value chains, although this will be a gradual change, as it first depends on companies engaging in component and product redesign. Product and component redesign will affect the production system, as companies simplify complex multi-component products into single component products. Another benefit is that the traditional model of disruptive innovation underlines that disruptive technologies typically have a lower unit price, despite having a higher usage cost. Moreover, industries such are Aerospace or Medical have higher unit costs, in particular incurred due to the nature of materials. The cost of parts will likely to be higher due to the greater complexity in the architectural design of these product systems, in comparison to predecessors. In contrast, disruptive innovations do not succeed in isolated environments. New markets are needed to increase the value of the potential disruptive innovation likely be caused by simpler, more affordable innovations that create a need in the customers, motivating them to transfer from the traditional mainstream market to the new one generated by innovators. Put differently, a disruptive innovation processes are based on long-term characteristics. These characteristics will only be visible after the existing products and existing firms have been replaced by new products and new entrants. Even though, literature reviews give insight into AM applications; this review cannot show the overall views of diffusion, therefore a vital role of AM technologies bring out with patent analysis.

SOLUTIONS AND RECOMMEDATIONS

Innovation Diffusion Models and Growth Curves

Diffusion is defined as a factor stimulates the adoption of technologies by product which an innovation is adopted. According to E. M. Rogers (1962) Diffusion theory, innovation is defined as an idea, object or practice that is perceived as new by individuals or organizations. Thus, diffusion is defined as the process which ''is communicated through some certain channels over time among members of a social system'' (Soffer, Nachmias, & Ram, 2010). In the technological diffusion view, technology diffusion is defined as the process by which innovations spread within and across economies. From a theoretical perspective, Epidemic, Rank and Game Theoretic models are major approaches that have been characterized diffusion process (Bocquet, Brossard, & Sabatier, 2007). The use of epidemic models which consider diffusion as resulting from the spread of information. An early model proposed by Mansfield (1961) suggests that diffusion follows a simple logistic penetration of the market, and depends on speed of diffusion based on the frequency of contacts between potential users and actual users. Consequently, the potential determinants of the speed of adoption depend on the process by which people become informed about technology (Kiesling et al., 2012).

However, different models have been proposed to take into account different situations. The "Diffusion Model" proposed by Bass (1969) is one of the well-known models in the field. The theoretical foundation of the Bass diffusion model (Equation 1) is combination of models of Fourt and Woodlock (1960) and Mansfield (1961) (Chang, Lai, & Chang, 2009). Bass's model estimates how a new idea or new behavior spreads across the market over time. It analyzes the diffusion process through consumers' behavior, under the basic assumption that the probability of initial purchase is related linearly to the number of previous buyers. Buyers or adopters may be innovators or imitators. The original Bass model is shown in Equation (1).

$$n\left(t\right) = \left(p + q\frac{N\left(t-1\right)}{m}\right)\left(m - N\left(t-1\right)\right) \tag{1}$$

In Equation (1) n(t) is the sales of a new product at time t; N(t) is the cumulative sales of a new product at time t; m is the market potential; p is the innovation coefficient and q is the imitation coefficient. In a pure innovation scenario (p > 0, q = 0), diffusion follows a modified exponential model; ia pure imitation scenario (p = 0, q > 0), it follows a logistic curve model (Meade & Islam, 2006). In patent related analysis, Bass model has been modified and the new form is given close approximation compared to the original (Cunningham & Kwakkel, 2011; You et al., 2017). The Bass-equivalent model, which is validated by its prediction for patent citation growth, is defined in Equation (2) (You et al., 2017).

$$n\left(t\right) = a + b\sum_{i=1}^{t-1} n\left(i\right) + c\left[\sum_{i=1}^{t-1} n\left(i\right)\right]^2 \tag{2}$$

In Equation (2), n(t) denotes the net patent amount at the time t while the symbols a, b and c are all fitting coefficients in model.

Another quantitative approach for understanding innovation is forecasting the diffusion of innovation. It has become a major research topic in the fields of innovation and technology research. Growth curve is an empirical model, originally developed in analogy to the growth of a living organism, designed to estimate the evolution of a quantity over time based on historical data (Meade & Islam, 2006; Young, 1993). Thus, they have been widely used to forecast analysis of technology lifecycles. Various studies have been conducted to identify the estimation of current development stage of technology domains with different growth curves (Meade & Islam, 2006; Yoon, Park, Kim, Lee, & Lee, 2014). For the patent data, the most frequently used models are The Gompertz model, the general Logistic model, the Pearl model and the Fisher-Pry model. The Pearl and Fisher–Pry models are generally used to forecast the cumulative growth of new adoptions (Cunningham & Kwakkel, 2011). Among those growth curves, Gompertz is considered appropriate for analyzing the life cycles of rapidly growing technologies (Bengisu & Nekhili, 2006; Martino, 2003; Yoon et al., 2014). The Gompertz model is shown in Equation (3)

$$Y_t = Le^{-ae^{-bt}} \tag{3}$$

In Equation (3) has three primary coefficients a, b, and L. The coefficients a and b, respectively, determine the location and shape of the curve, while L is its asymptotic maximum value (Meade & Islam, 2006; Yoon et al., 2014).

Patent Analysis

Patents, as a vital data source, are main outputs of research and development, and represent the characteristics of technology (Harhoff, Narin, Scherer, & Vopel, 1999; Trajtenberg, 1990). A vast amount of recent technical knowledge is available in patent documents, it has become increasingly important to exploit this knowledge since electronic patent database become accessible worldwide (Ernst, 2003). Hence, patent data have been considered as a key source in technology diffusion research (Cao & Zhao, 2013; Chang et al., 2009; Choi, Kim, & Park, 2007; Jun, Park, & Jang, 2015; Yan & Luo, 2016).

Patent analyses have been used to assess the knowledge diffusion and transfer processes in research and development which extract useful knowledge from databases. There are, of course, classification on using patent data as indicators of technology analysis (Karki, 1997). Patent analysis can be divided into three types of research level: Macro-level research of national or industrial analysis (Curran & Leker, 2011), micro-level research of particular technological diffusion analysis (Lee, Kim, & Cho, 2010) and forecasting (Chang et al., 2009). At the macro level, the major topics are technological innovation and evaluation of technological competitiveness of nations. At the micro-level, there are research activities such as identifying technological advantages and disadvantages of competitors (Choi et al., 2007; Cotropia, Lemley, & Sampat, 2013; Harhoff et al., 1999; Trajtenberg, 1990).

A patent contains "prior art'', which is to clearly explain the path of invention through citing linkages, such as patents, journals or other literature. In other words, the more a patent is cited, the wider the technology is diffused, and it implies a great number of applications and higher values (Lai & Wu, 2005). Patent citation analysis is widely used method for advanced analysis on technological diffusion through the examination of citation links among various patents. It is able to evaluate both the spill-over effects between industries and the importance of the technology-based transformation.

In this study, forward citation, a method for patent analysis, is applied. Forward citation in the patents is used to indicate the extent to which patents are closely related to technological breakthroughs and are considered technologically strong (Harhoff et al., 1999; Trajtenberg, 1990). In other words, a patent which is cited by other patents indicates its potential to create technological and economic value by making important contributions to the future path of technological diffusion (Karvonen & Kässi, 2013; Nemet & Johnson, 2012). In patents, there are key terms referred to as basic technologies which indicate the direction of information flow of the new technologies developed. The effectiveness of the technology is determined by the strength of the direction of information flow from key patents to newly filed ones. In typical patent evaluations, forward citation is used as "technological value" (Hu, Rousseau, & Chen, 2012). In order to compare each technology by in terms of technological strength, two performance indicators, Impact Index and Marketability index, are defined.

Impact Index (II)

The impact index is defined as the extent to which technology penetrate into other areas. In this case, number of citations to key patent can be determining as its impact. As each patent has different technological life time, these should be normalization for comparison. Impact index is shown in Equation (4).

$$II = \frac{C_i}{P_i} \tag{4}$$

e C_i denotes cumulative number of times i was cited and P_i. the number of patentin selected patent family. In this case, key patent has been already defined due to design of research. In order to compare patents citation periods are fixed.

Marketability Index (MI)

Marketability index (MI) is defined as the applicability of the products that utilizes a technology, and is evaluated using the number of citation over 4- digit CPC (Cooperative Patent Code) ratio in citing patents. This index can signal how technology diffuses into new technological territories and also perceived as qualifying technologies market size. MI is calculated is using Equation (5).

$$MI = \frac{F_i}{C_i} \tag{5}$$

where F_i denotes number of CPC codes in citing patents of key patents and C_i is number of times i was cited.

Analysis and Findings

Forward citation measure is frequently used for determining technological value of patents (Wang, Chiang, & Lin, 2010; Yasukawa & Kano, 2015). The definition of AM technologies covers a wide area of application. To understand the strength of technological breakthrough and growth based on time, forward citation analysis is mostly applied (Fallah, Fishman, & Reilly, 2011; Nair & Mathew, 2015). In this study, first commercial patents developed for liquid, powder, and solid-state materials, consisting SLA, SLS, FDM and LOM were selected as key AM methods (Coon et al., 2016; Han & Sohn, 2015; Huang et al., 2013). It is assumed that selected key AM methods are applied for different type of materials and application methods. Descriptive data for key methods are demonstrated in Table 1.

As can be seen in Table 1 the greatest number of patent citations was found for SLA method. Even though, FDM method is younger than others, it has more citation counts therefore it can be interpreted that solid-state materials have more impact than liquid and powder state materials. The change in the number of patent citations is a proxy measure of that technology's diffusion. In order to the understand diffusion

Table 1. Key AM patents and forward citation counts

Method	Patent No.	Patent Title	Patent Registration Date	Forward Citation Counts*[1]
SLA	4575330	Apparatus for production of three-dimensional objects by stereolithography	03/ 11 / 1986	555
SLS	4863538	Method and apparatus for producing parts by selective sintering	09/ 05 /1989	355
FDM	5121329	Apparatus and method for creating three-dimensional objects	06 /09 /1992	351
LOM	4752352	Apparatus and method for forming an integral object from laminations	06 / 21 / 1988	192

status of AM technologies, Bass equivalent model was applied using the patent data between 1993 and 2016. Bass model evaluates technological change based on innovation and imitation ratios, unlike other diffusion models (Daim, Iskin, & Ho, 2011; Radojičić & Marković, 2009). The diffusion of key patents based on Bass model is demonstrated in Figure 1.

To compare diffusion and growth of selected key patents, the 1993-2016 periods was chosen because citation periods are equal for each patent. In this period, diffusion of AM technologies was evaluated based on all citation counts. For the evaluation of diffusion, Bass equivalent model, proposed by Cunningham and Kwakkel (2011) was used. As can be seen in Figure 1, fit ratio of model (R^2=0.60134) can be interpreted as low and suggesting that AM technologies have not developed as much as expected. In previous innovation diffusion studies, fit ratios of Gompertz model were higher than our findings (Lim et al., 2012; Nam, Kim, & Lee, 2008; Turk & Trkman, 2012). In addition, growth curves were used to test the technologies to determining in which state of product life cycle. From the reviewed literature, it is clear that Gompertz curve is more appropriate for determining growth stage in technological forecasting (Bengisu & Nekhili, 2006; Chang et al., 2009; Martino, 2003; Suominen & Seppänen, 2014; Yoon et al., 2014). The findings based on Gompertz curve analysis are demonstrated in Table 2.

Based on the findings in Table 2, according to the Gompertz model, the growth curve of cumulative citations seems to be highly compatible with previous studies (Sood, James, Tellis, & Zhu, 2012; Suominen & Seppänen, 2014). As mentioned before, the Gompertz model is a growth curve used to describe the period in which

Figure 1. Diffusion trends of key patents

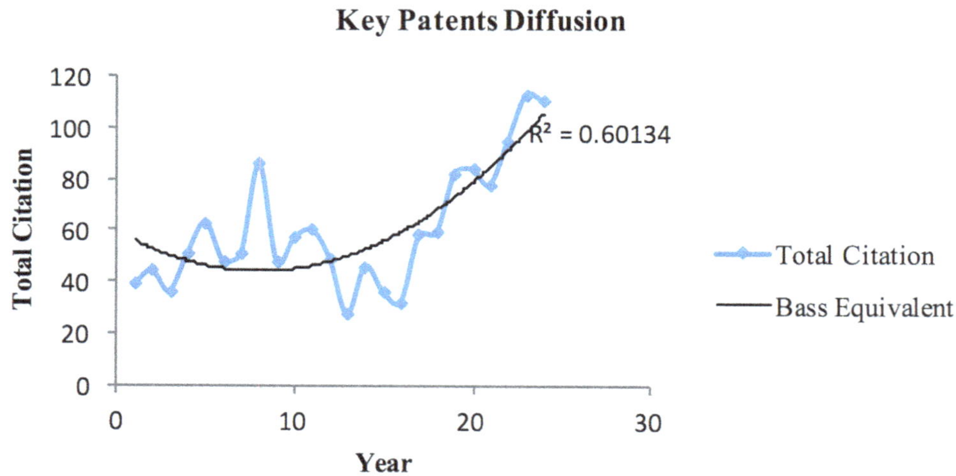

Table 2. Growth curve fit with Gompertz Model

Patent No	Coeff Det. (R^2)	Stand. Error
4575330	0,96	0,08
4752352	0,93	0,07
4863538	0,96	0,08
5121329	0,94	0,1

the growth that has not yet reached maturity continues in the technological life cycle. Although selected key patents have lost their preemptive right based on the year limitation for intellectual property rights, the four key patent citation curves reveal that for the technological strength that the patents represent, is still retained and has not yet reached the maturity. Individual citation performances of each key patent are shown in Figure 2, Figure 3, Figure 4 and Figure 5 respectively.

In Figure 2, there is only slightly fluctuation between 1989 and 2005. After this period, there is rapid growth in citation of patent 4575330. It can be interpreted that those patents approximately same period but attention to methods differs from each other.

Figure 3 should has same citation behavior as for the years 1990-2005. The main differences between Figure 2 and Figure 3 are ceiling slopes of peak years. In Figure 3 there is a non-continuous growth trend. In Figure 4, patent is rather new than compared to other key patents. Its citation trend has been increasing since 2008. The time lag of inflection point is same as Figure 2.

Figure 2. Citation distribution of patent 4575330

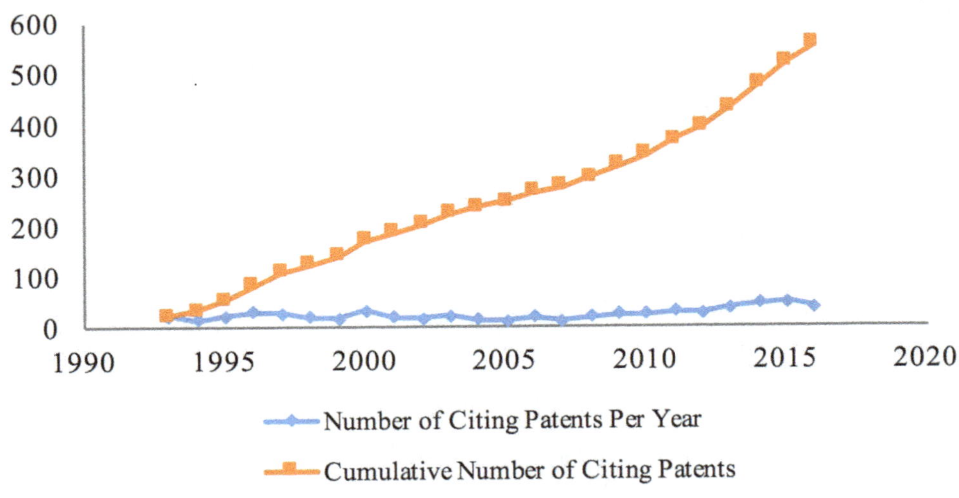

Figure 3. Citation distribution of patent 4863538

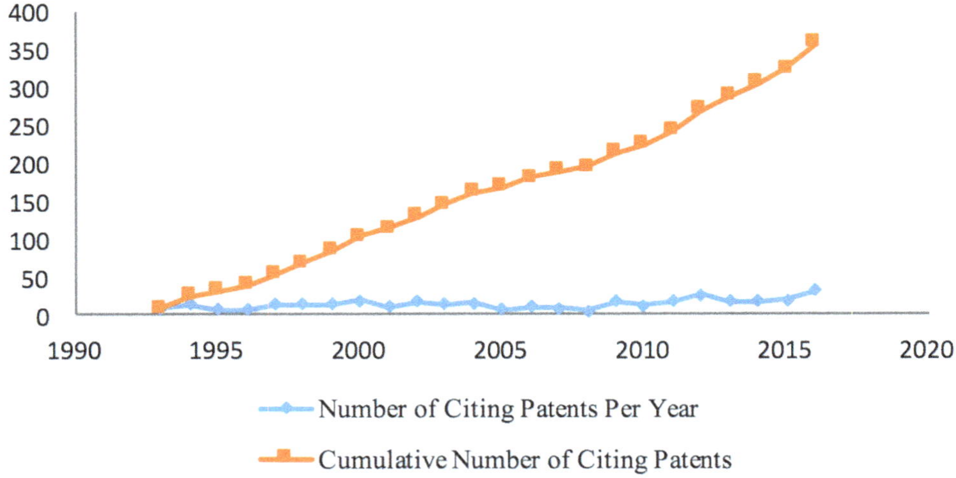

In comparison with Figure 3, Figure 4 has only slightly amount of forward citations. By contrast to filling dates of each patent, it is expected more differences than current status.

In Figure 5, it can be concluded that the yearly trend in patent citation is the most fluctuating patent among key patents, although, this patent was filed earlier than Figure 3, and Figure 4 citation count is lower than those patents. This implies

Figure 4. Citation distribution of patent 5121329

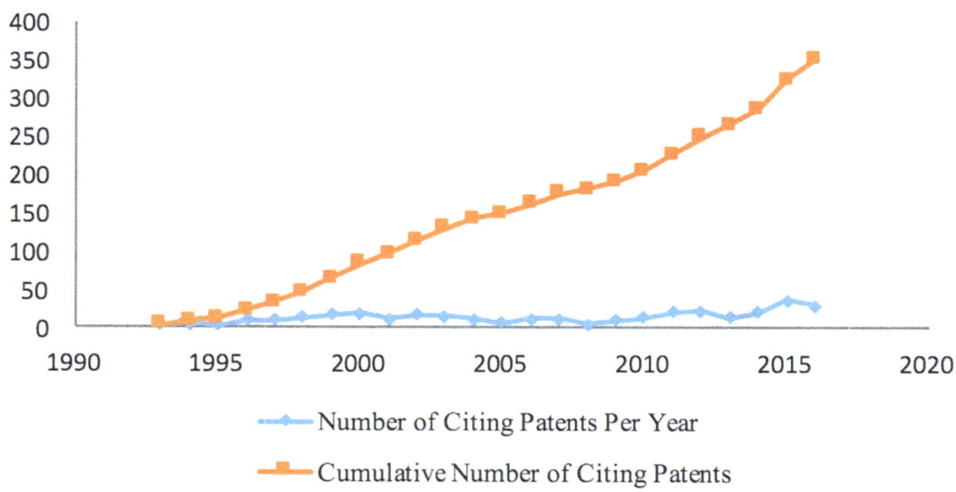

Figure 5. Citation distribution of patent 4752352

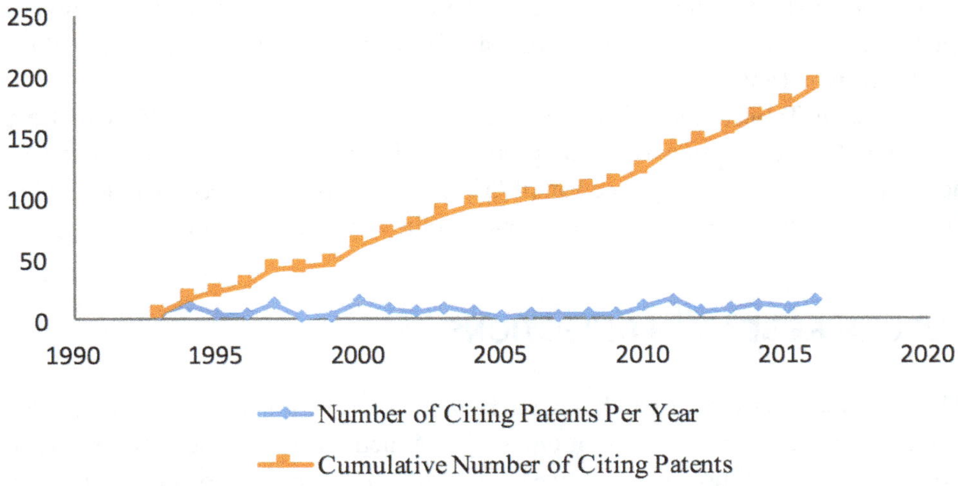

that this patent has smaller impact than others. Annually growing trends per patents indicate more attention to specific technology in terms of time activity pattern. However, this method of explanation suffers from limitations to provide how technology dependence paths evolve. To do so, performance indicator will help to make pair wise comparison between patents. The impact index values of each patent are shown in Table 3.

Patent 4575330 has the highest impact index. Patent 5121329 is relatively more recent than orthers, its impact indexed as high as patent 4863538. Those patents represent different AM methodology in terms of initial material state. Patent 4752352 has lowest impact index. Thus, it is in the same material family with Patent 4752352. This suggesting that time lag is not discriminative factor of technological impact and it may depend on other factors such as material, size, shape accuracy etc. Patent 4575330 is frequently applied, because it is liquid-based and often used for hobby activities (Bai, Liu, Wang, & Wen, 2017). In addition to technological

Table 3. Impact ındex of key patents

Patent No	Impact Index
4575330	0,38
4863538	0,25
5121329	0,24
4752352	0,13

impact, marketability impact is also analyzed based on the ratio of variety of CPC codes and citation count. Greater varied CPC codes showing that the method is found different application areas, and diffused into them. Marketability index values can be seen at Table 4.

As seen in Table 4, the smaller ratio of the MI indicates that technological effect more widely spread. Values reveal that commercial application of technology changes according to technology adaptation. 4752352 patent has the highest index value because, showing by the lower CPC code number compared to others.

FUTURE RESEARCH DIRECTIONS

The direction of further research is to study how researchers and industrial managers combine the selection of different types of AM methods or choose between them. For instance, to get more quantitative data, academic papers and patents can use as a proxy to measuring diffusion performance of each country or region. As a result of such research, both R&D managers and national and international policy makers get information on the ways of technological diffusion of 3D printing technologies which will be responding to the existing challenges of socio-technical issues at acceptable levels of risks and costs.

CONCLUSION

AM technologies have been considered as one of the most innovative technologies by both scholars and practitioners. Its potential to change the dynamics in many industries has made 3D Printing technologies a topic of great interest. In this study, 3D printing technologies have been examined from different standpoints.

Table 4. Marketability index of key patents

Patent No	MI	4-Digit CPC Number
4575330	0,22	120
4863538	0,26	98
5121329	0,28	98
4752352	0,44	84

In the literature reviewed, it can be seen that 3D printing technologies penetrate markets in which highly customized products are needed such as medicine and the automotive industry. Also, our review of literature has provided historical background of AM technologies and the four cores AM methods investigated in the analysis, Stereolithography (Patent no: 4575330), Selective Laser Sintering (Patent no: 4863538), Fused Deposition Modelling (Patent no: 5121329) and Laminated Object Manufacturing (Patent no: 4752352). In the analysis, forward citations have been used in order to understand the diffusion of the AM methods of choice.

The result of the conducted analysis indicates that AM technology has not yet reached the state of maturity. Result of the Bass Model suggests that the diffusion rate of the core AM technologies are not in accordance with the general fit of Bass model, indicating although there has been diffusion, it is not the expected level. The analysis on Impact Indices point out that despite emerging in slightly different timelines, the methods are very close in terms of extent of market penetration. This shows that the time between patents have limited effects compared to factors such as materials.

From the disruptive technologies and innovation points of view, this study points out that the current state of AM technologies is addressing to a niche customer. As it is known that disruptive technologies start with addressing the needs of niche customers, and later diffuse to a mainstream market, it can state that AM technologies have been slowly growing and evolving, and it can be considered disruptive technology with potential to creating major changes in the future. But it is also important to keep in mind that according to the results of the study, this is a gradual diffusion and it may take longer than scholars and practitioners predicted. Disruptive innovation is a process rather than an instantaneous event (Yu & Hang, 2010). Accordingly, in the case of 3D printing, this process is in motion, but is not as rapid as predicted. For this case, it is possible that 3D printing technologies are not yet in a position to be regarded as a revolution, but as we see its effects and growth in different sectors. We cannot say that it is an illusion either.

REFERENCES

Abernathy, W. J., & Clark, K. B. (1985). Innovation: Mapping the winds of creative destruction. *Research Policy*, *14*(1), 3–22. doi:10.1016/0048-7333(85)90021-6

Ahn, S.-H., Montero, M., Odell, D., Roundy, S., & Wright, P. K. (2002). Anisotropic material properties of fused deposition modeling ABS. *Rapid Prototyping Journal*, *8*(4), 248–257. doi:10.1108/13552540210441166

Bai, X., Liu, Y., Wang, G., & Wen, C. (2017). The pattern of technological accumulation: The comparative advantage and relative impact of 3D printing technology. *Journal of Manufacturing Technology Management*, *28*(1), 39–55. doi:10.1108/JMTM-10-2016-0136

Bak, D. (2003). Rapid prototyping or rapid production? 3D printing processes move industry towards the latter. *Assembly Automation*, *23*(4), 340–345. doi:10.1108/01445150310501190

Bass, F. M. (1969). A new product growth for model consumer durables. *Management Science*, *15*(5), 215–227. doi:10.1287/mnsc.15.5.215

Bechthold, L., Fischer, V., Hainzlmaier, A., Hugenroth, D., Ivanova, L., Kroth, K., . . . Sitzmann, V. (2015). *3D printing: A qualitative assessment of applications, recent trends and the technology's future potential.* Retrieved from https://www.econstor.eu/bitstream/10419/156627/1/StuDIS_2015-17.pdf

Bengisu, M., & Nekhili, R. (2006). Forecasting emerging technologies with the aid of science and technology databases. *Technological Forecasting and Social Change*, *73*(7), 835–844. doi:10.1016/j.techfore.2005.09.001

Berman, B. (2012). 3-D printing: The new industrial revolution. *Business Horizons*, *55*(2), 155–162. doi:10.1016/j.bushor.2011.11.003

Bertling, J., & Rommel, S. (2016). A Critical View of 3D Printing Regarding Industrial Mass Customization Versus Individual Desktop Fabrication. In S. Dickel, J.-P. Ferdinand, & U. Petschow (Eds.), The Decentralized and Networked Future of Value Creation (pp. 75-105). Springer.

Bocquet, R., Brossard, O., & Sabatier, M. (2007). Complementarities in organizational design and the diffusion of information technologies: An empirical analysis. *Research Policy*, *36*(3), 367–386. doi:10.1016/j.respol.2006.12.005

Bogue, R. (2013). 3D printing: The dawn of a new era in manufacturing? *Assembly Automation*, *33*(4), 307–311. doi:10.1108/AA-06-2013-055

Bourell, D. L. (2016). Perspectives on Additive Manufacturing. *Annual Review of Materials Research*, *46*(1), 1–18. doi:10.1146/annurev-matsci-070115-031606

Bourell, D. L., Beaman, J. J., Leu, M. C., & Rosen, D. W. (2009). A brief history of additive manufacturing and the 2009 roadmap for additive manufacturing: looking back and looking ahead. *Proceedings of RapidTech*, 24-25.

Bower, J. L., & Christensen, C. M. (1996). Disruptive technologies: Catching the wave. *Journal of Product Innovation Management*, *1*(13), 75–76.

Cao, Y., & Zhao, L. (2013). Analysis of patent management effects on technological innovation performance. *Baltic Journal of Management*, *8*(3), 286–305. doi:10.1108/BJOM-May-2012-0033

Caulfield, B., McHugh, P., & Lohfeld, S. (2007). Dependence of mechanical properties of polyamide components on build parameters in the SLS process. *Journal of Materials Processing Technology*, *182*(1), 477–488. doi:10.1016/j.jmatprotec.2006.09.007

Chang, S.-B., Lai, K.-K., & Chang, S.-M. (2009). Exploring technology diffusion and classification of business methods: Using the patent citation network. *Technological Forecasting and Social Change*, *76*(1), 107–117. doi:10.1016/j.techfore.2008.03.014

Chen, D., Heyer, S., Ibbotson, S., Salonitis, K., Steingrímsson, J. G., & Thiede, S. (2015). Direct digital manufacturing: Definition, evolution, and sustainability implications. *Journal of Cleaner Production*, *107*, 615–625. doi:10.1016/j.jclepro.2015.05.009

Cheng, Y., Huang, L., Ramlogan, R., & Li, X. (2016). Forecasting of potential impacts of disruptive technology in promising technological areas: Elaborating the SIRS epidemic model in RFID technology. *Technological Forecasting and Social Change*.

Choi, C., Kim, S., & Park, Y. (2007). A patent-based cross impact analysis for quantitative estimation of technological impact: The case of information and communication technology. *Technological Forecasting and Social Change*, *74*(8), 1296–1314. doi:10.1016/j.techfore.2006.10.008

Coon, C., Pretzel, B., Lomax, T., & Strlič, M. (2016). Preserving rapid prototypes: A review. *Heritage Science*, *4*(1), 40. doi:10.118640494-016-0097-y

Cotropia, C. A., Lemley, M. A., & Sampat, B. (2013). Do applicant patent citations matter? *Research Policy*, *42*(4), 844–854. doi:10.1016/j.respol.2013.01.003

Cotteleer, M., & Joyce, J. (2014). 3D opportunity: Additive manufacturing paths to performance, innovation, and growth. *Deloitte Review*, *14*, 5–19.

Crump, S. S. (1992). *Apparatus and method for creating three-dimensional objects*. Patent No. US5121329 A.

Cunningham, S. W., & Kwakkel, J. (2011). Innovation forecasting: A case study of the management of engineering and technology literature. *Technological Forecasting and Social Change*, *78*(2), 346–357. doi:10.1016/j.techfore.2010.11.001

Curran, C.-S., & Leker, J. (2011). Patent indicators for monitoring convergence–examples from NFF and ICT. *Technological Forecasting and Social Change*, *78*(2), 256–273. doi:10.1016/j.techfore.2010.06.021

Daim, T. U., Iskin, I., & Ho, D. (2011). Technology forecasting for residential energy management devices. *Foresight*, *13*(6), 70–87. doi:10.1108/14636681111179609

Danneels, E. (2004). Disruptive technology reconsidered: A critique and research agenda. *Journal of Product Innovation Management*, *21*(4), 246–258. doi:10.1111/j.0737-6782.2004.00076.x

Deckard, C. R. (1991). *Method and apparatus for producing parts by selective sintering*. Patent No. US5017753 A.

Desai, D. R., & Magliocca, G. N. (2013). Patents, meet Napster: 3D printing and the digitization of things. *Geological Journal*, *102*, 1691.

Ernst, H. (2003). Patent information for strategic technology management. *World Patent Information*, *25*(3), 233–242. doi:10.1016/S0172-2190(03)00077-2

Fallah, M. H., Fishman, E., & Reilly, R. R. (2011). Forward patent citations as predictors for patent valuation. *International Journal of Intellectual Property Management*, *4*(3), 165–178. doi:10.1504/IJIPM.2011.041081

Feygin, M. (1988). *Apparatus and method for forming an integral object from laminations*. Patent No. US4752352 A.

Fleming, L. (2001). Recombinant uncertainty in technological search. *Management Science*, *47*(1), 117–132. doi:10.1287/mnsc.47.1.117.10671

Fleming, L., & Sorenson, O. (2001). Technology as a complex adaptive system: Evidence from patent data. *Research Policy*, *30*(7), 1019–1039. doi:10.1016/S0048-7333(00)00135-9

Ford, S. (2014). Additive Manufacturing Technology: Potential Implications for U.S. Manufacturing Competitiveness. *Journal of International Commerce & Economics*.

Ford, S., & Despeisse, M. (2016). Additive manufacturing and sustainability: An exploratory study of the advantages and challenges. *Journal of Cleaner Production, 137*, 1573–1587. doi:10.1016/j.jclepro.2016.04.150

Forrest, E., & Cao, Y. (2013). Digital Additive Manufacturing: A Paradigm Shift in the Production Process and Its Socio-economic Impacts. *Engineering Management Research, 2*(2), 66. doi:10.5539/emr.v2n2p66

Fourt, L. A., & Woodlock, J. W. (1960). Early prediction of market success for new grocery products. *Journal of Marketing, 25*(2), 31–38. doi:10.2307/1248608

Gao, W., Zhang, Y., Ramanujan, D., Ramani, K., Chen, Y., Williams, C. B., ... Zavattieri, P. D. (2015). The status, challenges, and future of additive manufacturing in engineering. *Computer Aided Design, 69*, 65–89. doi:10.1016/j.cad.2015.04.001

Giannatsis, J., & Dedoussis, V. (2009). Additive fabrication technologies applied to medicine and health care: A review. *International Journal of Advanced Manufacturing Technology, 40*(1), 116–127. doi:10.100700170-007-1308-1

Gibson, I., Rosen, D., & Stucker, B. (2015). *Additive manufacturing technologies: 3D printing, rapid prototyping, and direct digital manufacturing.* Academic Press.

Gordon, R. (2015). Trends in commercial 3D printing and additive manufacturing. *3D Printing and Additive Manufacturing, 2*(2), 89-90.

Gross, B. C., Erkal, J. L., Lockwood, S. Y., Chen, C., & Spence, D. M. (2014). Evaluation of 3D printing and its potential impact on biotechnology and the chemical sciences. *Analytical Chemistry, 86*(7), 3240–3253. doi:10.1021/ac403397r PMID:24432804

Guo, N., & Leu, M. C. (2013). Additive manufacturing: Technology, applications and research needs. *Frontiers of Mechanical Engineering, 8*(3), 215–243. doi:10.100711465-013-0248-8

Han, E. J., & Sohn, S. Y. (2015). Patent valuation based on text mining and survival analysis. *The Journal of Technology Transfer, 40*(5), 821–839. doi:10.100710961-014-9367-6

Harhoff, D., Narin, F., Scherer, F. M., & Vopel, K. (1999). Citation frequency and the value of patented inventions. *The Review of Economics and Statistics, 81*(3), 511–515. doi:10.1162/003465399558265

Hu, X., Rousseau, R., & Chen, J. (2012). A new approach for measuring the value of patents based on structural indicators for ego patent citation networks. *Journal of the American Society for Information Science and Technology*, *63*(9), 1834–1842. doi:10.1002/asi.22632

Huang, S. H., Liu, P., Mokasdar, A., & Hou, L. (2013). Additive manufacturing and its societal impact: A literature review. *International Journal of Advanced Manufacturing Technology*, 1–13.

Hull, C. W. (1986). *Apparatus for production of three-dimensional objects by stereolithography*. Patent No. US4575330 A.

Hur, W. (2017). The patterns of knowledge spillovers across technology sectors evidenced in patent citation networks. *Scientometrics*, *111*(2), 595–619. doi:10.100711192-017-2329-7

Hutmacher, D. W., Sittinger, M., & Risbud, M. V. (2004). Scaffold-based tissue engineering: Rationale for computer-aided design and solid free-form fabrication systems. *Trends in Biotechnology*, *22*(7), 354–362. doi:10.1016/j.tibtech.2004.05.005 PMID:15245908

Jardini, A. L., Larosa, M. A., Maciel Filho, R., de Carvalho Zavaglia, C. A., Bernardes, L. F., Lambert, C. S., ... Kharmandayan, P. (2014). Cranial reconstruction: 3D biomodel and custom-built implant created using additive manufacturing. *Journal of Cranio-Maxillo-Facial Surgery*, *42*(8), 1877–1884. doi:10.1016/j.jcms.2014.07.006 PMID:25175080

Jeantette, F. P., Keicher, D. M., Romero, J. A., & Schanwald, L. P. (2000). *Method and system for producing complex-shape objects*. Patent No. US6046426 A.

Jiang, R., Kleer, R., & Piller, F. T. (2017). Predicting the future of additive manufacturing: A Delphi study on economic and societal implications of 3D printing for 2030. *Technological Forecasting and Social Change*, *117*, 84–97. doi:10.1016/j.techfore.2017.01.006

Jin, Y., Ji, S., Li, X., & Yu, J. (2017). A scientometric review of hotspots and emerging trends in additive manufacturing. *Journal of Manufacturing Technology Management*, *28*(1), 18–38. doi:10.1108/JMTM-12-2015-0114

Jun, S., Park, S., & Jang, D. (2015). A technology valuation model using quantitative patent analysis: A case study of technology transfer in big data marketing. *Emerging Markets Finance & Trade*, *51*(5), 963–974. doi:10.1080/1540496X.2015.1061387

Karki, M. (1997). Patent citation analysis: A policy analysis tool. *World Patent Information*, *19*(4), 269–272. doi:10.1016/S0172-2190(97)00033-1

Karvonen, M., & Kässi, T. (2013). Patent citations as a tool for analysing the early stages of convergence. *Technological Forecasting and Social Change*, *80*(6), 1094–1107. doi:10.1016/j.techfore.2012.05.006

Kellner, T. (2014). World's First Plant to Print Jet Engine Noozles in Mass Production. *General Electric Reports*. Retrieved from http://www.gereports.com/post/91763815095/worlds-first-plant-to-print-jet-engine-nozzles-in/

Kiesling, E., Günther, M., Stummer, C., & Wakolbinger, L. M. (2012). Agent-based simulation of innovation diffusion: A review. *Central European Journal of Operations Research*, *20*(2), 183–230. doi:10.100710100-011-0210-y

Kietzmann, J., Pitt, L., & Berthon, P. (2015). Disruptions, decisions, and destinations: Enter the age of 3-D printing and additive manufacturing. *Business Horizons*, *58*(2), 209–215. doi:10.1016/j.bushor.2014.11.005

Kostoff, R. N., Boylan, R., & Simons, G. R. (2004). Disruptive technology roadmaps. *Technological Forecasting and Social Change*, *71*(1), 141–159. doi:10.1016/S0040-1625(03)00048-9

Küchler, S. (2014). Additive technology and material cognition: A view from anthropology. *Journal of Cognition and Culture*, *14*(5), 385–399. doi:10.1163/15685373-12342133

Lai, K.-K., & Wu, S.-J. (2005). Using the patent co-citation approach to establish a new patent classification system. *Information Processing & Management*, *41*(2), 313–330. doi:10.1016/j.ipm.2003.11.004

Lee, M., Kim, K., & Cho, Y. (2010). A study on the relationship between technology diffusion and new product diffusion. *Technological Forecasting and Social Change*, *77*(5), 796–802. doi:10.1016/j.techfore.2010.01.003

Levy, G. N. (2010). The role and future of the laser technology in the additive manufacturing environment. *Physics Procedia*, *5*, 65–80. doi:10.1016/j.phpro.2010.08.123

Levy, G. N., Schindel, R., & Kruth, J.-P. (2003). Rapid manufacturing and rapid tooling with layer manufacturing (LM) technologies, state of the art and future perspectives. *CIRP Annals-Manufacturing Technology*, *52*(2), 589–609. doi:10.1016/S0007-8506(07)60206-6

Lim, J., Nam, C., Kim, S., Rhee, H., Lee, E., & Lee, H. (2012). Forecasting 3G mobile subscription in China: A study based on stochastic frontier analysis and a Bass diffusion model. *Telecommunications Policy*, *36*(10), 858–871. doi:10.1016/j.telpol.2012.07.016

Lyons, B. (2014). Additive manufacturing in aerospace: Examples and research outlook. *The Bridge, 44*(3).

Mansfield, E. (1961). Technical change and the rate of imitation. *Econometrica*, *29*(4), 741–766. doi:10.2307/1911817

Manyika, J., Chui, M., Bughin, J., Dobbs, R., Bisson, P., & Marrs, A. (2013). *Disruptive technologies: Advances that will transform life, business, and the global economy*. Retrieved from http://www.mckinsey.com/~/media/McKinsey/Business Functions/McKinsey Digital/Our Insights/Disruptive technologies/MGI_Disruptive_technologies_Full_report_May2013.ashx

Martino, J. P. (2003). A review of selected recent advances in technological forecasting. *Technological Forecasting and Social Change*, *70*(8), 719–733. doi:10.1016/S0040-1625(02)00375-X

Meade, N., & Islam, T. (2006). Modelling and forecasting the diffusion of innovation–A 25-year review. *International Journal of Forecasting*, *22*(3), 519–545. doi:10.1016/j.ijforecast.2006.01.005

Mueller, B., & Kochan, D. (1999). Laminated object manufacturing for rapid tooling and patternmaking in foundry industry. *Computers in Industry*, *39*(1), 47–53. doi:10.1016/S0166-3615(98)00127-4

Munz, J. O. (1956). *Photo-glyph recording*. Patent No. US2775758 A.

Nair, S. S., & Mathew, M. (2015). The dynamics between forward citations and price of singleton patents. *International Journal of Innovation and Technology Management*, *12*(03), 1540003. doi:10.1142/S0219877015400039

Nam, C., Kim, S., & Lee, H. (2008). The role of WiBro: Filling the gaps in mobile broadband technologies. *Technological Forecasting and Social Change*, *75*(3), 438–448. doi:10.1016/j.techfore.2007.04.008

Nemet, G. F., & Johnson, E. (2012). Do important inventions benefit from knowledge originating in other technological domains? *Research Policy*, *41*(1), 190–200. doi:10.1016/j.respol.2011.08.009

Paritala, P. K., Manchikatla, S., & Yarlagadda, P. K. (2017). Digital Manufacturing-Applications Past, Current, and Future Trends. *Procedia Engineering*, *174*, 982–991. doi:10.1016/j.proeng.2017.01.250

Park, J., Tari, M. J., & Hahn, H. T. (2000). Characterization of the laminated object manufacturing (LOM) process. *Rapid Prototyping Journal*, *6*(1), 36–50. doi:10.1108/13552540010309868

Pierrakakis, K., Gkritzali, C. D., Kandias, M., & Gritzalis, D. (2015). *3D Printing: A Paradigm Shift in Political Economy?* Paper presented at the 65th International Studies Association's Annual Convention, New Orleans, LA.

Pierrakakis, K., Kandias, M., Gritzali, C., & Gritzalis, D. (2014). *3D Printing and its regulation dynamics: The world in front of a paradigm shift.* Paper presented at the 6th International Conference on Information Law and Ethics, Thessaloniki, Greece.

Radojičić, V. D., & Marković, G. Z. (2009). *New technology forecasting using the bass model.* Paper presented at the Telecommunication in Modern Satellite, Cable, and Broadcasting Services, 2009. TELSIKS'09. 9th International Conference on. 10.1109/TELSKS.2009.5339534

Rayna, T., & Striukova, L. (2016). From rapid prototyping to home fabrication: How 3D printing is changing business model innovation. *Technological Forecasting and Social Change*, *102*, 214–224. doi:10.1016/j.techfore.2015.07.023

Rifkin, J. (2012). The third industrial revolution: How the internet, green electricity, and 3-d printing are ushering in a sustainable era of distributed capitalism. *World Financial Review, 1*.

Rogers, E. M. (1962). *Diffusion of innovations*. New York: Free Press.

Rogers, H., Baricz, N., & Pawar, K. S. (2016). 3D printing services: Classification, supply chain implications and research agenda. *International Journal of Physical Distribution & Logistics Management*, *46*(10), 886–907. doi:10.1108/IJPDLM-07-2016-0210

Sachs, E. M., Haggerty, J. S., Cima, M. J., & Williams, P. A. (1993). *Three-dimensional printing techniques*. Patent No. US5204055 A.

Samuelson, R. J. (2000). The internet and gutenberg. *Newsweek (Atlantic Edition), 135,* 2-2.

Schmidt, G. M., & Druehl, C. T. (2008). When is a disruptive innovation disruptive? *Journal of Product Innovation Management*, *25*(4), 347–369. doi:10.1111/j.1540-5885.2008.00306.x

Soffer, T., Nachmias, R., & Ram, J. (2010). Diffusion of Web Supported Instruction in Higher Education-The Case of Tel-Aviv University. *Journal of Educational Technology & Society*, *13*(3), 212–223.

Sood, A., James, G. M., Tellis, G. J., & Zhu, J. (2012). Predicting the path of technological innovation: SAW vs. Moore, Bass, Gompertz, and Kryder. *Marketing Science*, *31*(6), 964–979. doi:10.1287/mksc.1120.0739

Steenhuis, H.-J., & Pretorius, L. (2016). Consumer additive manufacturing or 3D printing adoption: An exploratory study. *Journal of Manufacturing Technology Management*, *27*(7), 990–1012. doi:10.1108/JMTM-01-2016-0002

Steenhuis, H.-J., & Pretorius, L. (2017). The additive manufacturing innovation: A range of implications. *Journal of Manufacturing Technology Management*, *28*(1), 122–143. doi:10.1108/JMTM-06-2016-0081

Suominen, A., & Seppänen, M. (2014). Bibliometric data and actual development in technology life cycles: Flaws in assumptions. *Foresight*, *16*(1), 37–53. doi:10.1108/FS-03-2013-0007

Swainson, W. K. (1977). *Method, medium and apparatus for producing three-dimensional figure product*. Patent No. US4041476 A.

Trajtenberg, M. (1990). A penny for your quotes: Patent citations and the value of innovations. *The Rand Journal of Economics*, *21*(1), 172–187. doi:10.2307/2555502

Turk, T., & Trkman, P. (2012). Bass model estimates for broadband diffusion in European countries. *Technological Forecasting and Social Change*, *79*(1), 85–96. doi:10.1016/j.techfore.2011.06.010

Uriondo, A., Esperon-Miguez, M., & Perinpanayagam, S. (2015). The present and future of additive manufacturing in the aerospace sector: A review of important aspects. *Proceedings of the Institution of Mechanical Engineers. Part G, Journal of Aerospace Engineering*, *229*(11), 2132–2147. doi:10.1177/0954410014568797

Vecchiato, R. (2016). Disruptive innovation, managerial cognition, and technology competition outcomes. *Technological Forecasting and Social Change*.

Walsh, S. T. (2004). Roadmapping a disruptive technology: A case study: The emerging microsystems and top-down nanosystems industry. *Technological Forecasting and Social Change*, *71*(1), 161–185. doi:10.1016/j.techfore.2003.10.003

Walsh, S. T., & Linton, J. D. (2000). Infrastructure for emergent industries based on discontinuous innovations. *Engineering Management Journal*, *12*(2), 23–32. doi:10.1080/10429247.2000.11415071

Wang, J.-C., Chiang, C., & Lin, S.-W. (2010). Network structure of innovation: Can brokerage or closure predict patent quality? *Scientometrics*, *84*(3), 735–748. doi:10.100711192-010-0211-y

Weeks, M. R. (2015). Is disruption theory wearing new clothes or just naked? Analyzing recent critiques of disruptive innovation theory. *Innovation*, *17*(4), 417–428. doi:10.1080/14479338.2015.1061896

Weller, C., Kleer, R., & Piller, F. T. (2015). Economic implications of 3D printing: Market structure models in light of additive manufacturing revisited. *International Journal of Production Economics*, *164*, 43–56. doi:10.1016/j.ijpe.2015.02.020

Yan, B., & Luo, J. (2016). Measuring technological distance for patent mapping. *Journal of the Association for Information Science and Technology*.

Yasukawa, S., & Kano, S. (2015). Comparison of examiners' forward citations in the United States and Japan with pairs of equivalent patent applications. *Scientometrics*, *102*(2), 1189–1205. doi:10.100711192-014-1466-5

Yoon, J., Park, Y., Kim, M., Lee, J., & Lee, D. (2014). Tracing evolving trends in printed electronics using patent information. *Journal of Nanoparticle Research*, *16*(7), 2471. doi:10.100711051-014-2471-6

You, H., Li, M., Hipel, K. W., Jiang, J., Ge, B., & Duan, H. (2017). Development trend forecasting for coherent light generator technology based on patent citation network analysis. *Scientometrics*, *111*(1), 297–315. doi:10.100711192-017-2252-y

Young, P. (1993). Technological growth curves: A competition of forecasting models. *Technological Forecasting and Social Change*, *44*(4), 375–389. doi:10.1016/0040-1625(93)90042-6

Yu, D., & Hang, C. C. (2010). A reflective review of disruptive innovation theory. *International Journal of Management Reviews*, *12*(4), 435–452. doi:10.1111/j.1468-2370.2009.00272.x

KEY TERMS AND DEFINITIONS

3D Printing: An innovative technology in the production and services industries that changes, substitutes, or supports the parts of/the whole chain of the traditional production paradigm through offering easier and cost-efficient solutions to businesses value or consumers.

Additive Manufacturing: Nontraditional production concept that is building objects up to one layer at a time from digital design without any mold or manufacturing background.

Disruptive Innovation: Is conceived as results from a combination of the emergent qualities of ideas based on observing the world differently, thinking the unthinkable and taking pre-emptive action to influence the environment.

Disruptive Technology: A bunch of scientific discoveries that changed ordinary product or technology capabilities into recent form in the basis of new competitive paradigm.

Forward Citation: The type of citation that receives citations from later papers or patents.

Growth Curve: Is an S-shaped curve that can frequently use in technology diffusion literature to describe performance of technology and diffusion over the life of technology.

Innovation Diffusion: Seeks to understand how new ideas, social learning styles, organized information dissemination and products spread throughout a society over time.

Patent Analysis: Is a method based on structured information of patent data such as filing dates, assignees, citations to analyze by bibliometric methods or data mining techniques.

ENDNOTE

1 In order to compare each patent citation period is fixed into 1993-2016.

Chapter 8
The Impacts of Peer-to-Peer Lodging Platform on the Traditional Lodging Industry:
California vs. Southern Europe

Anatoly Zhuplev
Loyola Marymount University, USA

Jonathan Dell
Loyola Marymount University, USA

DaVion Doby
Loyola Marymount University, USA

Joshua Tillipman
Loyola Marymount University, USA

ABSTRACT

The chapter presents a comparative analysis of trends in the peer-to-peer lodging and their impacts on the lodging industry of the metropolitan regions of the Greater Los Angeles, California, USA, and Southern Europe (Athens, Greece; Barcelona, Spain; and Vienna, Austria). The main research question is whether the emerging peer-to-peer lodging platforms, part of the exploding sharing economy, are competitive or complementary to the traditional lodging industry. Data was obtained from secondary research and collected in 24 field interviews. The study found variations in the complementarity/competitiveness equilibrium, depending on the market segment in hospitality industry and regulatory environment. The chapter draws recommendations for hotel competitiveness.

DOI: 10.4018/978-1-5225-4148-6.ch008

INTRODUCTION

International travel and tourism have been major contributors to the global economy since the 21st century (Tourism, 2016). Traditional patterns in lodging started to change as consumers gained access to the Internet through improved technologies (Hospitality Net – Hotels, 2016) enabling price comparison and stronger voice in the process. This paradigm has had profound impact on the traditional lodging platforms. This chapter aims to determine whether peer-to-peer (P2P) lodging platforms are competitive or complementary to the traditional lodging industry throughout Los Angeles, USA; Vienna, Austria; Athens, Greece; and Barcelona, Spain. The chapter also explores emerging strategies for traditional lodging organizations and regulatory entities.

BACKGROUND

Method

The North American Industry Classification System categorizes the general Accommodation and Food Sector under the industry code 721, with subsectors: "(1) traveler accommodation, (2) recreational accommodation, and (3) rooming and boarding houses." This study involves industry categories under NAICS codes 7211: Traveler Accommodation, 721110: Hotels (except casino hotels) and motels, 72119: Other Traveler Accommodation, and 721191: Bed-and-Breakfast Inns (United States Census Bureau, 2017).

Different types of accommodations in the lodging industry offer different consumer value propositions to different consumer segments. Among segments in the lodging industry as a whole, hotels clearly represent the dominant segment or mass market. They generate the most revenue and outnumber their counterparts worldwide. Niche accommodations such as bed and breakfast, hostels or cabins are complementary to the industry as a whole because they cater to a particular minority consumer segment of the market. In categorizing P2P lodging platforms as competitive or complementary to the traditional lodging industry it is reasonable to equate this phenomenon as representing a niche market (complementary) or having the capability of developing into the dominant segment (competitive).

Niche markets have their own dynamics. This study will be using the following variables supporting analysis of whether the dynamics in a particular location was competitive or complementary:

1. **Growth Rate of P2P Lodging Bookings:** Niche markets represent a minority of consumers. Their growth, as a result, will slow before gaining mass adoption. A declining trend in the bookings is an indicator that it is exhausting its potential to gain market share and thus will only complement the larger market.
2. **Available Supply of P2P Lodging Units:** A product cannot adequately fulfill the market needs if it is in limited supply. When government regulations, for example, reduce the number of legally rentable units, supply of lodging units experiences a downward pressure.
3. **Ratio Between Entire Place Rentals Versus Shared or Private Rooms:** Niche markets have unique features differentiating them from the mass market. Renting a shared or separate private room provides a much more immersive experience compared to entire place rentals. A higher mix of the former is a stronger indicator of the "live like a local" experience leading to a unique value proposition.
4. **Hotel Average Daily Rates (ADR) Premiums/Discounts:** Price can be a critical factor distinguishing one product from another. For example, in hostels (complementary product), consumers sacrifice some privacy and amenities in exchange for a lower room rate relative to hotels (mass market). Similarly, where P2P lodging platforms offer a steep discount to the traditional lodging industry in exchange for certain onsite services not offered by P2P lodging platforms, price becomes a distinguishing value proposition.
5. **Correlation Between Travel Data and Overnight Stays:** A weakening correlation between overnight travelers to a location and the number of overnight stays in hotels suggests that the dominant player is losing market share to another segment of the market. Where this trend persists and is amplified by strong growth in the P2P lodging market, there may be some potential for it to displace the dominant player.

Qualitative data was gathered through 24 personal, Skype interviews and surveys from March 2016 to June 2016 in the regions studied. The original study included the city of Rome, Italy, however under the space constraints in this publication it was excluded from analysis. Quantitative data was also incorporated, thus constituting an empirical-analytical methodology of the study. The quantitative P2P lodging data is based on Airbnb figures. Airbnb is by far the most popular among travelers and is the market leader in the P2P lodging sector: Airbnb has more than 2 million listings worldwide in more than 34,000 cities, in 191 countries and has serviced more than 60 million guests (About Us – Airbnb, 2016). Since its inception in 2008, Airbnb has reached a valuation of almost $30 billion, making it second to Uber as the highest valued start-up in the USA (Picker & Isaac, 2016). For these reasons, Airbnb data will be used as a proxy when evaluating P2P lodging platforms' overall impact on the lodging industry. This data analyzes Airbnb advertised listings by type, Airbnb vs. traditional lodging industry overnight stays, Airbnb vs. traditional lodging industry 2015 ADR & revenue per available room (RevPAR), and Airbnb year-over-year (YOY) growth in rooms available and rooms booked.

This chapter examines the current state of the lodging industry and sharing economy to explore whether P2P lodging platforms are competitive or complementary to the traditional lodging industry and draw recommendations for hotels and government influence.

GLOBAL LODGING INDUSTRY

Global Travel and Tourism Industry

The Travel and Tourism industry includes companies and agencies catering to the needs of the leisure or business travelers. Its major actors are airlines, charters, lodging organizations, foodservice establishments, tour providers, cruise lines, travel agencies and services. The industry is one of the largest in the world, generating $7.2 trillion GDP in 2015 and representing 9.8% of global GDP. The industry supports 284 million jobs, or 1 in 11 jobs worldwide (globalEDGE: Hospitality and Travel, 2017). This study focuses on the lodging industry segment within the Travel and Tourism industry.

Global Lodging Industry

The lodging industry includes companies providing traveler accommodations, such as hotels, motels, resorts, casino hotels, and Bed-and-Breakfast Inns (First Research, 2015). In 2014, the global lodging industry grew by 4% and generated $678.6 billion in revenue (Figure 1). The number of hotels and motels worldwide increased with a compound annual growth rate (CAGR) of 0.5% between 2010 and 2014, for a total of 444,244 establishments in 2014. The industry's volume is expected to rise to 472,227 establishments by the end of 2019, representing a CAGR of 1.2% for the 2014-2019 (MarketLine, 2015a).

Global Lodging Industry Competitive Landscape

The lodging industry is dominated by large international hotel companies. The success of individual companies depends on efficient operations and effective marketing. The

Figure 1. Global hotels and motels industry value
Source: MarketLine, 2015a

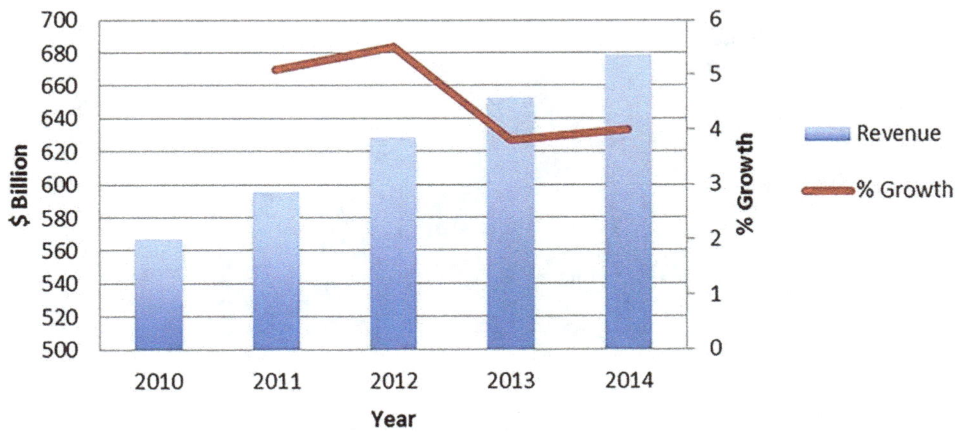

larger lodging companies have advantages in economies of scale, can easily raise capital, and have strong brand recognition. Small lodging companies can compete by offering personalization and exceptional customer service (First Research, 2015).

Major global market leaders include Marriott International, Inc., Hilton Worldwide Holdings Inc., AccorHotels group, Best Western International, Inc., Wyndham Worldwide, InterContinental Hotels Group PLC, and others.

Global Lodging Industry Analysis

The international lodging industry has been experiencing strong 4.3% annual growth over the past 5 years (IBISWorld, 2016a). This growth is driven by a global expansion of international tourism. More than 1.1 billion business and leisure tourists traveled the world in 2014 (First Research, 2016). The U.S. and Europe together account for 60% of the value in the global lodging industry with the Asia-Pacific region distant second at 24% of the market (MarketLine, 2015a), as seen in Figure 2.

Globalization, proliferation of the Internet and online travel agencies (OTAs) have facilitated global tourism, changed the travel industry, and made it easier for travelers to customize trips throughout the world. Major OTAs worldwide include Voyages, Expedia, Sabre, Opodo, the Priceline Group, Travelocity, Hotels.com, Travelgenio, Wotif.com, Webjet and others. These sites earned a reputation of searching across brands on one platform to find the best hotel at the best price. The hoteliers were slow to react, and as a result have become dependent on the OTAs. These relationships cost the hoteliers dearly, as OTAs can charge up to 33% of the booking (Speer, 2016). This depresses property's profit margin, and limits hoteliers'

Figure 2. Global hotels and motels industry geography segmentation: $ Billion, 2014
Source: MarketLine, 2015a

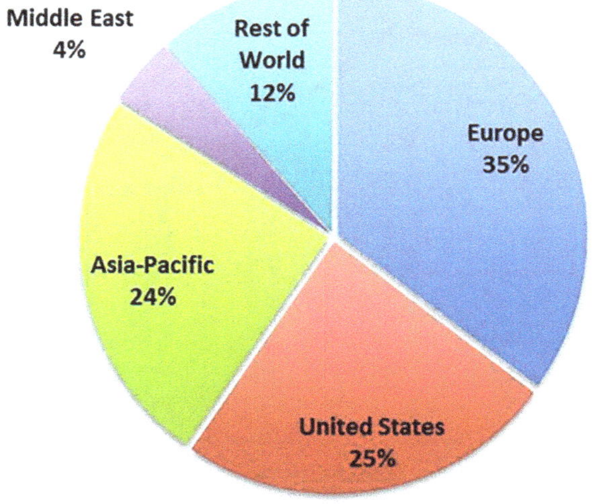

ability to compete with competitors, including the often lower priced P2P lodging platforms. "Expedia used to have a mandate that they would have the lowest price available. The hotel companies could not match or reduce their prices, but that's gone at this point, and the hotels are now trying to guarantee the lowest price available through their websites" (Baltin, 2016). This push to regain control over their own bookings makes good fiscal sense for hotels; however they are battling an entrenched consumer perception that the OTAs provide greater value. As long as this exists, a complete breakup between hotels and OTAs will be impossible, and their profitability will continue to be impacted.

The leisure segment in the global lodging industry was most lucrative in 2014, with total revenues of $509.3 billion, 75% of the industry's overall value (MarketLine, 2015a). The business segment contributed revenues of $167.8 billion in 2014, comprising only 25% of the industry's aggregate value (Figure 3). Both segments are affected by the strength of the economy: with more funds available for discretionary purchases comes more infusion of tourist dollars.

Global Lodging Industry Outlook

The industry is forecast to accelerate, with an anticipated CAGR of 5.6% for 2014-2019, which is expected to drive the industry to a value of $888.7 billion by the end

Figure 3. Global hotels and motels industry category segmentation: $ billion, 2014
Source: MarketLine, 2015a

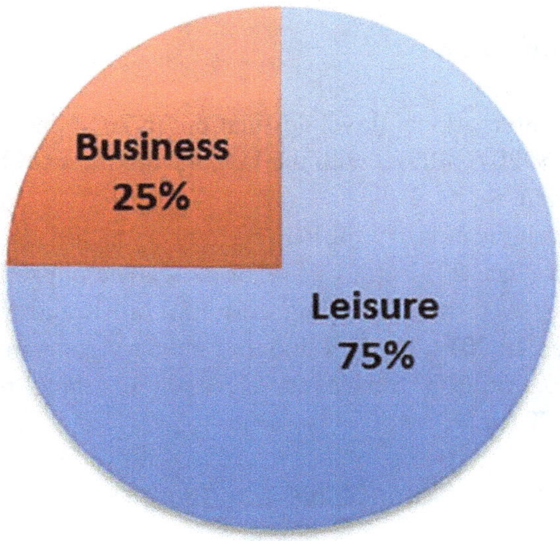

of 2019. The Asia-Pacific and U.S. segments of the industry in 2014-2019 will grow with CAGRs of 8.3% and 4.4% respectively to reach respective values of $244.5 billion and $206.3 billion in 2019 (MarketLine, 2015a).

Lodging companies will gain a future advantage by offering targeted products, services, amenities, and designs that cater to generational and cultural preferences based on consumer preferences for customized travel experience. Two emerging groups of travelers include the millennial generation (those born between 1980 and 2000) and Chinese tourists. As millennials enter their peak earning years over the next decade, they are expected to provide the majority of global spending for travel and leisure. Meanwhile, tourists from China spent nearly 30% more on international travel in 2014 compared to 2013 (First Research, 2015).

P2P lodging, now widely available through online platforms such as Airbnb, FlipKey and HomeAway, becomes a potential threat to the traditional lodging industry. These rentals give tourists the ability to stay with local people in their homes or apartments, either by renting a room or entire property (MarketLine, 2015a). The incumbents lobby for local and state laws that would restrict these P2P lodging platforms (First Research, 2015).

Globalization through technology will continue to drive growth in the lodging industry. Growing discretionary income levels among the rising global middle class, easing visa policies around the world, and the proliferation of travel information has heightened the desire to travel abroad. Lodging companies will need to customize and employ new business strategies to account for cultural and regulatory differences.

U.S. Lodging Industry

The U.S. is the top international travel destination: in 2014, international travelers spent an estimated $220.6 billion, with the U.S. having the world's largest share at 14.2% (AHLA, 2016).

Like the global lodging industry, the lodging industry in the U.S. has been thriving, growing every year since 2011 at a CAGR of 4.2%. The unparalleled growth has been driven by increases in travel spending, corporate profit, and consumer spending (IBISWorld, 2016a). In 2014, the U.S. lodging industry generated $176 billion in revenue and had 26.5 million overseas travelers stay in a hotel/motel during their U.S. visit (AHLA, 2016).

U.S. Lodging Industry Competitive Landscape

The lodging industry in the U.S. is dominated by several international players and numerous smaller independent players (MarketLine, 2015d). Approximately two-thirds of all rooms are chain affiliated (Freitag, 2016). The market leaders are Hilton Worldwide Holdings Inc. (10% market share), Marriott International, Inc. (10%), Wyndham Worldwide Corporation (9%), International Hotels Group (8%), and Starwood Hotels and Resorts Worldwide, Inc. (3% share) (Cederholm, 2014).

Larger chain brands in some cases own additional businesses, such as restaurants, bars, and clothing stores. To attract and sustain more business, operators try to offer more complex packages and value-added services, such as free breakfast, Wi-Fi, parking, or a free third night. Recently, a rising popularity of lifestyle hotels amongst major chains has been observed. These hotels are catering to the conscientious traveler's demands for eco-friendly practices, social responsibility, and affordability (MarketLine, 2015d).

U.S. Lodging Industry Analysis

The U.S. accounts for 25% of the global lodging industry and continues to thrive (MarketLine, 2015d). It is boosting with economic growth, marking five years of consecutive job creation. In 2014, the industry added more than 30,000 new hotel jobs and more than 100,000 new travel-related jobs, resulting in an increase of over $12 billion in travel-related wages and salaries (AHLA, 2016).

There were 53,432 properties (properties with 15 or more rooms) and 4,978,705 rooms by the end of 2014. Occupancy hit record 65.6% in 2015 (STIFEL, 2016), and RevPAR was up 6.3% last year from $74.12 in 2014 (AHLA, 2016). This solid performance and growth in the U.S. lodging industry is driven by robust economic performances in key feeder markets including Canada and the UK, as well as increasing tourism from emerging economics such as China (MarketLine, 2015d).

As seen in Figure 4, the leisure segment was the industry's most lucrative in 2014, with $116.8 billion in total revenues, constituting 70.3% of the U.S. industry's overall value. This business segment contributed $49.4 billion revenues in 2014, comprising 29.7% of the industry's aggregate value (MarketLine, 2015d).

U.S. Lodging Industry Outlook

Strong growth in international travel to the U.S. is expected to continue through 2020 with an average annual 4.3% growth rate in the volume of international visitors (AHLA, 2016). The lodging industry in the U.S. is forecast to rise to 56,208 establishments by the end of 2019, representing a CAGR of 1.1% for 2014-2019. The performance is forecast to reach a CAGR of 4.4% for 2014-2019, driving the industry to a value of $206.3 billion by the end of 2019 (MarketLine, 2015d).

Figure 4. U.S. hotels and motels industry category segmentation: $ billion, 2014
Source: MarketLine, 2015d

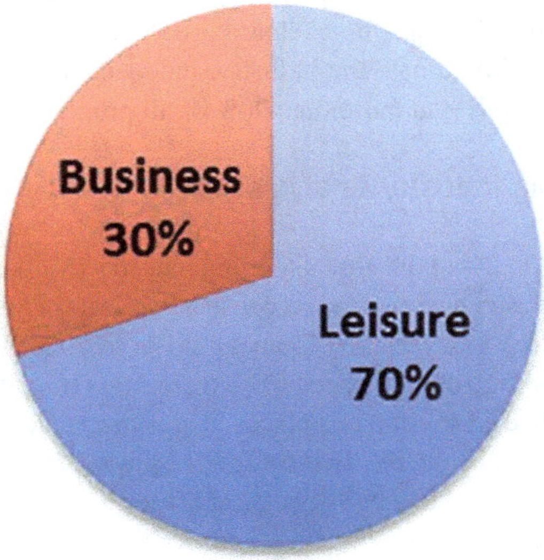

European Lodging Industry

2015 was a bumper year as record tourist arrivals propelled hotel performance for many European destinations. Tourists appeared to favor Europe over the Middle East and North Africa. City trips remain a top travel growth segment in Europe and strong demand, together with weak accommodation supply growth meant many hotels enjoyed a remarkable year. Europe, up 5% in tourism numbers, led global growth last year in absolute and relative terms supported by a weaker euro against the U.S. dollar and other main currencies. Arrivals reached 609 million, 29 million more than in 2014 (PWC, 2016b).

During 2005-2014, the number of nights spent in tourist accommodation establishments in the EU grew by 18%. Nearly 45% of these nights were spent in France (15%), Spain (15%), and Italy (14%) (Eurostat, 2016). In 2014, the European lodging industry generated $238.6 billion revenues, a 2.9% CAGR between 2010 and 2014 (MarketLine, 2015b). Continued economic recovery and a positive travel backdrop appear the main growth drivers for the European lodging industry.

European Lodging Industry Competitive Landscape

The European lodging industry consists of several large hotel operators; most of the leading players operate several brand chains. There are also a large number of independent players in the industry that create intense competition. The market leaders in Europe are InterContinental Hotels Group (686,000 rooms), AccorHotels (470,000 rooms), Best Western International (313,639 rooms), and Whitbread (50,000 rooms) (MarketLine, 2015b). In Europe, hotel chains are the minority, while independent lodging establishments dominate the industry. Independents constitute more than two-thirds of room supply and more than 90% of all properties (Sileo, 2015).

European Lodging Industry Analysis

Europe accounts for 35.2% of the global lodging industry (MarketLine, 2015a). The European lodging industry has reported six consecutive years of RevPAR growth since its double-digit decline following the financial crisis in 2008 (Figure 5). Europe's year-end 2015 RevPAR growth reached 6.9% to $83.94 (Hospitality Net, 2016a).

In 2014, the EU had 31 million bed places to accommodate tourists, spread over more than 570,000 establishments. In terms of bed places, France (5.1 million) and Italy (4.8 million) accounted for nearly one third of total available capacity (Eurostat, 2016). 2015 saw a 3.1% increase in demand for rooms, while supply growth saw a 0.8% gain: this imbalance is expected to continue. Major cities have been careful to

Figure 5. Europe hotels and motels industry value
Source: MarketLine, 2015b

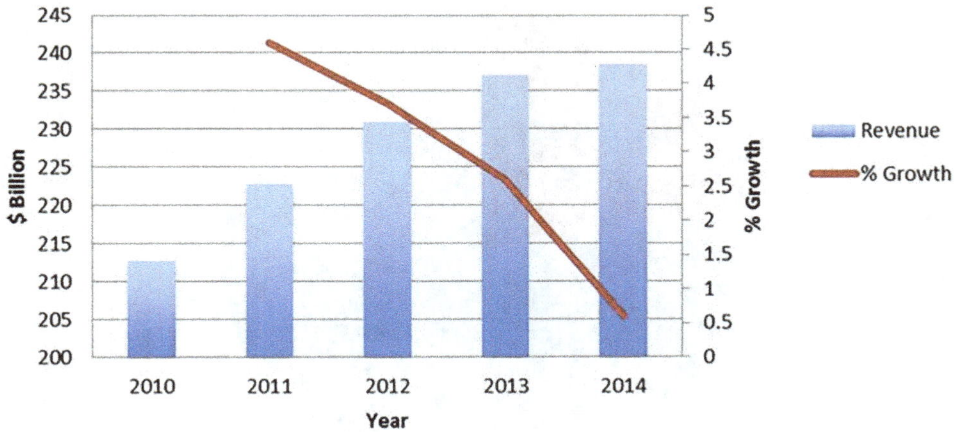

increase supply so to not damage the culture of the community. For most cities, the fast growth in the serviced apartment sector and shared space models like Airbnb adds to the competition facing hotels, as well creating challenges for those seeking to regulate them (PWC, 2016b).

As presented in Figure 6, the leisure segment, industry's most lucrative in 2014, generated $184.8 billion revenues, or 77% of the industry's overall value. The business segment in 2014 contributed $53.8 billion, or 23% of the industry's aggregate value (MarketLine, 2015b).

European Lodging Industry Outlook

The European lodging industry is poised to continue its recovery, after the 2008 financial crisis. Growth in the number of hotels in Europe is set to grow less rapidly; contributing factors here include the already-well developed nature of the market, continued economic uncertainty in the Eurozone, and a recently increased threat that traditional hotels face from the P2P lodging platforms. Industry performance is forecast to accelerate, reaching a 4.1% CAGR of between 2014 and 2019; that is expected to drive the industry value to $291.2 billion by the end of 2019 (MarketLine, 2015b).

Comparing the European and U.S. Lodging Industries

Europe and the U.S., the two largest segments in the global lodging industry, both experienced six consecutive years of RevPAR growth since 2009. 2015 marked the

Figure 6. Europe hotels and motels industry category segmentation: $ billion, 2014
Source: MarketLine, 2015

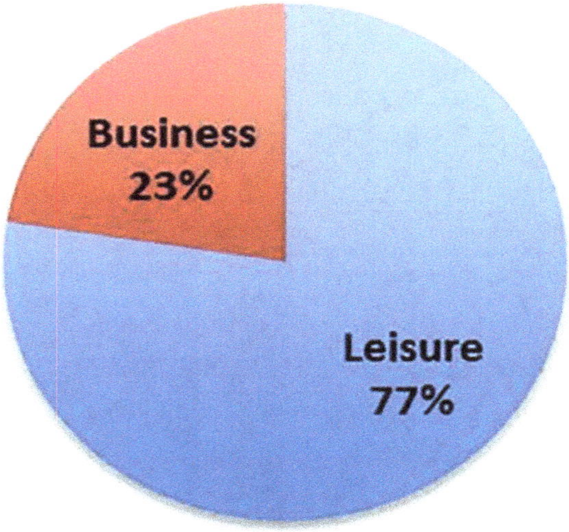

first year since 2010 in which the European hotel industry surpassed hotels in the U.S. in year-over-year RevPAR (Hospitality Net, 2016a).

Both the European and U.S. lodging industries are driven by international tourism growth, global travel rates, increasing room rates, the strength of the economy, and consumer discretionary spending (IBISWorld, 2016a). Global security concerns, like terrorism, also have an effect on the growth of both lodging industries.

As seen from figures 7 and 8, comparing hotel landscapes in the US and Europe, chain hotels dominate the U.S. lodging industry, comprising about two-thirds of the market (Freitag, 2016). Independent lodging suppliers operate at a disadvantage, lacking ample resources of the chain hotels. Independents have to rely more heavily on OTAs, like Expedia, in marketing their services. The European landscape is almost opposite to the U.S. Independent hotels comprise more than two-thirds of room supply in Europe. European properties are even more dependent on OTAs. OTAs represent 74% of European independent lodging online bookings in 2015 vs. 58% for the U.S. (Sileo, 2015).

Strengths of Traditional Lodging

Traditional lodging organizations in both the U.S. and Europe enjoy three major competitive strengths: continuity of service, breadth of service and loyalty programs.

Airbnb operates under a slogan "belong anywhere," which entices travelers choosing a dwelling rooted in the local flavor - to find hidden gems and experiences

Figure 7. U.S. hotel landscape
Source: STR

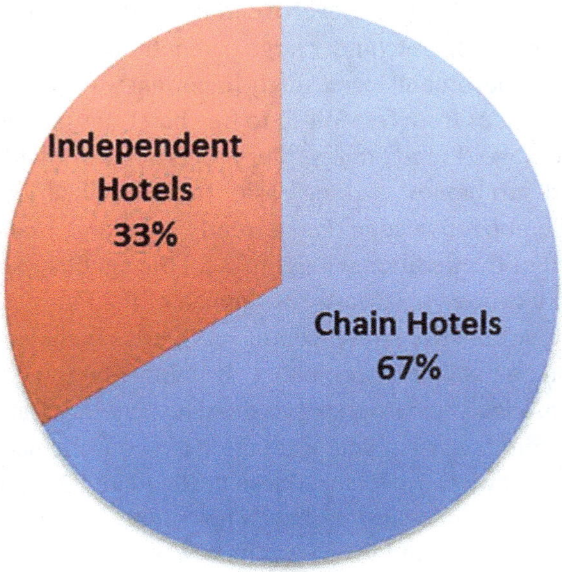

Figure 8. Europe hotel landscape
Source: Phocuswright Inc.

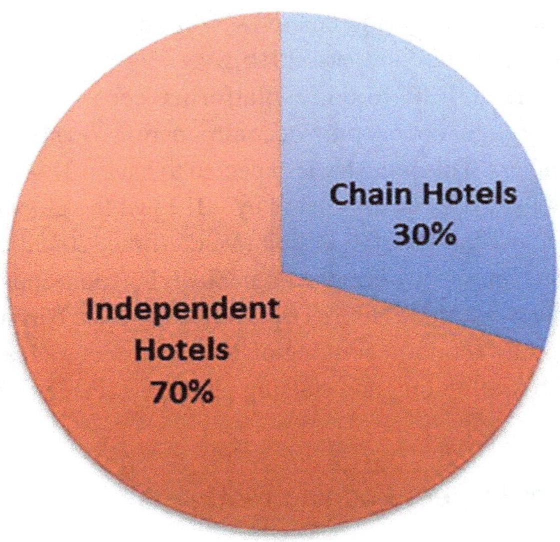

that would otherwise remain a mystery. In the franchise hotel model, mystery isn't always what travelers want. On the opposite, travelers in this category prefer consistency and predictability in the level of comfort and convenience (Speer, 2016).

As more hotel brands operate under large parent brands like Hilton or Marriott, loyalty programs have become an increasingly important way to retain customers. The programs offer incentives from free Wi-Fi to free hotel stays, all for simply staying loyal to the parent brand. Large brands offer properties worldwide across multiple known and trusted sub brands, making easier to accumulate loyalty points. This loyalty is a strong competitive edge that currently only large branded hotels enjoy.

Another benefit of the traditional lodging is all the ancillary services that make up a true hospitality experience. Matthew Lehman of the Fairmont Miramar Hotel explains: "We have a wonderful housekeeping service that includes three restaurants, a great bar out front, a pool with people there bringing you things – on and on. We have a fun lobby experience, a barber shop, a shoe shine guy, a great retail experience, room service... So, if that is what you want, there is a convenience to a hotel. There is a level of service that we can offer that doesn't exist [with P2P lodging]" (Lehman, 2016). The supporting infrastructure of such hospitality is not easily replicated in less formal lodging options, and for guests that want it, it is a powerful incentive. These include health and safety standards that serve as a strength of traditional hotels.

Sharing Economy and Peer-to-Peer Lodging Platforms

In 1995, eBay and Craigslist launched the concept of sharing economy (James, 2014). By 2025, the global sharing economy will generate revenues worth up to $335 billion (PricewaterhouseCoopers, 2016).

In this environment P2P lodging platforms center on sharing lodging accommodations through connecting accommodation owners with short-term accommodation seekers. This industry is expected to have a 31% revenue growth by 2025 (PricewaterhouseCoopers, 2016) and by 2019 just the global hotel and motel industry is expected to reach $890.6 billion (MarketLine, 2015a).

Accommodations can vary from entire homes to living room couches. Major actors in the P2P lodging sector include HouseTrip, HomeAway and Airbnb. Between 2013 and 2025, these sharing economy lodging platforms expect a 31% revenue growth, third highest after crowdfunding and staffing platforms (PricewaterhouseCoopers, 2016).

Strengths of Peer-to-Peer Lodging Platforms

One challenge for hotels is their inability in catering to larger groups of travelers who prefer to stay together. Booking several hotel rooms would still fail to provide a central living space for the family to be together. Some hotels offer larger multi-room suites, however it is often significantly costlier than renting a home on a P2P lodging platform. "The home-like aspect of it can be interesting, because it opens a way for different guest needs that probably are not completely served by the historic hotel-like accommodations," said Giulio Leporatti of the InterContinental Hotels Group (Leporatti, 2016).

The challenge of group accommodation for hotels appears one of P2P lodging platforms' greatest strengths. Rumyana Miteva of the HouseTrip states that many people looking for those features: "You get more space, perhaps better conditions and facilities that are available, such as a kitchen or a living room. You definitely get a lot more privacy. Those are some of the advantages that are a part of our industry and why they are so attractive to travelers" (Miteva, 2016). Additionally, P2P lodging platforms benefit of their ability to provide a wide variety of properties, allowing consumers to choose the exact amount of space or specific accommodation they prefer.

Perhaps the most important strength of the P2P lodging platforms is the price. While the ADR may be similar for hotels and hotel-comparable P2P lodging platforms, the final cost to the consumer is often lower for P2P listings. Many properties do not charge or collect Transient Occupancy Taxes levied by local governments to support tourist impact on local infrastructure. However, since sites like Airbnb do not share their property address data with local governments, legal enforcement can be difficult. Another reason for lower pricing is low entry barriers for P2P lodging platforms: property owner simply rent out their home while they're not using it, maximizing a resource that is already being paid for. The same argument extends to the professional peer-to-peer lodging listings as well: entrepreneurs wishing to convert apartment complexes into short-term rentals are not subjected to the same regulatory scrutiny that a hotel would. Lynn Mohrfeld of the California Hotel and Lodging Association recalled a story of a developer of a hotel in a historic building in Southern California:

He's gone through 2-3 years of city council approvals, development and public forums. Currently they're discussing what shade of beige best represents the historical nature that the photographs show. Now three blocks away, someone who has an apartment building has turned it into short-term rentals at almost the same rate. So the person down the street with no approval process, no health and safety, no regulation…s/he can open a hotel in a heartbeat. That commercialization is really what we have a problem with (Mohrfeld, 2016).

Being able to operate nimbly, especially in tightly controlled development markets, is a definite strength of P2P lodging platforms.

Another dimension of P2P lodging platforms is their ability to provide instantaneous supply to a market at periods of high demand. In 2015, Los Angeles hosted the Special Olympics World Summer Games, and the city was saturated with travelers. In these conditions, hotel rates rise significantly and hotels use the extra profit to support their labor-heavy services during the low season. For the Special Olympics event however, extra supply came online nearly instantaneously through P2P lodging platforms as residents-turned-opportunists recognized that they could leave for a few days and collect a premium for their dwellings (Baltin, 2916).

Peer-to-Peer Lodging Platforms Outlook

Some strengths of P2P lodging platforms may also be their weaknesses. While the lower regulatory burdens is a benefit for P2P lodging platforms now, it may change as the future regulatory regime is emerging. For example, the attention of lawmakers may be divided now, and enforcement funding may be hard to find, but if several serious health or safety issues arise in the P2P space, the spotlight will turn quickly. These consequences may extend beyond increased compliance measures to involve a violation of the Americans with Disabilities Act or stem from a lack of keyed access policy over the rented units. If real harm is done there is a question of liability that has yet to be addressed in many destinations. A series of precedent-setting and costly court battles would surely strain the segment's reputation and supply. As one hotelier put it, "There are a lot of red flags that the [P2P] industry has to come to grips with, primarily with security, safety, insurance coverage, and regulation." (O'Connor, 2016).

Another pressing issue for P2P platforms is limited control over the experience: a company like Airbnb has no control over the level of quality of its listings or hosts' responsiveness. An engaging review system is meant to self-police quality; however, similar systems have been effectively gamed to the point where fake reviews are a reality in most ecommerce spaces now, including Amazon's. In that case the retailer has seen legitimate sellers on its site fail after foreign-copies of their products infiltrate the site and garner sales reputability by utilizing faked reviews (Close, 2016). Some argue that reviews are less reliable today than they were five years prior. Airbnb does attempt to weight search results based on reviews and responsiveness of hosts (Airbnb Market Report, 2016), but once the booking is made Airbnb doesn't control what happens next. If a platform elects to police more fervently the offerings listed on their site, they risk trimming supply. For a company like Airbnb with a valuation close to $30 billion, cutting supply may be a risky proposition (Picker & Isaac, 2016). Conversely, if a platform elects not to control some standard of quality, they risk competition from others who will.

Table 1 highlights these driving forces and notes some possible threats for consideration as mentioned above.

City Findings: Los Angeles County

Los Angeles is both a city and a county. The city of Los Angeles incorporates 472 square miles and is home to 4 million residents. Los Angeles County includes the city of Los Angeles, and is spread out over 4,084 square miles with 10.1 million residents (Discover Los Angeles, 2016). This research focuses on Los Angeles County as a whole. Southern California at large has enjoyed a prosperous increase in travel, with San Diego attracting 34.3 million visitors (San Diego Tourism Authority, 2016) and Orange County, the home of Mickey Mouse and a booming convention circuit beckoning 47.3 million in 2015 (Martin, 2016).

Los Angeles is a major global destination for business and leisure travel. Aside from the weather and attractions, Los Angeles also serves as a gateway to Asia, with LAX Airport servicing many direct routes over the Pacific Ocean. In 2015, the Los Angeles area hosted the fifth consecutive year record breaking 45.5 million travelers, with 30.2 million staying overnight (Discover Los Angeles, 2016). Of the visitors to Los Angeles, 85% hailed from domestic locations, while 15% were international (Los Angeles Tourism, 2016).

International tourist arrivals, representing visitors remaining in the city for at least 24 hours, have grown considerably: over the past five years Los Angeles has seen that number grow from 4.96 million in 2011 to 5.55 million in 2015 (Figure 9). Viewed alongside hotel overnight stays, which grew from 25.5 million to 29.7 million over the same period, the two performance metrics have followed similar trajectories. The correlation coefficient over this period was .9576, indicating a strong upward trajectory (Doby et al., 2016).

Table 1. Driving forces, strengths and threats

Driving Forces of Sharing Economy	Strengths of Peer-to-Peer Lodging Platforms	Threats to Peer-to-Peer Lodging Platforms
• Advancements in Technology • Economic Realities • Consumer Shift in Asset Valuation •	• Group Accommodations • Price • Nimbleness • Lack of Barriers to Entry • Instantaneous Supply	• Security & Liability • Fake Reviews • Tax Regulations

Figure 9. Los Angeles tourist arrivals and overnight stays
Source: tourist arrivals per Euromonitor; hotel overnights per CBRE/PKF

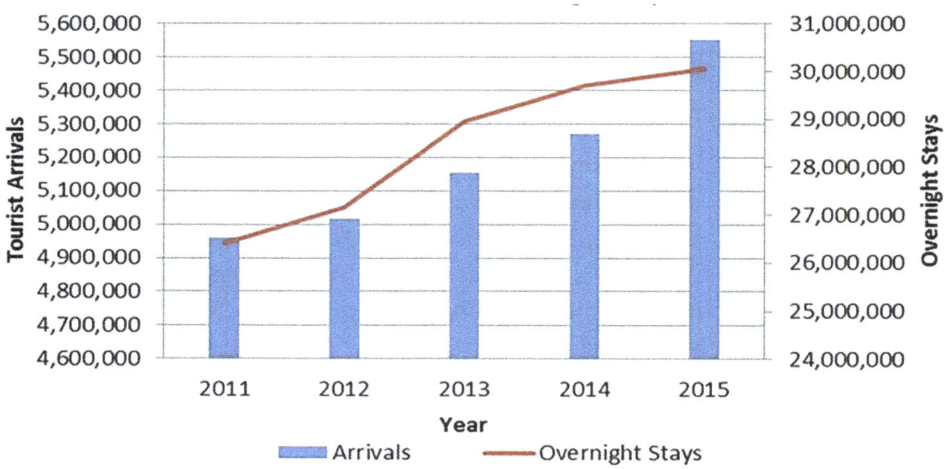

Los Angeles County Lodging Industry

Universally, the industry experts interviewed agreed that the lodging industry in Los Angeles County is currently in the mature stages of a semi-predictable cyclical peak. Patrick Speer, Sr. of Hilton Worldwide explained: "A typical cycle is about 7 [years] up and about 3 [years] down. We are somewhere in year 6 or 7." (Speer, 2016). Ellis O'Connor of MSD Hospitality agrees: "Lodging is still very good. We still see that there is a lot of runway left and don't foresee any downturn until probably 2018 at the earliest, in the context that no geopolitical or world events take place" (O'Connor, 2016).

Experts note economic recovery, new travelers from emerging middle classes in China and India, along with growing sentiment favoring experiences over goods. Giulio Leporatti of the InterContinental Hotels Group explains:

Travel demand grows as an effect of economic growth of course, but also if you think of the changing habits across generations, younger generations have more interest in traveling. And the growth is more than proportionate to the growth of the economy in terms of GDP, because you have more people. That goes beyond the threshold of disposable income, enabling them to travel - they have more access to travel (Leporatti. 2016).

Hotel Room Supply

Los Angeles County is home to 98,135 hotel rooms, ranked 6th in the U.S. These rooms are spread across 997 properties, registered as hotels within their respective municipalities. These hotels have benefitted from increasing demand, resulting in over 28.5 million room nights sold in 2015. While this represented a 1.2% increase year over year, the growth has slowed compared with 2014 and 2013, which saw demand rise 3.2% and 2.1% respectively (Discover Los Angeles, 2016).

Hotel Occupancy

Occupancy rates in Los Angeles County have flattened in the last three years, due partly to new supply. In 2015, net supply of traditional lodging units increased by 157 rooms, a 0.2% increase. New traditional supply has been unsteady in recent years, resulting in a total net contraction of 166 rooms since 2011. This contraction, however, varies across all hotel classifications. Inventory in the upper end of the market including luxury, upper upscale and upscale classes grew by 2% (939 rooms), while inventory in the midscale, economy and independent classes contracted by 2% (1,238 rooms) (Market Pipeline Report: Los Angeles County, 2016).

Ellis O'Connor of Santa Monica-based MSD Hospitality notes: "What we have seen in the 7 years of recovery is that hotels in the pipeline in 09, 10 and 11 ceased, and so you had this strong recovery of demand and no increase in supply. Now we have record levels of growth, but we're now getting closer to our 20-year average of supply increase. But with all of this coming online at once and while the industry is growing, individual properties are seeing diminished year over year increases, because of the new supply" (O'Connor, 2016).

ADR and RevPAR

This growth in the upper end of the market may be responsible for some positive data trends: ADR and RevPAR numbers are growing by healthy margins: ADR grew from roughly $147 to over $158 in 2015, and RevPAR jumped from $116 to $126 (Discover Los Angeles, 2016). This, coupled with the 79.7% occupancy rate, seems encouraging news. Los Angeles outpaced the top 25 markets in the U.S. by a healthy margin in all three metrics.

Peer-to-Peer Market Segment

Airdna reported 29,654 Airbnb property listings in the Los Angeles metropolitan area as of May 2016. Figure 10 shows: 40% of these listings were for a shared or

Figure 10. Airbnb listings by type: Los Angeles May 2016
Source: Airbnb per AirDNA, 2016

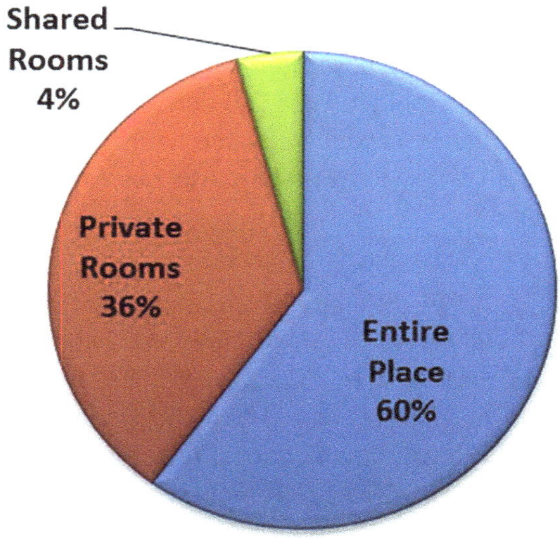

private room within a residence, and 60% were for full use of an apartment or home (Doby et al., 2016). This ratio is important in the regulatory context since some municipalities forbid full home rentals. With that product comprising 60% of the short-term rental market, the effects are significant. Without the tools and resources to enforce the regulations, P2P hosts continue to violate the laws unchecked and without assistance from the platforms, the lack of compliance only continues to grow.

Airbnb continues to experience exponential growth in the Los Angeles area. While bookings follow the typical seasonal pattern for the lodging industry, the total number of bookings for any month in 2016 has far exceeded the same total a year prior. This spread reached its height in April. Total Airbnb bookings for this month were 9,214 in 2015 compared to 16,874 in 2016, an increase of 83%. Bookings in May also peaked at 18,061 (Figure 11).

Mainstream popularity, strong ADR, and low entry barriers have contributed to a surge in new P2P lodging listings worldwide. Airbnb properties increased 42% in Los Angeles County between June 2015 and January 2016 (Martin, 2016). This rapid growth has opened the door for part-time proprietors and many others dedicating properties full-time to the short-term rental market. Another estimate shows that nearly 45% of Airbnb listings in Los Angeles County are listed by a host that has multiple properties on the platform (Inside Airbnb, 2016).

This increase in supply may provide incoming travelers with choices for lodging; however, it also depresses an already tight permanent rental market. In the event

Figure 11. Airbnb booked listings: Los Angeles
Source Airbnb per AirDNA, 2016

Financially, the ADR of hotel-comparable Airbnb properties reached $157.38 and RevPAR averaged $94.87 in 2015 (Doby et al., 2016). While the ADR for the hotel-comparable properties was very close to hotels, the RevPAR is 27.5% lower (Figure 12). This is due to a major difference in occupancy. There were around 1.4 million Airbnb nights listed in Los Angeles County during 2015, however only 842,809 Airbnb nights were booked. As a result, average occupancy for Airbnb properties was 60.2% - decidedly lower than 79.7% at traditional lodging facilities (Doby et al., 2016).

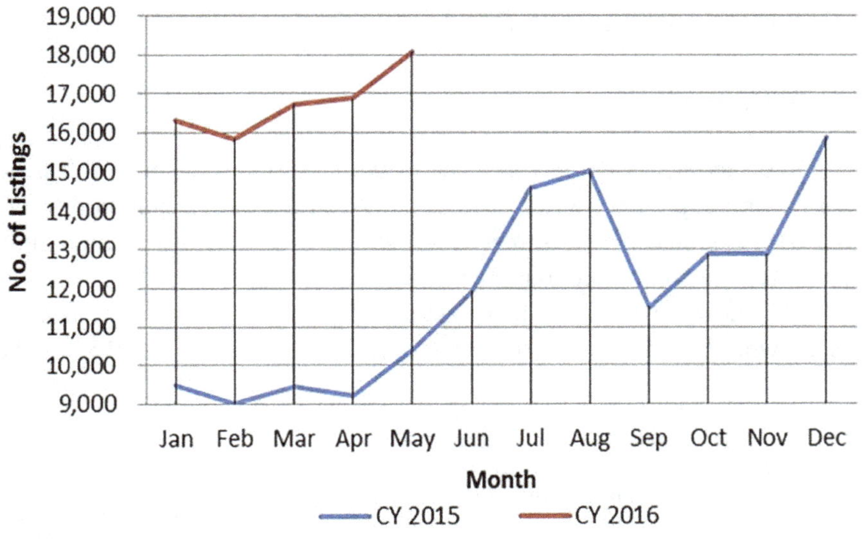

Figure 12. Los Angeles ADR/Rev par hotel vs. Airbnb***
Source: Airbnb data per AirDNA; hotel data per LA Tourism/Convention Board

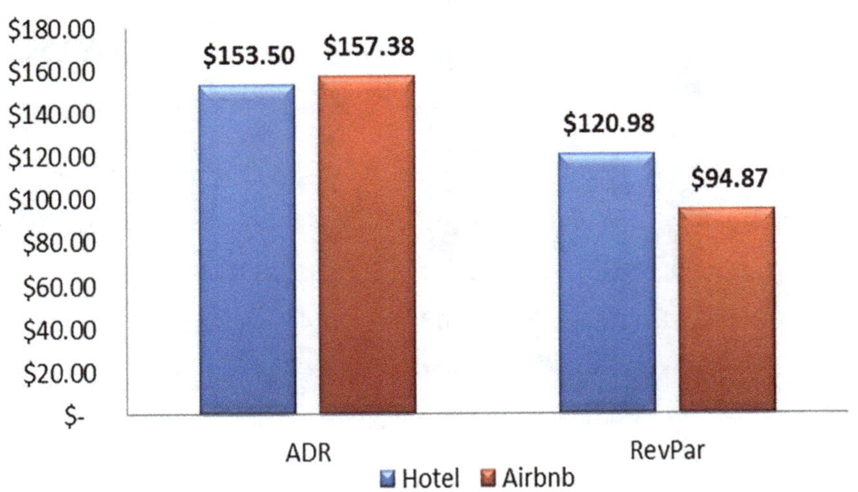

that an apartment is rented out full time, or when property owners decide to convert former long term rentals to short-term rentals, that inventory comes off the market and constrains supply for permanent residents. Rents in Los Angeles County are expected to rise 8.9% between 2015 and 2018 (Khouri, 2016), an indication that demand continues to outpace supply. While this increase cannot be fully attributed to the cannibalization caused by the increase in short-term rentals, experts have judged it as a significant and possibly governable factor in terms of the short-term rental market on the local community.

City Findings: Vienna

Vienna is the capital, largest city and one of the nine states of Austria. It is home to 1.83 million people accounting for 21% of Austria's total population (Statistic Austria, 2016). Vienna also hosts many international organizations. Vienna's unique charm and liveliness continue to attract international tourists.

Figure 13 shows that Vienna recorded a 5.8% increase in international tourist arrivals, growing from 5.4 million travelers in 2014 to 5.72 million in 2015 (Flughafen

Figure 13. Vienna tourist arrivals and overnight stays
Soucre: tourist arrivals per Euromonitor, overnight stays per Wien: Hotelangebot & -auslastung

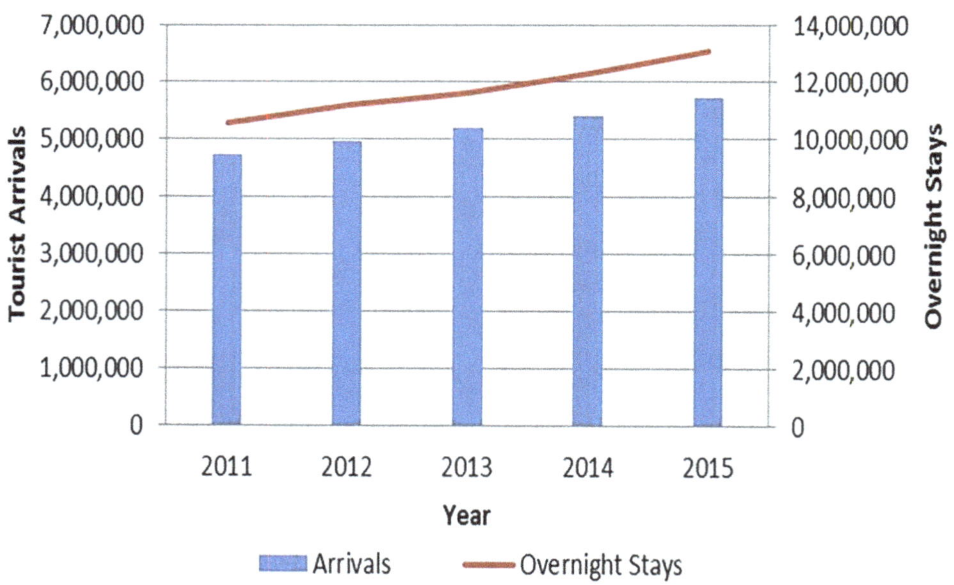

Wien, 2016). It ranks as 9th top destination to visit in Europe and is popular for both summer and winter vacation. In fact, Austria leads Europe's market for winter holiday travel with a 50% market share. Vienna is globally ranked as the most prosperous city in the world and is number one for having the highest quality of life (H1-2014-Hotel-and-Tourism-Markt-Report, 2016). The growth in tourism to the city has been felt by the hotel industry. Hotel overnight stays slightly outpaced international tourist arrivals, growing from 12.3 million in 2014 to 13.1 million in 2015, a 6.48% increase. Over the past 5 years, these two performance measures have followed each other closely. The correlation coefficient for the two measures from 2011 to 2015 is .998, indicating a strong linear relationship (Doby et al., 2016).

Vienna Lodging Industry

Hotel Room Supply

The number of hotel beds rose from 50,000 in 2005 to 70,000 beds today, and is expected to reach 80,000 beds by the year 2020. This 3% average increase in hotel room matches the 3% average increase in hotel demand (Penz, 2016). Despite the economic crisis in Europe tourism in Vienna continues to grow, and the lodging industry, including both traditional hotels and P2P lodging platforms, continues to thrive.

Hotel Occupancy

The demand of today's travelers seeking urban areas is on the rise and consequently the lodging industry in Vienna is growing. The city has continued to add many hotel beds over the last few years to meet demand, as building housing and hotels was cheap. As a result, the number of hotel rooms supplied continues to increase upward while the hotel occupancy rates grew only slightly from 72% in 2014 to 73% in 2015 (PWC, 2016b).

ADR and RevPAR

Another sign of Vienna's continued growth was an increase in ADR from €95 in 2014 to €100 in 2016, the highest since 2007. During the peak of the economic crisis, the higher end of the market suffered the most. The 4.4% YOY increase can be attributed primarily to a reversal in the downward trend of 4 and 5 star hotels. RevPAR, driven by both the increases in ADR and occupancy rates, similarly increased from €69 to €72 (PWC, 2016b). As the economy gradually recovers, so too does the image of fine luxury and culture to Vienna.

Peer-to-Peer Market Segment

Similar to the traditional lodging industry, P2P rentals have also increased over the past year (Pwc-European-Cities-Hotel-Forecast, 2016). In May 2016, there were 3,720 Airbnb listings in the city, compared to just 1,607 one year prior - a robust 132% growth. While only 28% of the units last May were for shared or private rooms, 72% were for use of an entire place (Figure 14). This larger share suggests that the majority of supplied listings provide for a more hotel comparable experience, than it would to stay in a shared or private room.

Airbnb experienced tremendous growth in Vienna throughout 2015. Bookings reached peak in December 2015 at 4,103 (Doby et al., 2016). However, this growth seems to be levelling off (Reitterer, 2016). Some experts have suggested that the P2P lodging market has reached its peak in the city and may even be headed for a downturn. Visitation to Vienna is highest in the winter months, so the fact that no month this year has eclipsed December's total isn't a strong indicator that these bookings have come to a halt. The YOY growth, however, has slowed (Figure 15). There are several causes that may be contributing to the slowdown in bookings, but the supply of available units may be the top factor.

Figure 14. Airbnb listings by type: Vienna May 2016
Source: Airbnb per AirDNA

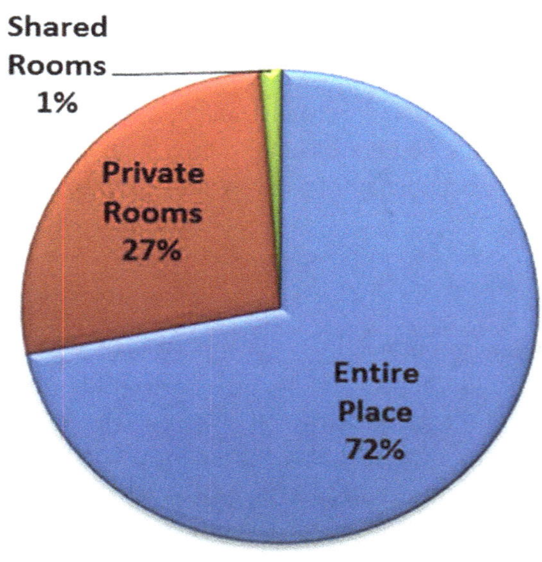

Figure 15. Airbnb booked listings: Vienna
Source: Airbnb via AirDNA

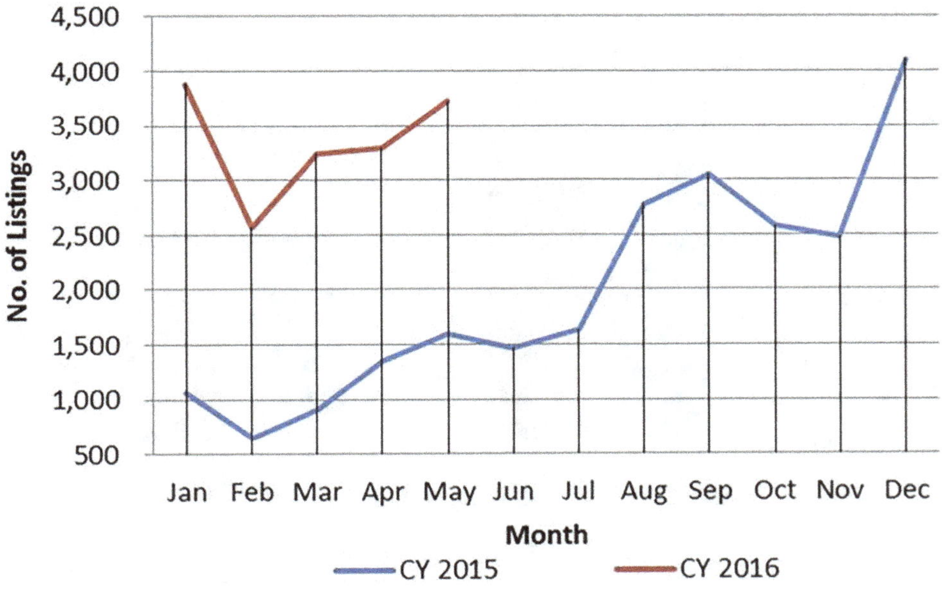

On average, staying at a hotel in Vienna is more expensive than renting a P2P unit on Airbnb. In 2015, the ADR for a hotel-comparable unit on Airbnb was €85.33 and the €100 ADR for the traditional lodging industry represents a premium of 17.2%. Certain demographics don't mind paying the extra cost. For example, 80% of business travelers choose hotels over P2P lodging accommodations in Vienna (Penz, 2916). The difference in RevPAR is even more pronounced: in 2015, this figure for Airbnb units was just €44.58 compared to €72 for hotels (Doby et al., 2016). This differential was affected by the occupancy rate, which was 73% for hotels and only 53% for Airbnb (Figure 16). Another contributing factor is that the traditional lodging industry benefitted from a recovery of high-end hotels. At the height of the recession, they had to significantly reduce their room rates. Under some recovery, the higher room rates charged by these accommodations have brought up the ADR.

One factor slowing down the growth of P2P lodging platforms in Vienna is the regulatory constraints on rentable units within the city. Almost 60% of Viennese residents live in subsidized housing. They are legally forbidden from renting out their place of residence on Airbnb or any other P2P lodging platform (Wieser, 2016). However, if the residence is not government subsidized, sharing or renting a room

Figure 16. Vienna ADR/RevPar hotel vs. Airbnb
Source: Airbnb via AirDNA, hotel data per PWC Hotel forecast

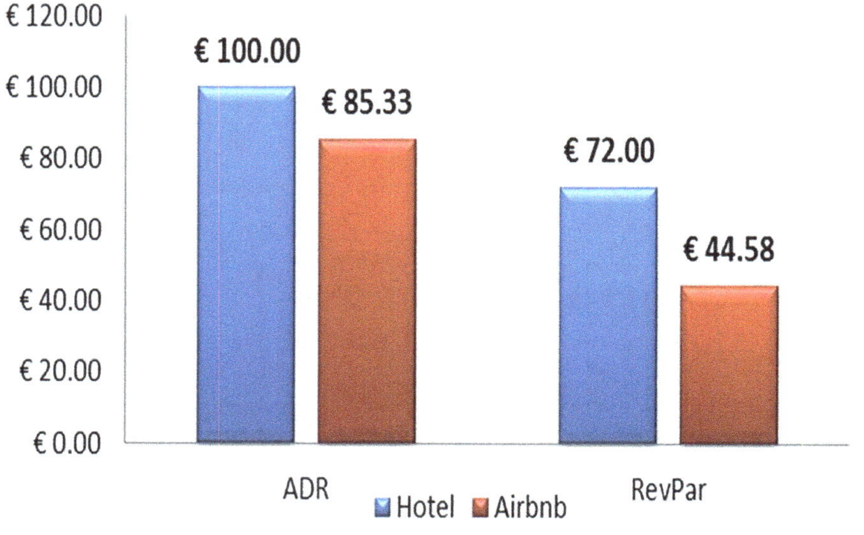

is legal in Austria as long as the host meets the following requirements: lives in the same building, does not offer more than 5 rooms or 10 people at any one time and only serves breakfast (Vienna Official Website, 2016). However, today most of the units listed in Vienna on Airbnb's website are in districts where tourism is common and over 70% are for use of an entire home or apartment. It suggests that only the 28%, which represent shared and private rooms, are conforming to the "live like a local experience" advertised by Airbnb. With the limited supply of legally rentable rooms tied up by opportunistic hosts, the housing market is being compressed and the pressure is on city to respond to a growing housing crisis (Reitterer, 2016).

Government and Regulations

Vienna's upscale reputation has aided hotels by allowing premium room rates commensurable with the stature of the city. The lower room rates in the P2P lodging sector erode that image. As a result, city has not actively supported these alternative accommodations. In fact, the tourism boards, including the Austrian Hotel Association, which are funded in part by overnight taxes collected by traditional hotels, have continually criticized P2P platforms offering short-term rentals and have called on the city government to take action and enforce upon these companies the same rules hotels have to comply with and ensure an even playing field. Currently, unlike traditional hotels, Airbnb hosts do not operate under the same city rules in regard to taxes and safety. Complying with existing regulations increases the overhead and operating costs for traditional hotels, causing the city government to address these inequalities by trying to build a working relationship with Airbnb to gather host data. This data will be used by the government to collect city tourism taxes and enforce city laws. However, these efforts, particularly with Airbnb, have not yet succeeded and P2P lodging platforms have had some success under European data protection laws, which legally prevent them from providing host information (ktv_adittrich, 2016).

The tourism industry in Austria is largely regulated on a provincial level and not on a national level. Vienna can therefore make independent decisions affecting the city and its citizens. The City of Vienna appears cognizant of the presence of P2P lodging platforms and acknowledges the sharing economy is here to stay and grow, while technological possibilities are still far from being fully exploited. However, government looks to regulate this new and developing industry in a way that protects the interest of its citizens. The City of Vienna endeavors to turn the sharing economy absolutely into a fair economy (Turning the Sharing Economy, 2016). The city wants entrepreneurs to act within the rules and regulations. Vienna is committed to fair competition and the principle of "good jobs" (good pay and fair working conditions), which means the same rules apply to all businesses, no matter whether business is done online or offline.

In order to even the playing field between Airbnb hosts and traditional hotels, State legislators are expected to pass regulations by December 2016 (Turning the Sharing Economy, 2016). The city has already started the information campaign and plans to launch online resources of the city laws to facilitate awareness of the current rules and regulations among people hoping to rent their apartments aware. The city also introduced a new reporting requirement for platform operators as accommodation agencies. When the city government began to monitor Airbnb hosts, raise fines (up to €2000), and request that hosts provide exact addresses, Airbnb coincidentally began cooperating with city officials. As a result, several European governments are considering creating registry numbers to list units so governments can track and collect taxes with relative ease.

City Findings: Athens

Athens is the capital and largest city of Greece, with a population of 3.7 million people. The Athens metropolitan surface area covers approximately 2,928 square kilometers. The city itself has a population of approximately 665,000 people, which usually doubles during business hours. Athens is a national business hub and plays an important business role in the entire Balkan region (Papadimitriou, 2014).

After five rough years of the market correction in key metrics following Greece's economic downfall since 2008, 2013 saw a turnaround in tourism, displaying an increased number of arrivals as well as a significant increase in hotel occupancy rates.

Athens has one airport, Athens International Airport. Piraeus is the main seaport of the city and is the largest port in Greece (second largest in the Mediterranean Sea). Between 2011 and 2013, international tourist arrivals to Athens have dropped from 2.9 million to 2.6 million. This trend has reversed climbing to 4.16 million arrivals in 2015 (Doby et. al., 2016). Over the same period, overnights for hotels and similar accommodations took a similar path to recovery (Figure 17). Hotel overnight stay data for 2015 was not readily available at the time of this writing, but given its relationship with international arrival figures, there is likely to be a year over year gain. From 2011 to 2014, these two metrics had a positive correlation coefficient of .932, suggesting a strong dependence of overnight stays on tourist arrivals (Doby et al., 2016).

The Athens International Airport recorded an increase of 10.1% year over year in international arrivals in Q1 2016. However, this increase is not reflected in the occupancy of the Athens hotel sector as an improvement of only 3.5% year over year was recorded in Q1 2016. The Athens Hotel Association attributes this difference to the booming sharing economy in Athens, where the supply on platforms like

Figure 17. Athens tourist arrivals and overnight stays
Source: tourist arrivals per Euromonitor, hotel overnights per Greek Statistical Authority

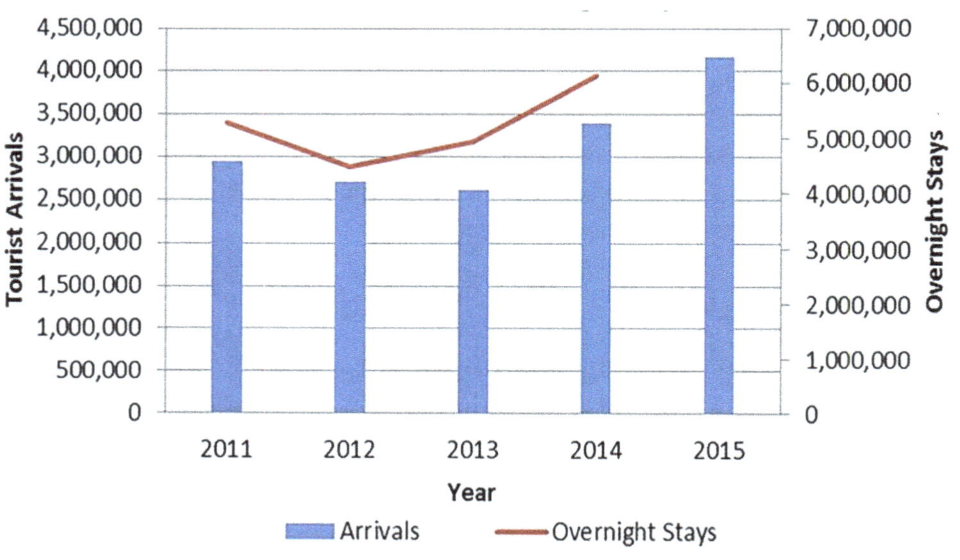

Airbnb is increasing at a rapid pace. In winter, the difference cannot be attributed to tourists leaving directly to the islands (2016Q1 Greek Hospitality Industry Performance, 2016).

Depressed prices have made Athens, and the country in general, a more desirable destination for tourists visiting from abroad. Foreign travelers have accounted for more than 65% of Athens overnight stays for nearly every year over the last decade (Lodging in Greece, 2017). Northern Europe, in particular, has been particularly important to the industry's recovery. With Germany and UK representing 22% of all arrivals in 2015, the 28 EU countries were responsible for 15 million arrivals, representing 63% of the total (2016Q1 Greek Hospitality Industry Performance, 2016).

Athens Lodging Industry

Summer is the peak tourist time in Athens. Most tourists visiting Greece during summer only spend a few days in Athens and continue on to the Greek islands for the remainder of their trip. During winter the Greek islands are not visited by tourists. As a result, many of the hotels there have to shut down operations. About 50% of 5, 4 and 3-star hotels in Greece operate on a seasonal basis and can be characterized as 'resort' hotels while the other half operate all year round and are categorized as 'city' hotels. The seasonality of the industry has an even greater effect on hotels categorized with less stars. About 60% of these hotels only shut down during the off season (NAIHellas, 2015).

Hotel Room Supply

The supply of hotels in Athens has fluctuated significantly over the last couple of years, and larger hotels that closed were replaced by smaller ones. Per GBR Consulting, for the period 2007 through 2013, 37 hotels, representing 2,156 rooms, shut down. Over the same period, 36 hotels opened, but with just 671 rooms (2016Q1 Greek Hospitality Industry Performance, 2016). This compression countered by an influx of tourism, facilitated growth in the P2P lodging sector. Meanwhile, the traditional hotel industry has recovered some of this lost supply: at the end of 2014, the city of Athens, including its suburbs, had 15,049 rooms, 52% of which are 4 star or 5 star hotels (2016Q1 Greek Hospitality Industry Performance, 2016).

Hotel Occupancy

Despite the significant increases in international tourist arrivals throughout 2015, several months during the latter part of the year saw an occupancy rate decrease.

During the fall, month over month occupancy rates dropped by 3.8%, 6.1%, and 4.6%, respectively (GTP Editing, 2015). The gap seems unexplainable, if not for the impact of P2P lodging sector. In total, the occupancy rate for the year increased from 72.2% in 2014 to 74.9% in 2015, a 3.6% gain. Over the same period, international tourist arrivals grew from 3.39 million to 4.16 million, a 22.65% gain (Doby et al., 2016). The occupancy rates stated exclude the islands, but that still does not account for the difference during the winter months, when visitations to islands are low.

ADR and RevPAR

In 2015, Athens saw significant increases in both ADR and RevPAR. ADR grew from €85.96 in 2014 to €90, a 4.7% gain. RevPAR grew from €62.11 to €67.38, an 8.5% gain (Doby et al., 2016). This is an interesting dynamic. Travelers are being drawn to the city because of their common perception that the Greek economic crisis makes things inexpensive. Yet the influx in tourism is actually driving prices up under the supply and demand dynamics. Staying in less expensive P2P lodging units, however, better aligns consumer perception with reality.

Peer-to-Peer Market Segment

In May 2016, Athens had 3,540 listings with 2,210 bookings, 84% of which was for use of an entire home or apartment. This disproportion in listings types has caused the Hellenic Chamber of Hotels to dub much of the emerging P2P lodging market as the "shadow economy (Christidou, 2016). Entire place rentals act as a much more direct competitor to the traditional lodging industry, and opportunistic Greeks are seizing in on the opportunity to cash in on this growing phenomenon.

Hotel room rates in Athens far exceed those of Airbnb units. In 2015, the ADR for hotels in Athens was €90 against €58.88 for comparable Airbnb units. This represents a 52.9% premium and provides economic travelers strong incentives to select P2P lodging platforms. RevPAR for hotels in 2015 was €67.38, against €32.43 earned by Airbnb hosts (Doby et al., 2016). On average, hotels are earning higher rental rates per room, but much of this gain is lost once the labor and maintenance costs are included. The citywide accommodation and financial dynamics are presented in figures 18-20.

City Specific Dynamics

One factor contributing to the success of P2P lodging platforms in Athens is the plentiful supply of homes and apartments to rent (Profile of Greece, 2016). With a

Figure 18. Airbnb listings by type: Athens May 2016
Source: Airbnb per AirDNA

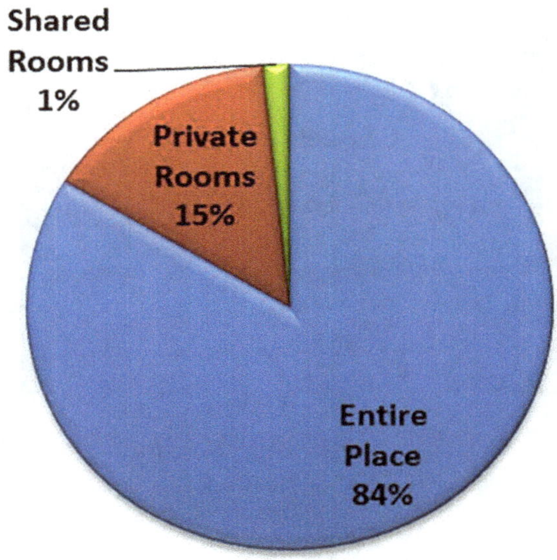

Figure 19. Airbnb booked listings: Athens
Source: Airbnb per AirDNA

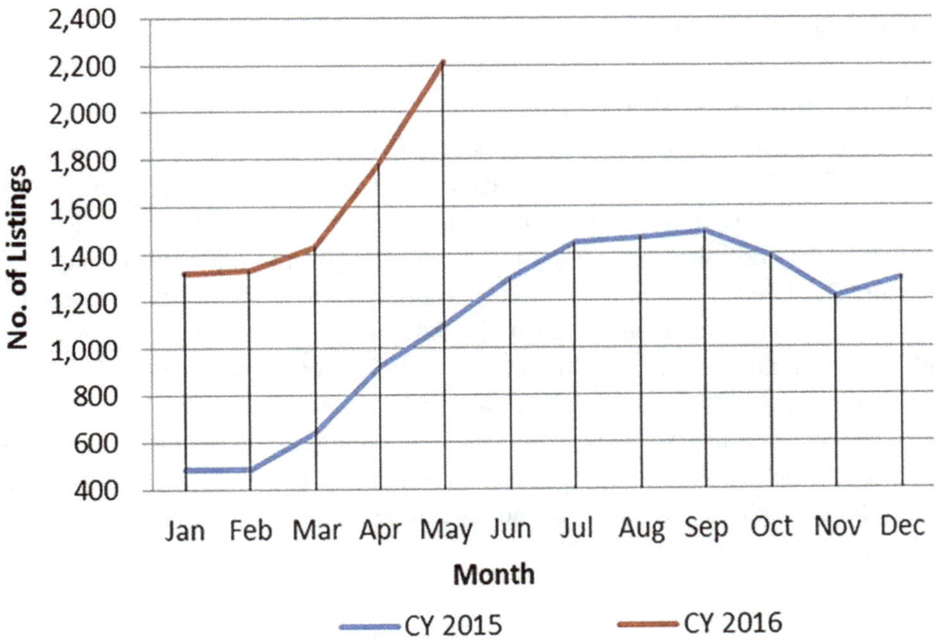

*Figure 20. Athens ADR/RevPar Hotel vs. Airbnb**
Source: Airbnb data per AirDNA, hotel data per GBR Consulting

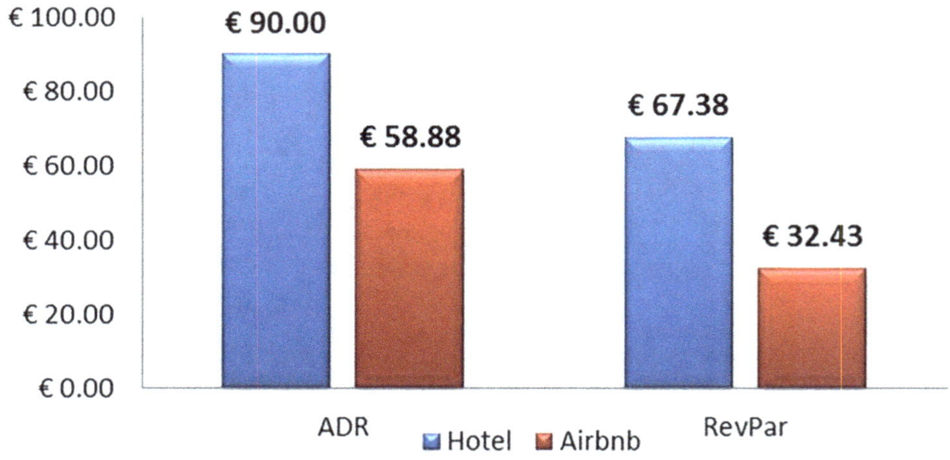

hot summer and because parents used to live in a village or an island or somewhere outside Athens, many Greek families still have a house "there" (Theofanopoulou, 2016). In fact, about 50% of Greece's population owns a second home (Profile of Greece, 2016). An abundance of supply, financial hardship faced by homeowners, and a heavy influx of tourists hoping to benefit from lower prices, act in support of P2P lodging platforms.

Government and Regulations

P2P lodging platforms in Greece are neither regulated, nor controlled by government agencies. They do not fit in any legislative framework and, generally, are not supervised by an (independent) mechanism. Due to the increasing popularity of P2P lodging platforms, they definitely inflict losses on both the government revenue and hotel revenue. The Greece economy misses taxes adding up to €350 million per year due to the operation of illegal tourism accommodations. This equates to hotels losing 12 million overnight stays, translating into €554 million of lost revenue (GTP, 2015) that could potentially create an additional 15,000 jobs annually in the traditional hotel sector. An industry report states that P2P lodging platforms cause an annual loss to the Greek state between €260 and €276 million. If now a similar tax applies to P2P lodging platforms, public revenues can be further strengthened by €322 to €353 million annually (GTP, 2015).

City Findings: Barcelona

The 1992 Olympic Games, hosted by the city of Barcelona, put it on the map as one of the top global tourist destinations and since then it has experienced exponential growth in the travel and lodging industries. Barcelona is the second largest city in Spain and the leading international tourist destination in the country. In the last few years, Barcelona has experienced phenomenal tourist growth and is now one of the five most visited cities in Europe (PWC, 2016b).

International tourist arrivals, in particular, have been strong, increasing over 2014 to 2015 from 6.26 million to 6.61 million, a 5.68% gain. In part, this growth has been facilitated by new international flights: in 2015, the Barcelona airport achieved a 5.7% increase in passenger arrivals against 2014. This has also greatly contributed to the growth in hotel overnight stays which have gone from 15.6 million in 2011 to 18.4 million in 2015 (Figure 21). The correlation coefficient for these two performance measures was .9949, suggesting a strong uphill linear relationship (Doby et al., 2016).

Barcelona Lodging Industry

Hotel Room Supply

The lodging industry in Barcelona has experienced strong growth in hotel room supply. In 2010, Barcelona counted 31,776 hotel rooms, distributed over 328 hotels;

Figure 21. Barcelona tourist arrivals and overnight stays
Source: tourist arrivals per Euromonitor, overnight stays per Instituto Nacional de Estadistica

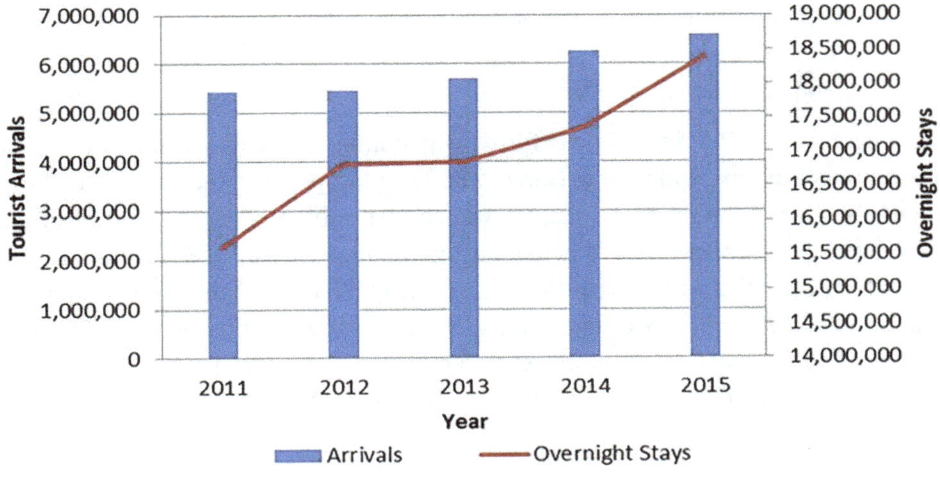

by 2015, this supply has grown to 40,379 hotel rooms, distributed over 639 hotels. While over this period, the total number of hotels increased by 95%, despite this vast increase in supply, demand continues to grow at a faster relative pace, resulting in the rising hotel occupancy rates (PWC, 2016b).

Hotel Occupancy

Barcelona continues to attract more tourists, driving a strong demand in the lodging industry. Hotel occupancy rates grew from 68% in 2010 to 75% in 2015 (PWC, 2016b). In addition to tourist attractions, Barcelona is also a leading MICE (meetings, incentives, conventions, and exhibitions) destination, hosting nearly 2,000 events a year. Forty percent of tourists visiting Barcelona in 2014 cited a convention or exhibition as their prime motivation to visit. The GSMA Mobile Congress alone, the world's largest exhibition, conference and networking event for mobile operators, brings more than 87,000 people to Barcelona (PWC, 2016a).

ADR and RevPAR

Barcelona's growth in tourism has propelled a strong financial performance in the lodging industry. Hotel ADR has consistently increased since 2010, from €110.20 to €124 in 2015. Additionally, Barcelona hotels have experienced a substantial growth in RevPAR, increasing from €74.61 in 2010 to €94 in 2015 (Doby et al., 2016). Last year's RevPAR ranks 9th in all of Europe and is forecast to move to the 8th highest RevPAR by 2017. The strong growth in RevPAR was driven in part by marginal hotel room supply growth and healthy occupancy levels. The flow of events and conferences planned for 2016 and 2017, as well as a hotel moratorium, are expected to drive growth in RevPAR, largely through price increases. RevPAR is forecast to increase to €97 in 2016 and €103 by 2017, and ADR is expected to grow 3.2%, taking it to €129 by the end of 2016 (PWC, 2016b).

Peer-To-Peer Market Segment

The recent rise of P2P lodging platforms is becoming increasingly threatening to the traditional lodging industry in Barcelona. The most popular and successful these services in Barcelona in recent years have been Airbnb, HomeAway, and FlipKey. For example, there were 18,474 active Airbnb listings in May 2016. From August 2014 – May 2015 to August 2015 – May 2016, Airbnb increased its room availability from 675,855 to 1,497,774, a staggering 121.61% increase (Doby et al., 2016). The number of active listings in Barcelona also reveals a more balanced distribution of the different types of units than some of its peers. During May 2016, 54% of the active listings were available for use of the entire place and 46% were for private and shared rooms (Figure 22). In order to determine if P2P lodging platforms are a threat to hotels, it is important to examine their performance compared to traditional lodging companies' performance in Barcelona. In 2015, Airbnb had a 60% occupancy rate, with 759,812 listing nights booked out of 1,267,099 listing nights available. The traditional lodging industry stated occupancy levels at 75% in 2015. During that time, 3.8% of total overnight bookings in the lodging industry were from Airbnb overnights (Doby et al., 2016).

In 2015, Airbnb's ADR stood at €96.29 against €125 for hotels' ADR. This represents a 29.8% pricing premium for hotels. The lower prices is an advantage that P2P lodging platforms have over hotels, and tourists often choose the lower price when looking for a place to stay in Barcelona. In 2015, Airbnb had a RevPAR of €57.36, which was also lower than hotels' RevPAR of €94 (Figure 23). This suggests that hotels were able to generate 63.9% more revenue per room against Airbnb hotel comparable listings (Doby et al., 2016).

Airbnb's growth in 2016 appears to have stalled somewhat in Barcelona compared to the prior year. The summer months are definitely the most popular time to visit the city, so the numbers appear poised to climb again. However, while the number of

Figure 22. Airbnb listings by type: Barcelona May 2016
Source: Airbnb per AirDNA

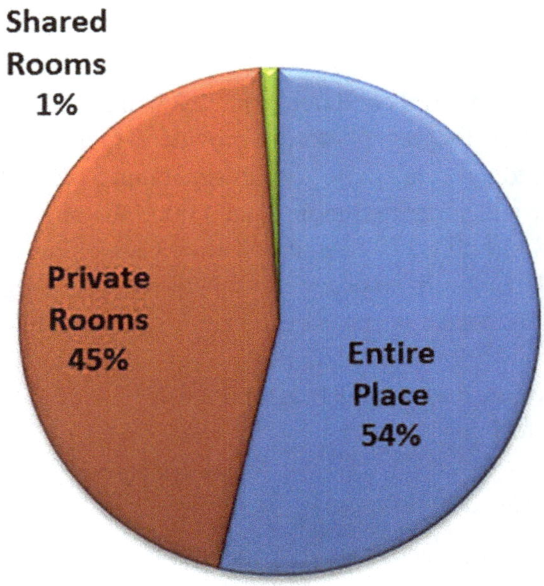

*Figure 23. Barcelona ADR/RevPar hotel vs. Airbnb**
Source: Airbnb per AirDNA, Hotel data per PWC Hotel Forecast

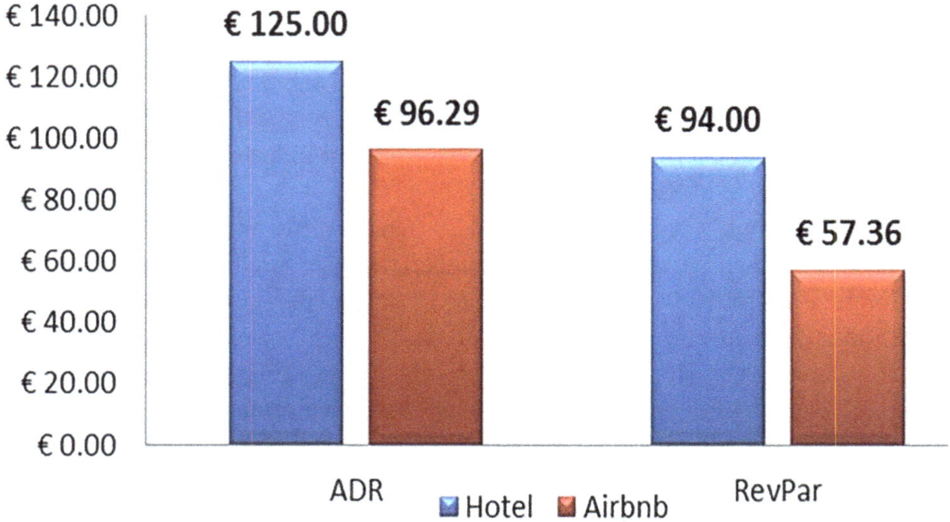

bookings in 2016 is likely to exceed 2015, the current trajectory is not as steep. As Figure 24 shows, the number of bookings actually began to level out at the beginning of the year and fell slightly from December 2015 to January 2016.

City Specific Dynamics

With high unemployment levels in Barcelona, 22.3% in 2015 (PWC, 2016b), locals need income, and P2P lodging platforms provide a great opportunity for extra income. According to a recent study, Airbnb generated $175 million in economic activity in one year and supported more than 4,000 jobs in Barcelona. The study found that 75% of Airbnb hosts earn at or below Catalonia's average household income (Airbnb, 2016). From an economic standpoint, P2P lodging platforms are an ideal match for Barcelona because of the city's high unemployment consistent with the growing consumer trend of wanting to "live like locals" with an authentic experience. As a result, Barcelona's lodging industry has seen a dramatic growth in P2P lodging platforms.

Government and Regulations

Tourist revenue generating activities in Catalonia benefit the area in which they provide business, resulting in a regional legislation in place to tax hotels and tourist

Figure 24. Airbnb booked listings: Barcelona
Source: Airbnb per AirDNA

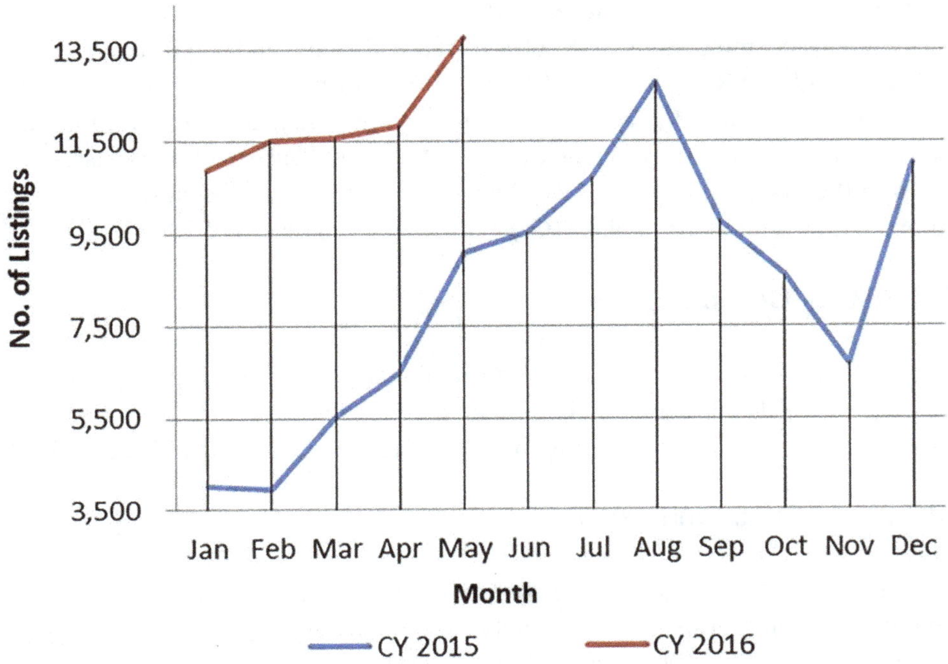

accommodations. Since 2012, tourist accommodations are required to pay a VAT of 10% of their overall revenue (Agencia Tributaria, 2016). Apartment and house rentals are exempt from this VAT; however there is a massive amount of units and rooms, 202,530 total Airbnb listings in 2015 alone (Doby et al., 2916) that are rented out for tourism purposes, thus bringing in revenue like a normal tourist accommodation. There is a discrepancy in Barcelona between tax contribution by traditional lodging accommodations and P2P lodging platforms. The International Hotel and Restaurant Association describes P2P lodging platforms as unfair to competition (MarketLine, 2015c), in part because they haven't been taxed the same way as traditional tourist accommodations, creating an uneven playing field.

In 2012, the Catalonian government initiated legislative initiatives toward regulating apartment establishments and homes for tourist use. The regulation requires all hosts of tourist rental accommodations to register their unit as an official tourist rental accommodation and obtain a government license. Hosts must also register with the Tourism Register of Catalonia (RTC). They are then given an NIRTC, which is the identification number required for all advertised tourism accommodations in

Catalonia. The government is then able to collect the data from all the registered vacation rental accommodations and appropriately tax them.

Progress has been made in leveling the playing field amongst tax regulation of tourist rental accommodations: 81,100 tourist rental homes and 446,051 individuals registered with the RTC between 2013 and 2015. Hosts have been providing their data in order to be taxed as a travel accommodation, although there are still many who don't register their homes. The Catalonia Strategic Tourism Plan estimates that half a million beds in Catalonia are illegal accommodations, because they aren't following the regulations in place (Generalitat de Catalunya, 2016).

Cross-City Comparisons

This section presents comparative environmental analysis of the four cities, specifically looking into the legal/regulation issues and taxes, licenses/permits, social/cultural aspects, and economics. Key findings of this analysis are presented in Figures 25-30.

Legal/Regulation and Taxes

Tourist and value added tax (VAT) contribute to local economies. This tax differs per city and country. Los Angeles County has a transient occupancy tax of 12%. In Barcelona tourist accommodations pay a VAT of 10% of their overall revenue (Agencia Tributaria, 2016). Vienna requires a similar tax, where tourist accommodations pay 3.2% of the accommodation cost (Brownlow, 2016). Similar to Los Angeles, Vienna, and Barcelona, Athens tourist accommodations must pay a tax of 24% of their revenue (Theofanopoulou, 2016).

Airbnb and other P2P lodging platforms have not been held to the same standards of paying taxes as hotels have been. P2P lodging platforms have been reluctant to share data with governments that would be used for taxing them, thus complicating governments' effort to enforce tourist tax regulations. Los Angeles and Santa Monica are setting a new precedent, by establishing an agreement with Airbnb to collect transient occupancy taxes on their behalf. In Barcelona, Vienna, and Athens it is the host's responsibility to collect those taxes on their own and then submit them to the government.

Barcelona, Vienna, and Los Angeles County, specifically Santa Monica, have all started fining hosts of P2P lodging platforms for not disclosing their units and paying taxes to their governments. Vienna fines hosts up to €2,000, while Barcelona hosts pay up to €3000 for illegally renting out their flats (País, 2015). Similarly, Santa Monica recently fined a host $3,500 for illegally operating a vacation rental business without city permits (Ganezer, 2016). Barcelona has even gone so far to fine Airbnb directly. They administered a €30,000 fine to Airbnb to send a message, although the minimal fine was really too small to make a significant impact thus far (Abad, 2016). In contrast, cities like Athens have done little to combat tax issues with P2P lodging platforms, as they haven't started administering fines to hosts.

Licenses/ Permits

Barcelona and Santa Monica have led the way in implementing processes to track and collect taxes from P2P lodging platforms. Barcelona requires hosts to register with the government to obtain a specific tourist accommodation license. This has been somewhat successful in helping the government to track and collect tourist taxes from P2P lodging platforms. Santa Monica, and soon to be Los Angeles, require hosts to register with the city to obtain a Transient Occupancy Registration certificate in order to collect and pay taxes. Vienna is not far behind Barcelona and Santa Monica in working with Airbnb to gather hosts' addresses and collecting data directly from hosts. Additionally, Vienna has a private accommodation law that says residents cannot rent out more than 10 beds without a license from the government (Penz, 2016).

Unlike Barcelona, Vienna, and Los Angeles County, Athens has done little to facilitate tax collection from hosts of P2P lodging platforms. Their governments are putting their trust in the hosts, hoping they will pay the tax on their own without any regulations and oversight. The government in Athens used to require hosts using P2P lodging platforms for 30 days to be professional organizations, but have recently abolished that (Christidou, 2016). Athens has introduced processes to collect the appropriate taxes from P2P lodging platform, resulting in millions of euros missing in tax revenue.

Social/Cultural

Comparing cross-country/city analysis of the social/cultural aspects of P2P lodging platforms generated interesting findings. Most notable were the residents of Vienna (Balint & Indra, 2016), and Athens (Benetou, 2016), who generally view P2P lodging platforms as a necessary means to supplement their current incomes (Guido, 2016). Financially, residents of these cities feel they must have additional sources of income to maintain a standard of living, and renting their homes through P2P lodging platforms provides this opportunity. An owner in Athens explained that most Athenians have an apartment in the city and a second home in the countryside that has been passed through family generations. Due to recent government mandated increases on property taxes, Athenians are now having difficulty paying for these second homes (Benetou, 2016): the establishment of P2P lodging platforms has brought some relief for the Athenians. One would argue however, why these individuals don't just lease their properties out long term and move away from the P2P lodging platforms. Benetou and Guido both mentioned that long-term renters have a plethora of rights that protect them from many requests of landlords (Benetou, 2016; Guido, 2016). Landlords do not want to bother with unreasonable renters or be forced into a situation where a renter does not pay rent but due to renter rights, they cannot be removed from the unit (Benetou, 2016). The P2P lodging platforms however, prevent any of this from happening, providing a more attractive alternative.

While residents of Vienna and Athens may view P2P lodging platforms as a necessary tool to supplement their incomes, Barcelona and Los Angeles appear to be using the platforms the way they were intended to be used. The brand positioning of P2P lodging platforms has traditionally been focused on making connections, "living like the locals" and sharing space. Therefore, listings of private and shared rooms would fit this proposition. In Barcelona and Los Angeles, the percentage of private and shared rooms listed exceeds those of other countries, allowing for the conclusion that residents of these cities view P2P lodging platforms closer to its intended purpose.

Figure 25. Airbnb listings by type
Source: Airbnb per AirDNA

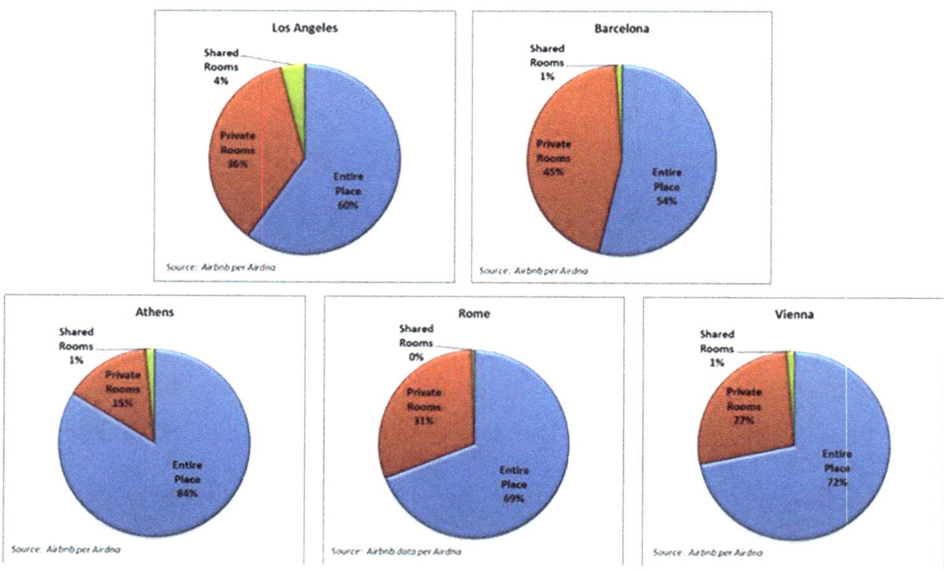

It is also interesting to cross-compare the growth of available Airbnb listings from December 2014 to December 2015 (Figure 26). Each of the cities experienced growth, but Athens saw the largest growth with 115.4%, followed by Vienna (91%), Barcelona (47.5%), and Los Angeles with 42.8% (Doby et al., 2016). This is significant growth for each city, apparently reflective of residents' social acceptance of P2P lodging platforms.

Economics

ADR Premiums

A comparison of the various performance measures reveals different economic dynamics across the countries/cities. In the US, for the 12-month period ending September 2015, the ADR for Airbnb units was $29.31 or 25% higher than traditional hotels over the same periods (Lane & Woodworth, 2016). This contradicted the widespread perception of Airbnb being a less expensive alternative to hotels. Furthermore, it dispels the notion of price as a distinguishing value proposition for P2P lodging platforms in the U.S. This is true even in Los Angeles where this study found that hotels actually offer a 2.5% discount to Airbnb units.

Figure 26. Hotel ADR premiums 2015 (includes hotel comparable Airbnb listings)
Source: Airbnb per AirDNA, Hotel ADR per PWC Hotel 201602917 Forecast Reports, Athens Hotel ADR per GBR Consultings, Metro Hotel ADR Weighted based on counties' overnight stays

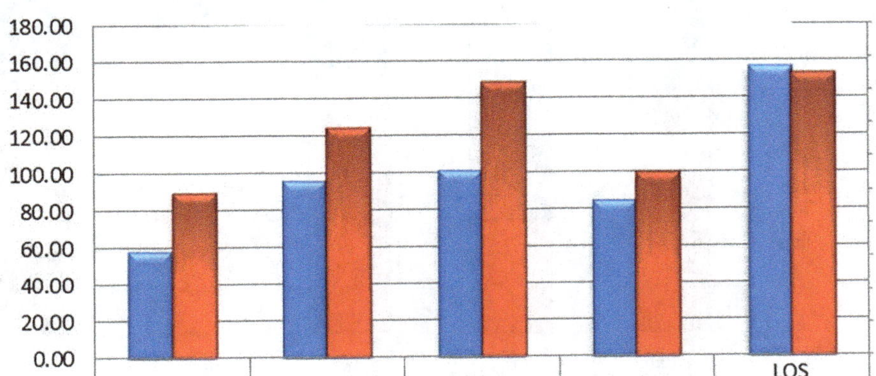

	ATHENS	BARCELONA	ROME	VIENNA	LOS ANGELES*
Airbnb	€ 58.88	€ 96.29	€ 101.53	€ 85.33	$157.38
Hotels	€ 90.00	€ 125.00	€ 149.00	€ 100.00	$153.50
Premiums	52.9%	29.8%	46.8%	17.2%	-2.5%

For the European locations under this study, however, the opposite is true. In fairness, comparing the ADR for a shared or private room to a hotel room may not be equitable. However, even when the analysis is limited only to Airbnb hotel comparable listings, defined as entire place listings with 3 or fewer bedrooms, the comparison still reveals substantially higher hotel premiums: for the targeted European locations they range from 17.2% in Vienna to as much as 52.9% in Athens.

Airbnb units' premium amenities may be the cause of their higher ADRs. In particular, 90% of the units in the U.S. offer access to a kitchen and 68% offer access to a washing machine. By contrast, only 60% of hotels in the U.S. even offer a microwave (Lane & Woodworth, 2016). This logic however does not explain the relationship between hotel and P2P lodging platforms' ADRs in the targeted European locations. The percentage of Airbnb properties with access to a kitchen are all similarly greater than 90% and access to a washer/dryer ranges from 72% in Athens to 82% in Vienna (Doby et al., 2016). In fact, Athens with the lower percentage of washer/dryer access actually has a more substantial Hotel ADR premium against Vienna.

Figure 27. 2015 Overnight Stays; Airbnb share of the market
Source: Airbnb per AirDNA, hotels ADR per....

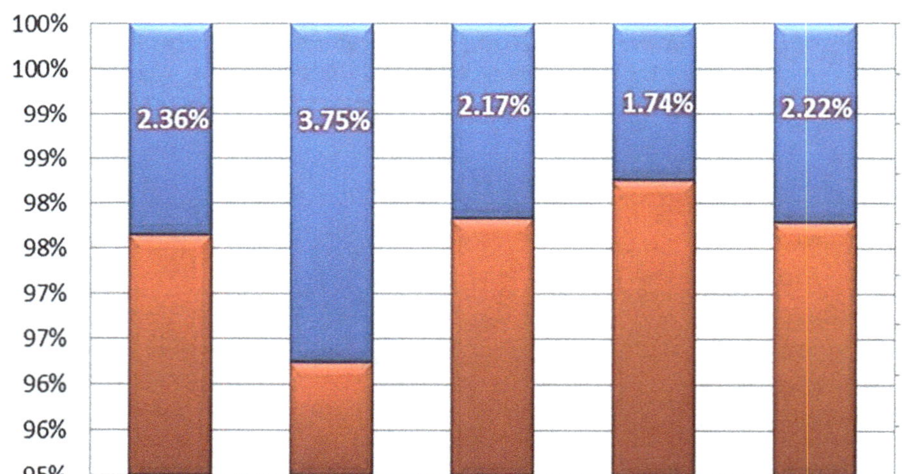

	ATHENS*	BARCELONA	ROME	VIENNA	LOS ANGELES**
Airbnb	148,187	759,812	587,170	231,738	1,400,354
Hotels	6,136,739	19,478,755	26,420,620	13,091,903	61,743,687

Table 2. Properties with amenities, %, April 2016

	Athens	Barcelona	Vienna	Los Angeles
Kitchen	96	92	97	91
Washer/Dryer	72	80	82	78
Air Conditioning	80	53	11	74
Wi-Fi	95	95	97	97
Free Parking	11	4	7	63
Breakfast	14	8	5	8

Source: Airbnb data per AirDNA.
* Los Angeles metropolitan area

So what is causing the hotel premium in Europe? One difference between the hotel markets in the U.S. and Europe is the distribution of independent versus branded hotels. While approximately only one-third of the hotel supply in the U.S. is comprised of independent hotels (Worgull, 2016), this percentage jumps to nearly 60% in Europe (Clampet, 2016). What makes this dynamic important is OTAs' influence on hotels in terms of setting room rates. Independent hotels have a higher reliance on these agencies, and as a result pay higher fees. These rates have only grown over the years. Back in the 1990s when OTAs were growing in popularity, their commissions were around 5%. Today they range from 15% to 30%, and hotel companies with less than 50 properties pay an even higher rate (Clampet, 2016). Companies of this smaller size comprise the vast majority of the European hotel market. When coupling the fact that tourism is on the rise with the cost pressures created by using OTA's, a better sense of why hotels in Europe are not adjusting their prices in response to the P2P lodging platform phenomena is apparent. More people visiting these cities increases the demand for lodging and increased costs from OTAs eat into their margins.

Furthermore, the city of Athens, where hotel premiums are highest, is home to some of the most culturally iconic sites in Europe. Hotels in these culturally rich locations, particularly those in close proximity to these historical sites, are able to maintain and even increase the amount they are able to charge their guest, further widening the gap with P2P lodging platform rates.

Overnight Stays

Two important measures in assessing P2P lodging platforms' threat to the lodging industry are its concentration of the market and its rate of growth. While renting homes to tourists and travelers has been around for decades, the proliferation of this practice is still in its infancy. Therefore, while these single digit percentage rates may seem non-threatening, they are merely the starting point. Furthermore, an understanding of the growth rate in these different locations may provide insights on what this segment of the lodging industry may become.

Barcelona is an interesting case in terms of P2P lodging platform's concentration of the lodging industry. At 3.75% of total overnight stays, it is the highest of the locations under this study. Tourism for the city has experienced exponential growth. Becoming the fifth most visited city in Europe (Lane & Woodworth, 2016) has created pressure on the infrastructure of the city to support that growth. The supply of hotels hasn't been able to keep up. P2P lodging platforms can and likely do step in and absorb the excess supply of overnight travelers. A counter argument might suggest that if that were the case, occupancy rates should rise until the point where hotels are turning customers away. This may well be the case during peak seasons, but this phenomenon likely occurs along different segments of travelers. For example, when budget hotels meet their capacity, the overflows of tourists are not going to naturally move into the premium end of the market, flowing instead into similarly priced P2P accommodations. This may materialize by sacrificing some of the services and conveniences offered by a hotel, but for some travelers, access to a kitchen and washer/dryer may be a welcomed trade-off.

Available and Booked Rooms

Traditional hotels are generally considered special purpose assets, with a fixed inventory of saleable rooms (PWC, 2016b). Of course, new hotel construction will periodically add to supply, but this does not happen overnight. In contrast, the supply of P2P lodging accommodations can fluctuate widely as hosts with minimal barriers to entry add and remove their listings on a whim. While the number of booked Airbnb rooms follows the typical seasonality of a peak during the summer months and a trough thereafter, the growth in the number of available rooms for this sample has steadily risen, irrespective of the time of year. This consistency of supply acts as a sort of assurance to travelers that there will be available supply when they need it. Remarkably, this trend appears to hold true for all of the locations included in this study.

City Results

A detailed analysis of the current lodging industry status across cities and a discussion of the competitiveness variables delineated above create a basis for addressing the main

Figure 28. Airbnb hotel comparable available rooms
Source: Airbnb per AirDNA

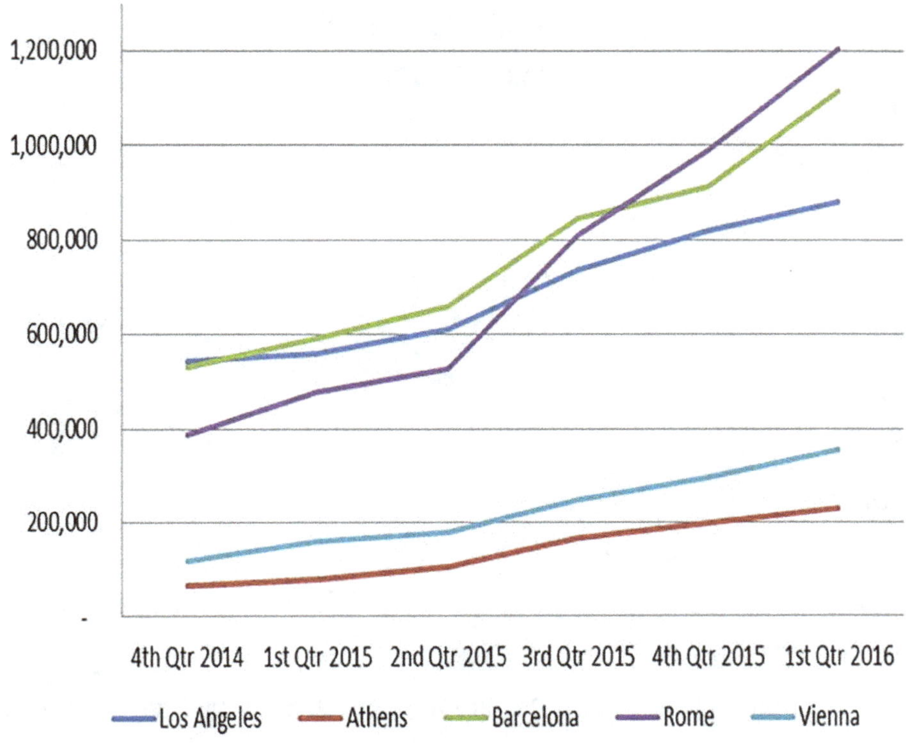

research question on whether P2P lodging platforms competitive or complementary to the traditional lodging industry and what strategies can be forecast for traditional lodging organizations and regulatory entities. Below, city decisions are discussed based on the competitiveness variables.

Los Angeles

The study in Los Angeles based found that P2P lodging platforms are competitive to traditional lodging organizations.

Hotel ADR Premiums/Discounts

Price is not the primary value proposition utilized by P2P lodging platforms to appeal to consumers in Los Angeles. An economic traveler can find inexpensive lodging in a shared or private room, but A-list celebrities easily spend $10,000 a night for

Figure 29. Airbnb hotel comparable booked rooms
Source: Airbnb per AirDNA

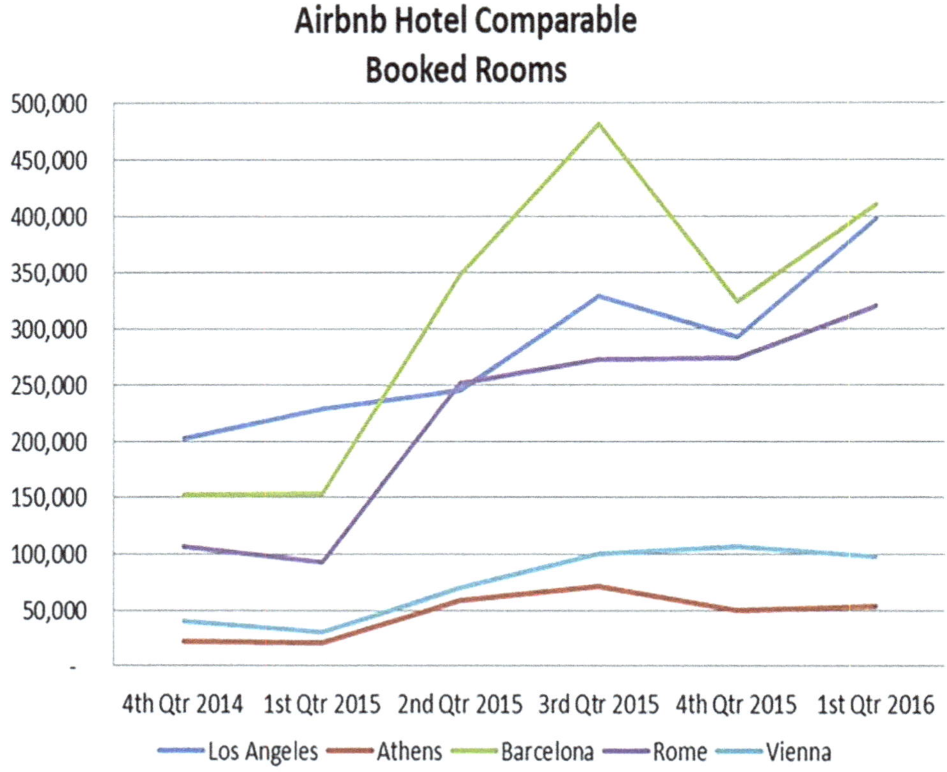

Source: Airbnb data per Airdna

an entire home advertised on Airbnb (Chung, 2016). The implication, therefore, is that this alternative lodging option does not only appeal to a budget focused niche market or any specific market at all. Travelers to Los Angeles utilizing P2P lodging platforms come from a wide social and economic spectrum. These services are competitive to the traditional lodging industry as a whole because both draw from a largely overlapping consumer base representing the majority of the market.

The ADR for hotels in Los Angeles is close to hotel comparable Airbnb listings. In fact, it was $3.88 higher against hotels in 2015 (Doby et al., 2016). Airbnb and mainstream P2P lodging platforms have only been exposed to positive demand growth in Los Angeles (Mohrfeld, 2016). As a result, their ADR may be artificially high.

Figure 30. Hotel ADR premiums 2015 (includes hotel comparable Airbnb listings)
Source: airbnb per AirDNA

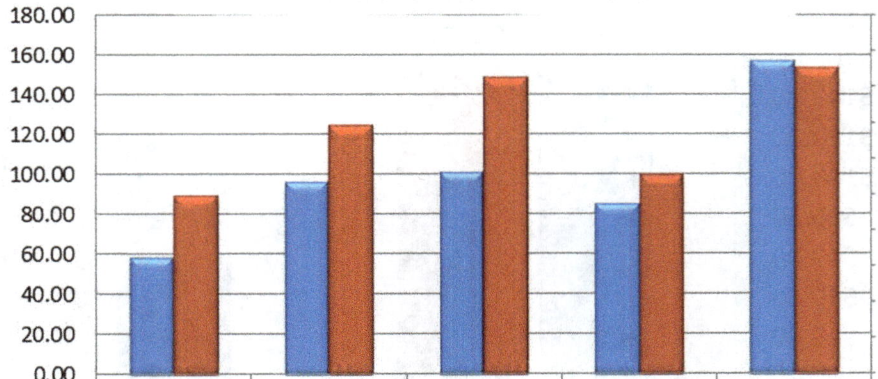

	ATHENS	BARCELONA	ROME	VIENNA	LOS ANGELES*
Airbnb	€ 58.88	€ 96.29	€ 101.53	€ 85.33	$157.38
Hotels	€ 90.00	€ 125.00	€ 149.00	€ 100.00	$153.50
Premiums	52.9%	29.8%	46.8%	17.2%	-2.5%

Additionally, when factoring in the additional taxes, resort fees, and parking charges, staying in a hotel may be even more expensive. Ultimately, while differences in regulatory treatment and labor costs affect the profitability of operating a hotel versus renting out a home or apartment via a P2P lodging platform, both are competing for the same customers across the full spectrum of price (Figure 31).

Available Supply of P2P Lodging Units

It is another significant factor in the Los Angeles market unique to the sharing economy. One strength of P2P lodging platforms was the ability to generate supply from residential units when demand at a destination peaks. Net supply of traditional lodging facilities stayed relatively flat in 2015. Airbnb properties however grew robustly. There were nearly 7,000 more private room and full home Airbnb listings added during 2015, a 7% addition to the Los Angeles County supply, albeit with a previously mentioned caveat; unlike hotels, a portion of P2P lodging units are not available for rent at all times. This means that the true supply of P2P lodging units fluctuates daily. Bookings however are a more appropriate indicator of market share: full home bookings on Airbnb rose 61% while private room bookings rose 91% during the same period. In terms of completed sales, Airbnb properties alone

Figure 31. Los Angeles ADR/RevPar hotel vs. Airbnb*
Source: Airbnb per AirDNA, hotel data per La tourism/convention board

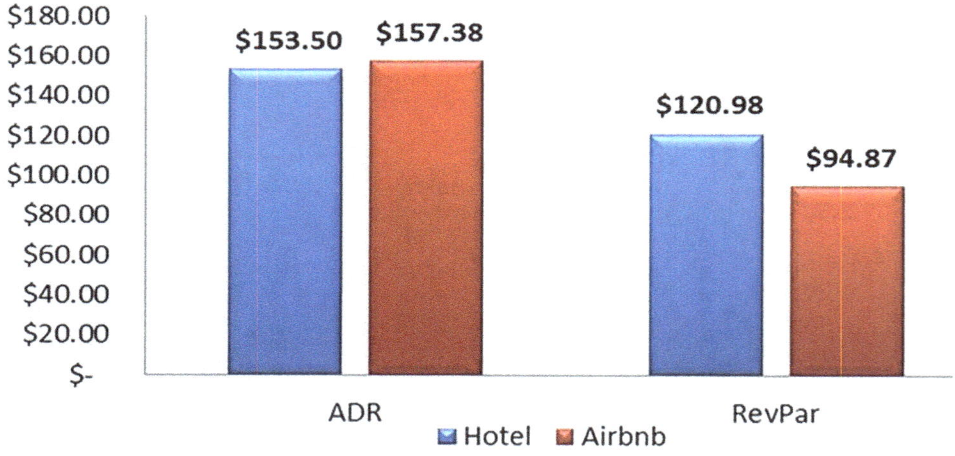

accounted for 1.7% of all listing nights booked in Los Angeles County (Doby et al., 2016). In an industry where even the largest brands only make up 5-6% of a given market (Leporatti, 2016), the rapid integration of a player at 1.7% of the market is a significant competitive factor (Figure 32).

Correlation Between Travel Data and Overnight Stays

The correlation between ancillary travel data for 2015 and traditional lodging overnight stays further supports the competitive nature of P2P lodging platforms. The number of airline passengers arriving in Los Angeles increased 6% in 2015. Total visitors to LA County increased 2.5%. Overnight rooms booked rose 1% (Figure 33). These data suggest that traditional lodging bookings are increasing at a slower rate than overall travel figures in Los Angeles. ADR has risen 7% year over year, however occupancy only grew 1% (Discover Los Angeles, 2016). This resulted in RevPAR gains; however, the increase is more a product of ADR growth rather than occupancy growth. With an increase in travelers to Los Angeles County and a stagnant occupancy, it is plausible that new supply of lodging inventory is being utilized.

Vienna

The study found that P2P lodging platforms are complementary to traditional lodging organizations in Vienna based on the following variables.

Figure 32. Airbnb advertised listings: Los Angeles

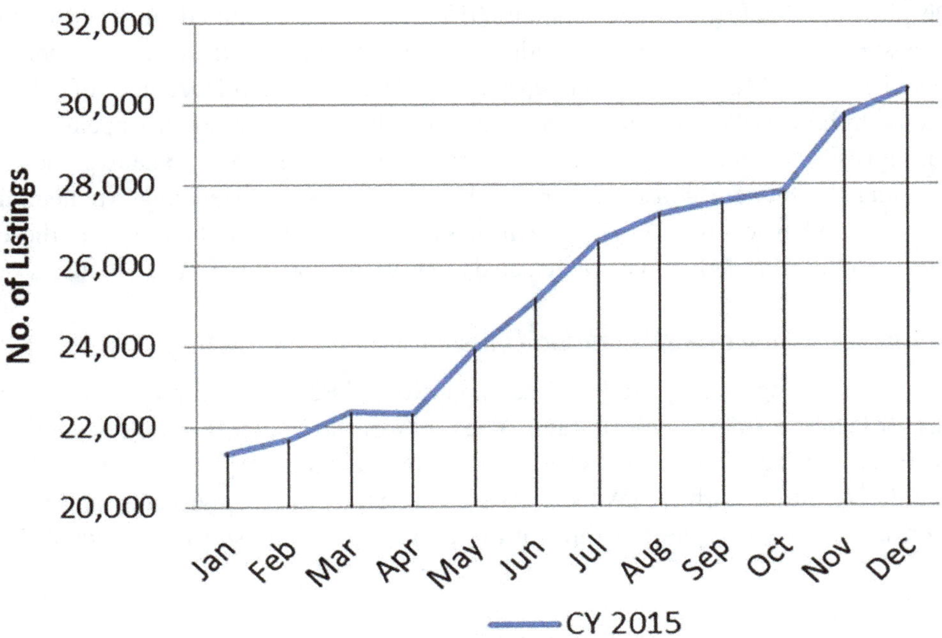

Figure 33. Los Angeles tourist arrivals and overnight stays
Source: tourist arrivals per Euromonitor, hotel overnights per CBRE/PKF

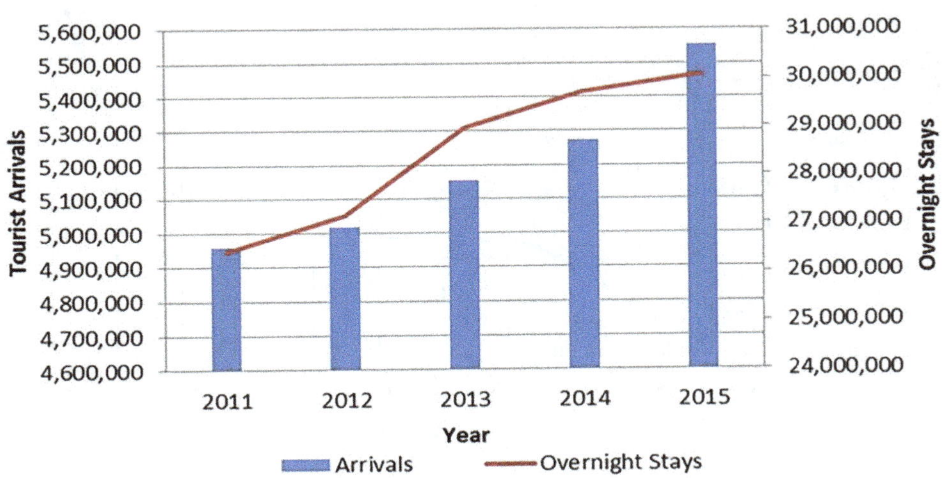

Growth Rate of P2P Lodging Bookings

Despite remarkable growth throughout 2015, the count of Airbnb listings receiving at least one booking during the month appears to be leveling off. In fact, no month this year (as of May 2016), has eclipsed the 4,103 total from December of 2015. Such a trend is typical for niche products when they enter a market. Appealing to a minority segment, they grow for some time, but soon become exhausted never reaching the broader consumer base. While Airbnb is experiencing exponential growth in other locations, a slowdown in Vienna may be an indication that these alternative accommodations have reached their full potential in the city (Figure 34).

Available Supply of P2P Lodging Units

One contributing factor curtailing the expansion of P2P lodging platforms is the lack of legally available units to rent. When 60% of the Viennese people are living in social housing, that is somewhat financed by the state, they are not allowed to rent this room via Airbnb" (Wieser, 2016). Officials try to stop violators and have even launched a campaign to inform the public. Opportunistic renters won't be

Figure 34. Airbnb booked listings: Vienna
Source: Airbnb per airDNA

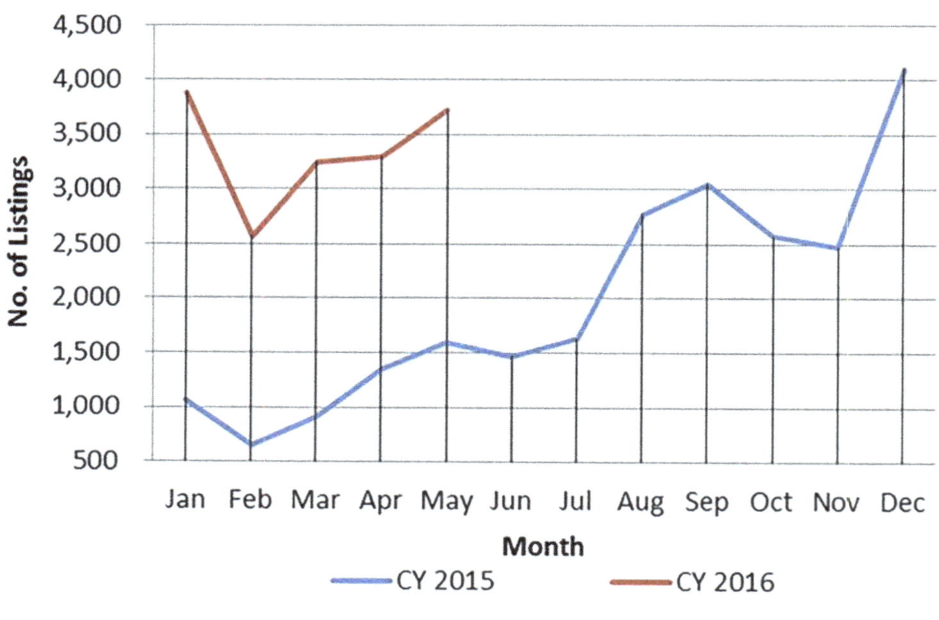

allowed to turn their apartments, partially subsidized with tax payer money, into their own rental property. The number of Airbnb advertised listings has fallen this year (Figure 35), a sign that they are slowly getting the message. With limited supply to accommodate the mass market, P2P lodging platforms will be relegated to compete amongst themselves for a smaller group of consumers.

Hotel ADR Premiums/Discounts

P2P lodging platforms are poised to take advantage of a "low price" value proposition. This is a common strategy used by niche products to attract budget conscience consumers. Measured by ADR, to stay at a hotel visitors to Vienna are paying a 17.9% premium over the cost of an average Airbnb hotel comparable unit. In addition to the savings from room rates, visitors can also take advantage of the fact that 82% of the Airbnb units in Vienna have a washer/dryer, 97% have free Wi-Fi, and 97% have a kitchen (Dobt et al., 2016). Taking advantage of these amenties provides a unique experience different from staying in a hotel. Conversely, tourists are even providing the locals with new experiences. Meanwhjile, while the "live like a local" niche market is alive in Vienna, but in a city known for classical music and luxurious palaces, hotels will continue to dominate the market.

Figure 35. Airbnb advertised listings: Vienna

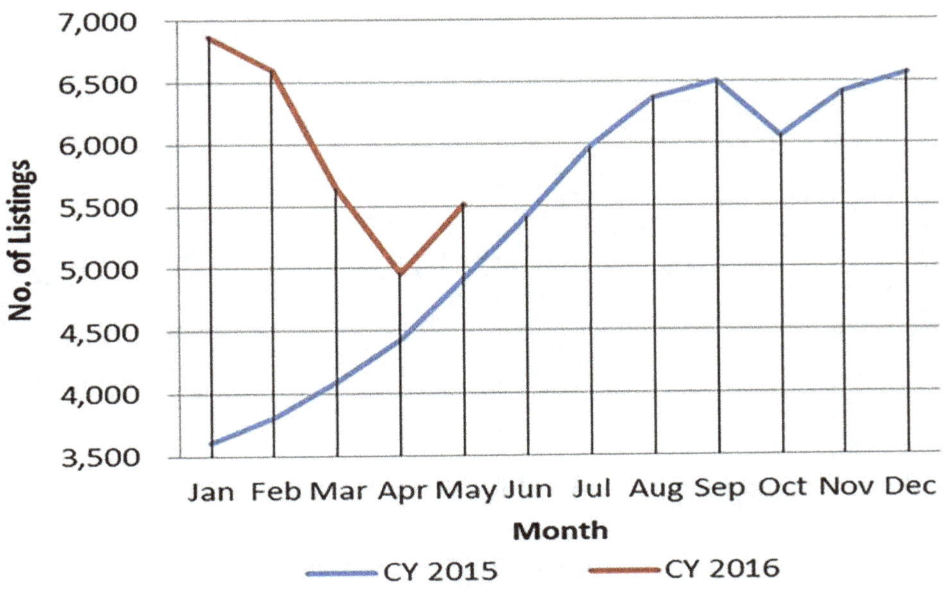

P2P lodging platforms face a challenge in Vienna. Public opposition within the city is strong, with officials going so far as to fine violators breaking regulations. They have even succeeded in getting Airbnb to add a link to the city rules on their website. Growth in tourism has coincided with new development, allowing new hotels to compete for every tourist visiting the city. Airbnb and others must to contend for a smaller segment of the market.

Athens

The study found that P2P lodging platforms are competitive to traditional lodging organizations in Athens based on the following variables.

Growth Rate of P2P Lodging Bookings

The rate of growth in the Airbnb listings, receiving at least one booking during the month, has shown no indication of slowing down. In comparing the upward trajectory for the first 5 months in 2016 to the same period in 2015, it appears to be growing at an even faster rate (Figure 36). The economic crisis in Greece has created the tourist perception that Athens is a cheap place to visit right now. As room rates within the

Figure 36. Airbnb booked listings: Athens
Source: Airbnb per AirDNA

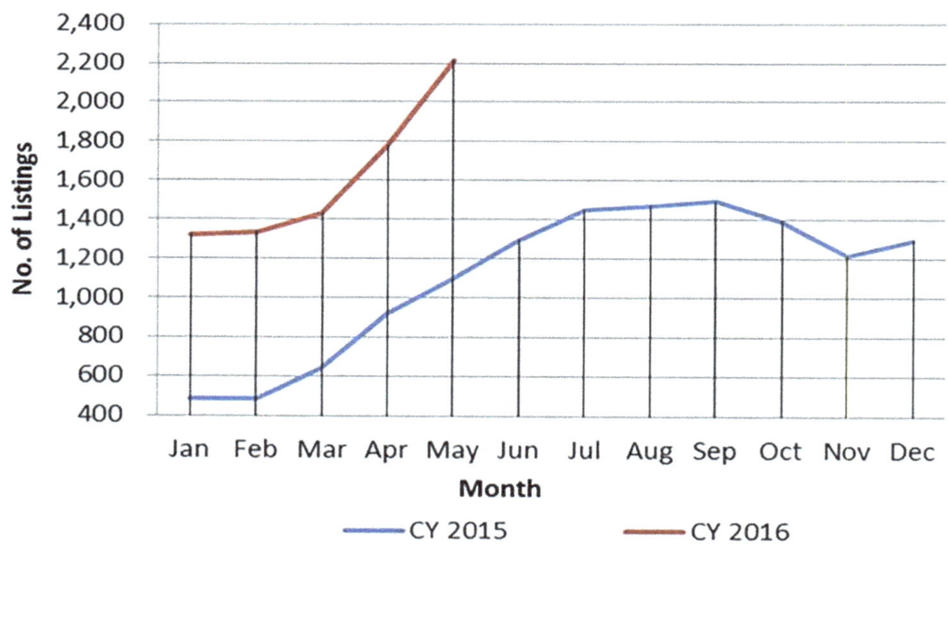

hotel industry increase in response to the influx of tourism, this perception slowly dissipates. Relatively inexpensive P2P lodging units, however, bring that perception back in view. Meanwhile, there is a concern that before long these rentals could be indistinguishable from hotels and are going to be like a hotel because of the trends toward all-encompassing standardization and a diminishing feeling of staying in a Greek home: in the end, it will be not a Greek home, but just a special apartment for a tourist (Christidou, 2016).

Available Supply of P2P Lodging Units

One contributing factor to the success of P2P lodging platforms in Greece is the bountiful supply of rentable units. The country has one of the highest rates of homeownership in the EU and about half of the country's citizens own a second home. This creates a vast pool of potential listings for P2P lodging platforms and enough supply for becoming the dominant market segment. The hardships of economic crisis have pushed many Greeks toward using their home to generate extra income and may be the only option some to pay their mortgage and respond to an increasing property tax burden. Considering that, many of them also have a second home, and subsequently two tax payments to make, renting one of them out to pay the bills looks rational.

Ratio Between Entire Place Rentals vs. Shared or Private Rooms

Many Greeks have a second home and over 84% of them are for use of an entire place, home or apartment (Figure 37). They need not worry about where to go if they rent out their home; they can just go to their other one. With that, the P2P market segment appears very capable of direct competition with hotels. Sharing a room or renting a private room in someone's home is a niche market. For some, the lack of privacy would be too much to bear. Entire place listings provide a much more comparable experience to what one would expect from a traditional accommodation. Furthermore, as complementary services emerge, like Laundrapp, which provides "hotel-quality laundry hire and cleaning services for Airbnb hosts," visitors to Athens may soon not even know the difference (Laundrapp, 2016).

Correlation Between Travel Data and Overnight Stays

Despite the country's economic woes, the travel and tourism industry in Athens is booming. At the rate tourists are visiting the city hotels should have to turn consumers away. Instead, for three months in late 2015, they actually experienced a decline in

Figure 37. Airbnb listings by type: Athens May 2016
Source Airbnb per AirDNA

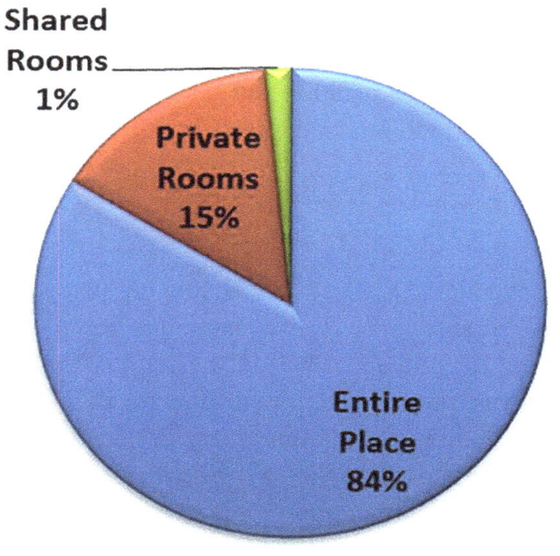

their occupancy rates. This significant development suggests that given the choice to stay in either a hotel or a P2P lodging unit, enough visitors chose the latter, that despite an uptick in arrivals, the overall overnight stays in hotels fell.

P2P lodging platforms pose a serious threat to the traditional lodging industry in Athens. These alternative accommodations have the potential to overthrow the hotel industry to become the dominant segment of the market. International hotel brands do not have a strong presence in the city. Instead, the market is dominated by family owned businesses whose proprietors often "do not have a degree that supports the operation of such lodging" (Theofanopoulou, 2016). They are, as a result, not properly equipped to suppress the threat posed by this emerging phenomenon. In fact, they are at a complete disadvantage. Labor costs, taxes, and the expense associated with the city regulations compliance, all create a burden not borne by their emerging competitor.

Barcelona

The study found that P2P lodging platforms are complementary to traditional lodging organizations in Barcelona based on the following variables.

Growth Rate of P2P Lodging Booking

The slowdown in growth of P2P bookings in Barcelona is indicative of a market exhausting the supply of its potential customers: the 2015 booking volume increased rapidly in the beginning of the year and peaked during the summer months. The growth in 2016 has shown only modest gains. The market for P2P lodging platforms has not been exhausted yet, but it may be approaching that point. P2P lodging platforms allow consumers to select units with a variety of amenities and locations that can differentiate them from hotels. The ability to cook your own meal in a foreign place is important for travelers. Location is also very important, and travelers want to be in a unique neighborhood that they might not enjoy by staying at a hotel. For these reasons, P2P lodging in Barcelona provides a unique value proposition to a particular type of traveler. These travelers, however, may not represent the majority of market and thus P2P bookings have slowed down as it reaches the point of exhaustion (Figure 38). Time will tell if P2P lodging in Barcelona has settled down completely, but for now, the significant drop suggests that, these alternative accommodations will remain on the fringe of the market, complementing the lodging industry as a whole.

Figure 38. Airbnb booked listings: Barcelona
Source: Airbnb per AirDNA

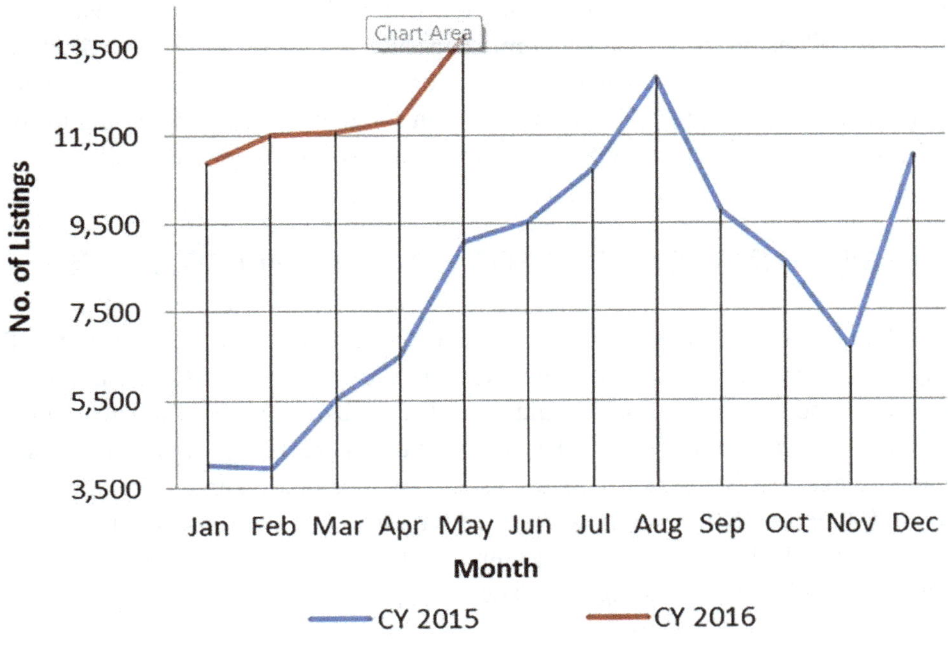

Ratio between entire place rentals versus shared or private rooms - 57% of travelers in Barcelona booked an entire place and 57% is more than half, the lowest percentage across the cities studied. This is a significant finding as the percentage of travelers who stayed in private or shared rooms almost matches that of entire places. This is a more balanced distribution of the different types of units. Barcelona P2P lodging platforms offer social benefits and a "local experience" differing from a hotel. Consumers want the authentic local experience when they travel now and the most preferred and best way to accomplish that is to live in a local's house, just like a local would, and Barcelona P2P lodging platforms are achieving this. Travelers are able to get insider recommendations from their host, meet and interact with someone who is part of the Barcelona community, and cook and do laundry like an ordinary member of the community. This consumer experience is completely different from a hotel stay. This leads to conclude that P2P lodging platforms are complementary because travelers are using them in Barcelona for a unique value proposition.

Hotel ADR Premiums/Discounts

There is a significant difference in the ADR between hotels and P2P lodging platforms in Barcelona. Airbnb ADR was €96.29 in 2015, which was lower than hotels' ADR of €125 during the same time. This represents a 29.8% pricing premium for hotels. The lower price point is an advantage that P2P lodging platforms have over hotels, and tourists often choose the lower price when looking for accommodation in Barcelona. Consumers face an economic benefit when choosing to stay with P2P lodging platforms instead of hotels in Barcelona. Travelers often decide on where to stay based on the lower price. Therefore, in Barcelona, P2P lodging platforms providing a different value proposition appear to play complementary role.

STRATEGIC IMPLICATIONS FOR HOTEL COMPETITIVENESS

Despite numerous challenges hoteliers must consider due to the emergence of P2P lodging platforms, there are opportunities that these lodging organizations can capture to stay competitive in the shifting industry. The industry will continue to change because this change is an organic part of evolution where the consumers' needs are changing and lodging organizations must adjust. The controversies around these new P2P lodging platforms will be resolved with time, but the consumer will never stop evolving. If the issue were not centered on the current P2P lodging platforms, then it would be on something else. The most productive mindset they can have

*Figure 39. Barcelona ADR/RevPar Hotel vs. Airbnb**
Source: Airbnb per airDNA, hotel data per PWC forecast

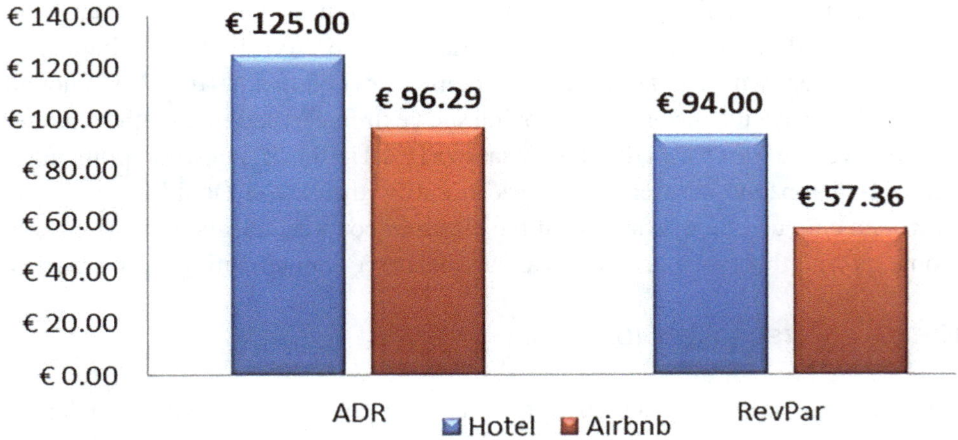

is focusing on making themselves better (Pertesis, 2016). Having a positive and proactive outlook will aid hoteliers in facing new industry challenges in stride. This offers the following recommendations on how hoteliers can stay competitive as the lodging industry and consumers evolve.

Brand Positioning

Positioning is one of the most critical elements of a company's marketing strategy. How a consumer compares a brand versus the competition and what value they assign to a brand on recall, is critical for business. For this reason, it is incredibly important to know the targeted consumers and understand their changing needs. The lodging consumer is now more than ever, well educated, well researched, price sensitive, socially aware, technology savvy and accessible among many other developments. To maintain a strong position with their consumers, hotels should:

Create "Instagramable" Moments

The establishment of P2P lodging platforms is the result of the consumer demanding a different lodging experience. They want to "live like the locals," be adventurous and share their experiences through social media. Therefore, hoteliers should understand the motivation of the guest's visit and provide those "instagrammable" moments for them. Matthew Lehman of the Fairmont Miramar in Santa Monica spoke on how his hotel has been able to collaborate with extreme adventure tour companies to create "instagrammable" activity packages for their guests that are unique to their hotel experience. Additionally, upon request, guests can have a stylist visit their bungalow to provide them with the latest trends in swimwear (Lehman, 2016). While not all hotels might have the resources to provide such extreme experiences, other smaller moments can be created. At the H10 Casanova Hotel in Barcelona, a complimentary glass of champagne is offered at check in and at the Hilton DoubleTree a fresh chocolate chip cookie is available at the check in counter. Guests will share these moments with their social networks and associate your hotel with great memories.

Create a Personality/Story

Hoteliers may have no control over the changing consumer or evolving industry, but they do have the power to control their "house." To withstand the instability of any industry, a reason for why a business exists is important to portray and project. How do you differentiate yourself from your competition? The COCO-Mat Hotel Athens provides a great example. Their philosophy is to help their guests live well by providing them a place to eat well and of course, sleep well (Pertesis, 2016). Exceptional sleep systems and furniture, all handcrafted from sustainable materials create COCO-Mat's attractive personality (COCO-MAT, 2016). Hilton Hotels and Resorts recently established a new brand of hotel entitled, Tru. Offering a unique personality, Tru positions itself with the desires of today's new traveler and becomes a category disrupter. It's built from a belief that being cost conscious and having a great stay don't have to be mutually exclusive. Tru focuses on what matters most to guests, with a hotel that is more than just a place to sleep, it's a true travel experience" (Tru by Hilton, 2016). Consumers will know that when they stay at a Tru location, they will receive a safe, affordable and great experience (Speer, 2016).

Be Flexible

A major strength of P2P lodging platforms, besides their lower average daily rate compared to most hotels, is their ability to satisfy an array of custom requests. Such requests can be recommendations for local restaurants and events, laundry services, space to fit a large family or kitchen facilities. This personalized attention is what keeps consumers of all demographics coming back to P2P lodging platforms. Therefore, a hotelier should be open and willing to accommodate the changing requests of consumers. Leaving a personalized card in guest rooms, offering late check-out when available and ensuring the staff provides the best customer service are just some examples of being flexible (Socialdraft, 2017). Michaela Reitterer of the Austria Hotel Association emphasizes the importance of catering to families. As groups and families are among the majority of P2P lodging platform users, in order to stay competitive and attractive hotel facilities in her Association always strive to accommodate families to the best of their ability by reserving adjoining rooms and even offering discounts for children (Reitterer, 2016). Matthew Lehman of the Fairmont Miramar Hotel in Santa Monica summarizes this concept: "we can fight it all we want, but instead why don't we accept it as our new reality and try to understand where it is going to take us and figure out how to be successful within it" (Lehman, 2016).

Internal Business Operations

While altering the mind of a consumer may not be the easiest task, a hotelier can alter internal business operations with a greater success rate. From a small adjustment on a website to a large implementation of new systems, this may be what a hotelier needs to stay current in the ever-changing lodging industry. Below are a few internal business operation recommendations that can be optimized for greater industry success.

Strategize for More Website Direct Bookings

As mentioned earlier, online travel agencies (OTAs) emerged with the arrival of the Internet and progressed with today's new traveler. Scanning the Internet, platforms such as Priceline, Travelocity, Expedia and Hotels.com claim they can provide the best hotel rates when booking through them. These OTAs though can charge up to a 33% booking fee, which must be paid by the hotels (Speer, 2016). This fee can make a significant impact on a hotel's profit. Additionally, these platforms can lead to consumer confusion, "sometimes consumers think they are booking reservations that can be canceled without charge, only to find out later that is not the case - or they arrive at a hotel to discover that they do not actually have reservations" (Tugend, 2016). This confusion can instill a negative customer perception of a hotel, regardless if the hotel is responsible or not. Therefore, while today's traveler might view these OTAs as providing a stronger value than booking directly with a hotel, it behooves hoteliers to strategize how they can pull away from OTAs and gain more control over their bookings. Marriott Hotels and Hilton Hotels and Resorts have already begun campaigns encouraging consumers to book direct with their hotels. Offering incentives, such as free Wi-Fi, special rates and fast check in are enticing consumers to book direct and place trust in a hotel's website (Tugend, 2016).

Invest in Consumer Reviews

Historically, when a consumer would search for a hotel they would filter their options based on location, luxury ranking and price. Today's traveler however, is more likely to first look at a hotel's consumer reviews before considering location or price (Di Baia, 2016). Consumers trust each other more than industry driven categorization, such as luxury rankings. Di Baia, owner of the Blue Hostel, he states that often hotels that would be considered a 5-star hotel in the industry will have lower average consumer reviews than his hostel. This, Di Baia claims, is his competitive advantage. Their mission is guest satisfaction, not money, and it shows, with an average consumer review of 4.5 stars out of 5 and a "Certificate of Excellence" on TripAdvisor (Blue Hostel, 2016). To achieve such high ratings, The Blue Hostel provides personalized touches from the moment a guest books to the moment that guest departs. Gestures such as travel recommendations, customized city maps, multi-country outlet adapters, coffee machines and updated accommodations keep The Blue Hostel ranking as a favorite among guests (Di Baia, 2016). Consumer reviews can also increase a hotel's search engine ranking. When search engines, such as Google, perform their search engine optimization, they crawls the web for mentions of a specific search term and verified links from other websites. The more reviews a hotel has on trusted review sites such as TripAdvisor, the higher chance that hotel will appear in a consumer's search results. Ultimately, leading to a higher chance a consumer will book with that hotel. Consumer reviews are gaining more power in the purchasing process and are proving to have a high return on investment.

Establish Brand Loyalty

Brand loyalty is the "Holy Grail" of marketing (Ayyar, 2014). Consumers who are loyal to a brand will spend more than first-time consumers. They are less price sensitive and are brand advocates (Vides, 2014). Patrick Speer of Hilton Worldwide explains how Hilton Hotels and Resorts has created a brand portfolio for their consumers. By offering accommodations in the lower end market all the way to the highest end market, Hilton can capture consumers at an early stage and let them progress through their brands (Speer, 2016). Establishing the Hilton HHonors program has encouraged consumers to stay loyal to Hilton through perks such as discounts, rewards and special offers. Speer shared that Hilton HHonors members will travel an extra 20 minutes to stay at a Hilton property just so they can receive their perks or points and something catastrophic has to happen to break their loyalty (Speer, 2016). These types of loyalty programs are a great way to inspire repeat business and have been proven effective, but for boutique or family-run hotels, these resources might not be available. To maintain loyalty with limited resources, Entrepreneur Magazine suggests 3 practices. First is constant innovation. For a hotel, this might be something simple to installing Wi-Fi throughout the hotel or implementing keyless room entry technologies. Being open to new trends in the industry and listening to consumers' requests will make a positive impact on consumers. The second practice is hiring positive and capable employees and keeping them happy. The staff is responsible for portraying the mission and goals of the organization. It is the staff's hands to determine how enjoyable a consumer's stay will be. This leads to the third practice of, excellent customer service. How a guest is greeted or how an issue is resolved will influence a consumer's perception of a hotel. If their experience has exceeded their expectations, they are more likely to return to that hotel and become a loyal guest (Ayyar, 2014). Since repeat consumers can be the largest revenue generators, it is imperative to invest in brand loyalty practices.

Embrace Technology

A survey by YouGov found that 89% of hotel guests want hotels to adopt "stay-enhancing technology." Hotel guests concluded that technologies such as mobile check in, keyless room entry and the capability to order room service all through smartphones would save them time and enhance their stay experience at a hotel. Of the 89% of participants who yearn for new technologies, 39% said that they would be more likely to book with a hotel that had these technologies, over a hotel that did not (Worgull, 2016). This is significant in exerting strong impacts on revenues. These numbers and the demand for technologies will continue to grow as the consumer and industry forever evolve. It would be foolish not to mention social media as a technology that hotels must embrace as well. Social media includes, "blogs, forums, business networks, photo-sharing platforms, social gaming, microblogs, chat apps, and last but not least social networks." By 2018, it is expected that there will be 2.5 billion users of social media, a third of the Earth's entire population and Facebook alone, captures 1.5 billion users. Globally, the average daily time spent on these networks is almost two hours, creating an opportunity for companies to interact with their consumers on a deeper personal level than ever before (cycles and Text, “Topic,” 2016). Consumers use these networks to connect with peers, share their opinions, research topics and stay informed with world events. The Pew Research Center found that, “half of all adult Facebook users have more than 200 friends in their network...and the average (mean) number of friends is 338” (Cocco, 2016). Meaning, when a user shares their opinion on Facebook, it will reach 200-338 other impressionable individuals. If a user posts about their experience with a company, it could lead to a positive or negative impact for that company. Matthew Lehman of the Fairmont Miramar Hotel views social networks as a free marketing opportunity (Lehman, 2016). Of course, he would prefer posts about his hotel be positive as he understands the impact of P2P recommendations, so he managers his hotel aiming to provide guests with positive content to post on social media. The numbers for Facebook alone, provide reason for why a company must invest in social media. Through social media, a company can monitor consumer opinion of their business, respond to negative and positive feedback, connect with potential consumers and share promotions, special offers and their company values.

Optimize Current Strengths

Throughout this research a myriad of hotel strengths were discovered. Strengths that resonated the most were safety, consistency, quality service, cleanliness, amenities and standards. When a consumer reserves a hotel room these attributes are expected, however, when reserving a P2P lodging accommodation, these attributes are not necessarily what come to mind first. Hotels offer a wider range of benefits for the consumer and this should be optimized. These strengths are what will keep consumers staying at hotels. Therefore, it is essential for a hotel to promote them, ensure they are performing at their highest capacity and update them when necessary. Also, facilitate feedback from guests and adjust when consumers are demanding a change. Another under-recognized strength of hotels and a benefit hotel guest receive daily is housekeeping. P2P lodging platforms typically do not provide this benefit, ultimately, creating an opportunity for hotels. Depending on the situation, it might befit a hotel to utilize their housekeeping staff and provide daily housekeeping for P2P units.

The lodging industry will continue to face new challenges, new competition and an evolving consumer, regardless of enforced regulations. The recommendations above will however, help hoteliers be open to consumer feedback and stay relevant in the lodging industry. Studying the consumer, understanding motivations and following through on requests will allow the hotelier to stay ahead of their competition.

FUTURE RESEARCH DIRECTIONS

Since the early 2000s, when the sharing economy started to proliferate, P2P lodging platform has been part of a stream of technological innovations and socio-economic changes with disruptive impacts. This study is among the first comparative research projects involving multiple countries, stakeholders, and subsectors in the lodging industry. The study found that P2P lodging is emerging as a strategic alternative complimentary to the traditional lodging industry in some countries; for other countries, the study found it as a competitive factor to the traditional lodging industry. Countries wary in their attractiveness to consumers in the global lodging industry, their political-economic differences and cultural attitudes toward the sharing economy. In this context, the study revealed that the emergence and growth of P2P lodging has affected different groups of stakeholders in the lodging industry differently with dissimilar cumulative country-specific and stakeholder-specific effects.

The study offers the following directions for future research.

1. Competitiveness vs. complementarity of the P2P lodging platform against traditional lodging platform across countries, global regions, and market segments.
2. Positive and adverse impacts of the P2P platform on the key stakeholder groups in the lodging industry (travelers seeking accommodation, traditional hotel operators, municipal government agencies, short-term property renters/ landlords, local community interest groups).
3. Socio-economic impacts of the P2P lodging platform on adjacent industries (individual and business travel, car rentals, etc.).
4. Social impacts of the P2P lodging platform: availability of affordable housing for different demographic segments.
5. Economic impacts of the P2P lodging platform: travelers (pros and cons comparative to traditional lodging) and renters/landlords.
6. Municipal impacts of the P2P lodging platform: municipal taxes, budgets, law enforcement, noise levels, crime, water, waste management, energy, transportation and other infrastructure.

CONCLUSION

The Travel and Tourism industry is booming worldwide. Hotels are seeing a rise in occupancy rates and are generating record ADR and RevPAR. As a result, the true disruptions caused by P2P lodging platforms are masked within the growth of the hospitality industry as a whole. The outperformance of these various metrics may actually be underperforming relative to where they could be if P2P lodging platforms did not exist. As Thomas Caton, Chief Revenue Officer for Airdna, noted

there is this conviction that we see a lot in the hotel industry that Airbnb people, if there wasn't Airbnb, [would] be sleeping on a couch, sleeping on the street, [or] not going on the trip. If you accept that some proportion of people would have made that trip otherwise, then you have to accept that those people would have been staying in a hotel and that that has in some way affected either your occupancy or the rate you can charge in different areas (Caton, 2016).

It is undeniable that some percentage of travelers who have chosen to stay in P2P accommodations would have otherwise stayed in a hotel. Whether that percentage continues to grow will largely depend on how the industry and its regulators respond to this phenomenon. The dynamics differ from location to location, as hoteliers must decide whether or not to view P2P lodging platforms as competitive or complementary and legislators must decide whether or not to create laws that curtail its growth or promote its future. Hotels can only be successful by adjusting to changing consumers trends and regulation can only be effective when enforcement creates compliance. The latter may poise the more serious threat, as it is the nature of a "shadow economy" to elude governmental control. Among all of the uncertainty, city officials, academics and hoteliers across the globe can agree on one thing; P2P lodging platforms are making a major impact and they are here to stay.

ACKNOWLEDGMENT

The original study was conducted in 2015-2016 by a group of MBA students from Loyola Marymount University as part of their Comparative Management Systems program. The group included Jonathan Dell, DaVion Doby, Brittany Dupuy, Biniam Fassehatsion, Weixing Li, Bonnie McAfee, Axel Messager, Joshua Tillipman, and Anatoly Zhuplev (group faculty advisor). The authors would like to express appreciation for the group members' contribution to the study.

REFERENCES

2016Q1 Greek Hospitality Industry Performance. (2016). GBR Counsulting.

21.10.010 - Title. | Municipal Code | Long Beach, CA | Municode Library. (n.d.). Accessed July 27, 2016. https://www.municode.com/library/ca/long_beach/codes/municipal_code?nodeId=TIT21ZO_CH21.10GEPR_21.10.010TI

21.15.360 - Bed and Breakfast Inn. | Municipal Code | Long Beach, CA | Municode Library. (n.d.). Accessed July 27, 2016. https://www.municode.com/library/ca/long_beach/codes/municipal_code?nodeId=TIT21ZO_CH21.15DE_21.15.360BEBRIN

721: Accommodation. (n.d.). Accessed July 29, 2016. http://naics-codes.findthedata.com/l/107/Accommodation

7211: Traveler Accommodation. (n.d.). Accessed July 29, 2016. http://naics-codes.findthedata.com/l/408/Traveler-Accommodation

721110: Hotels (except Casino Hotels) and Motels. (n.d.). Accessed July 29, 2016. http://naics-codes.findthedata.com/l/1738/Hotels-except-Casino-Hotels-and-Motels

72119: Other Traveler Accommodation. (n.d.). Accessed July 29, 2016. http://naics-codes.findthedata.com/l/638/Other-Traveler-Accommodation

721191: Bed-and-Breakfast Inns. (n.d.). Accessed July 29, 2016. http://naics-codes.findthedata.com/l/1740/Bed-and-Breakfast-Inns

Aaron Smith92. (2014). 6 New Facts about Facebook. *Pew Research Center*. Retrieved from http://www.pewresearch.org/fact-tank/2014/02/03/6-new-facts-about-facebook/

Airbnb. (2016). *Airbnb Economic Impact*. Retrieved on February 28, 2016 from http://blog.airbnb.com/economic-impact-airbnb/

Airbnb Market Report - April 2016 (AirDNA). (2016). AirDNA.

Airbnb. (n.d.). *About Us - Airbnb*. Retrieved on August 2, 2016 from https://www.airbnb.com/about/about-us

Airbnb. (n.d.). Who We Are, What We Stand for. *The Airbnb Blog - Belong Anywhere*. Retrieved on October 3, 2013 from http://blog.airbnb.com/who-we-are/

American Hotel & Lodging Association (AHLA). (2015). *Lodging Industry Trends 2015*. Retrieved from https://www.ahla.com/sites/default/files/Lodging_Industry_Trends_2015.pdf

Anastasia. (2015). An Introduction to Sharing Economy. *Cleverism*. Retrieved from https://www.cleverism.com/introduction-to-sharing-economy/

Ayyar, R. (2014). The 3 Characteristics of Brands That Command Customer Loyalty. *Entrepreneur*. Retrieved from https://www.entrepreneur.com/article/237364

bornintheusa617. (2013). *Vienna - City of Dreams*. Accessed July 27, 2016. https://www.youtube.com/watch?v=AFZBOTgL_Hk

Botsman, R. (2013). The Sharing Economy Lacks A Shared Definition. *Co.Exist*. Retrieved from http://www.fastcoexist.com/3022028/the-sharing-economy-lacks-a-shared-definition

Brownlow, M. (2016). *Is There a Hotel or Tourism Tax in Vienna?* Retrieved February 4, 2016 from http://www.visitingvienna.com/accommodation/hotel-tax-vienna/

Cable News Network (CNN). (2015). *World's Most Visited Cities*. Retrieved on July 30, 2016 from http://edition.cnn.com/2015/01/29/travel/gallery/most-visited-cities-euromonitor/

Cederholm, T. (2014). *Hilton's Market Share in the Domestic and International Market - Market Realist*. Retrieved from http://marketrealist.com/2014/10/hiltons-market-share-domestic-international-market/

Chandler, J. (2016). Los Angeles Is One Step Closer to Regulating Airbnb. *Curbed LA*, Retrieved from http://la.curbed.com/2016/6/23/11985210/los-angeles-airbnb-law-regulation-legal

Chung, G. (2016). *Justin Bieber Stays at Beyoncé's $10,000-a-Night Airbnb | Celebuzz*. Retrieved on August 11, 2016 from http://www.celebuzz.com/2016-03-18/justin-bieber-beyonce-airbnb-home-house-mansion/

Clampet, J. (2016). *Everything You Wanted to Know About the Hotel Industry's Gripes Against OTAs – Skift*. Retrieved on July 23, 2016 from https://skift.com/2016/04/25/everything-you-ever-wanted-to-know-about-hotel-industrys-complaints-against-otas/

Close, K. (2016). *Amazon Is Cracking Down on Fake Product Reviews - Yahoo Finance*. Retrieved from http://finance.yahoo.com/news/amazon-cracking-down-fake-product-141618091.html

COCO-MAT. (2015-2018). *About Us*. Retrieved July 18, 2016 from http://www.coco-mat.com/us_en/company/

Crowe, P. (2016). *It's A Deal: Marriott Is Buying Starwood Hotels for $12.4 Billion*. Retrieved from http://www.businessinsider.com/marriott-and-starwood-shareholder-vote-2016-4

Desks Near Me | Find a Space to Work. (n.d.). *Desks Near Me*. Accessed August 1, 2016. https://desksnear.me/en

Discover Los Angeles. (2016). *Facts About Los Angeles*. Retrieved on March 30, 2016. from http://www.discoverlosangeles.com/content/ares-header-nav.html

Discover Los Angeles. (2016). *Research*. Accessed July 3, 2016. http://www.discoverlosangeles.com/tourism/research

Doby, D., Fessehatsion, B., & Tillipman, J. (2016). *LMU MBA Hotel & Airbnb Data Tables*. Academic Press.

Euromonitor International. (2017). *Lodging in Greece*. Retrieved from http://www.euromonitor.com/lodging-in-greece/report

Euromonitor International. (n.d.). *Top 100 City Destinations Ranking*. Accessed June 30, 2016. http://go.euromonitor.com/rs/805-KOK-719/images/TCD-presentation_FINAL.pdf

Eurostat. (2016). *Tourism Statistics - Annual Results for the Accommodation Sector - Statistics Explained*. Retrieved July 17, 2016 from http://ec.europa.eu/eurostat/statistics-explained/index.php/Tourism_statistics__annual_results_for_the_accommodation_sector

FEVITUR. (n.d.). Salvetti & Llombart, and ESADE. *Study of Holiday Rentals*.

First Research. (2015). *Lodging Industry*. Author.

Flughafen Wien - Passagiere. (2018). Retrieved on July 26, 2016 from http://www.viennaairport.com/

Freelancer - Hire & Find Jobs. (2016). Retrieved August 1, 2016 from https://www.freelancer.com/?t=a&utm_expid=294858-448.vN25f6toQF6x9f65mvAcgg.1&utm_referrer=https%3A%2F%2Fwww.google.com%2F

Freitag, J. (2016). Non-Branded Hotels Outperforming the Competition. *Lodging*. Retrieved from http://lodgingmagazine.com/non-branded-hotels-outperforming-the-competition/

Ganezer, D. (2016). Rent AIRBNB, Go To Jail: Subletting Five Santa Monica Apartments, Gets Man Misdemeanor Conviction. *Santa Monica Observer*. Retrieved from http://www.smobserved.com/story/2016/07/11/news/rent-airbnb-go-to-jail-subletting-five-santa-monica-apartments-gets-man-misdemeanor-conviction/1594.html

Generalitat de Catalunya. Departament d'Empresa i Ocupacio. (2015). *European Tourism Forum - Luxembourg 2015*. Author.

globalEDGE. (n.d.). *globalEDGE: Hospitality and Travel*. Retrieved from http://globaledge.msu.edu/industries/hospitality-and-travel

Gozzer Arias, S. (2015). Barcelona Cracks down on Illegal Subletting through Airbnb. *EL PAÍS*. Retrieved from https://elpais.com/elpais/2015/09/01/inenglish/1441115926_651764.html

GTP. (2015). The 'Sharing Economy' in Tourism Deprives Greece of Millions of Euros. *GTP Headlines*. Retrieved from http://news.gtp.gr/2015/10/08/sharing-economy-tourism-deprives-greece-millions/

GTP Editing. (2015). Athens: Hotel Occupancy Rates Decline Despite Rising Visitor Arrivals. *GTP Headlines*. Retrieved from http://news.gtp.gr/2015/12/08/athens-hotel-occupancy-rates-decline/

Hilton Worldwide. (n.d.). *Hospitality Company | About Hilton Worldwide*. Accessed July 29, 2016. http://www.hiltonworldwide.com/about/

History of Rail Transport. (2016). In *Wikipedia, the Free Encyclopedia*. Retrieved from https://en.wikipedia.org/w/index.php?title=History_of_rail_transport&oldid=729427314

HNN. (n.d.). Accessed July 21, 2016. http://www.hotelnewsnow.com/Articles/26920/Infographic-Stay-enhancing-tech-at-hotels

Horwath HTL. (2014). Retrieved from H1-2014-Hotel-and-Tourism-Markt-Report-Vienna-Salzburg-Austria (1).pdf

Hospitality Net. (n.d.). Hospitality Net - Europe Hotel Industry Passes U.S. in Growth for the First Time since 2010. *Hospitality Net*. Accessed July 17, 2016. http://www.hospitalitynet.org/news/4074750.html

Hospitality Net - Hotels | A Brief History - By Jacques Levy-Bonvin. (n.d.). *Hospitality Net*. Accessed August 2, 2016. http://www.hospitalitynet.org/news/4017990.html

Hospitality_Report_2015_NAIHellas. (2015). Naihellas.

IBISWorld. (2017a). *Global Hotels & Resorts Market Research | IBISWorld*. Retrieved on July 11, 2016 from http://www.ibisworld.com/industry/global/global-hotels-resorts.html

IBISWorld. (2017b). *Hotels & Motels in the US Market Research | IBISWorld*. Retrieved from http://www.ibisworld.com/industry/default.aspx?indid=1661

In Focus- Athens, Greece. (2014). HVS.

Inside Airbnb: Los Angeles. Adding Data to the Debate. (2016). *Inside Airbnb*. Retrieved on July 6, 2016 from http://insideairbnb.com/los-angeles/

InterContinental Hotels Group. (2015). InterContinental Hotels Group PLC. Retrieved from http://www.ihgplc.com/files/results/prelims15/downloads/prelims15_results.pdf

InterContinental Hotels Group. (n.d.). *InterContinental Hotels Group PLC : About Us - Overview*. Accessed July 29, 2016. http://www.ihgplc.com/index.asp?pageid=16

Khouri, A. (2016). Southern California Apartment Rents Are Expected to Continue Rising through 2018. *Latimes.com*, Retrieved from http://www.latimes.com/business/realestate/la-fi-casden-forecast-20160412-story.html

ktv_adittrich. (n.d.). *Vermieten von Wohnungen für touristische Zwecke*. Retrieved on July 26, 2016 from https://www.wien.gv.at/arbeit-wirtschaft/privat-vermieten.html

Labate, V. (2016). *Ancient Journeys: What Was Travel Like for the Romans? | Ancient Origins*. Retrieved on July 29, 2016 from http://www.ancient-origins.net/artifacts-ancient-technology/ancient-journeys-what-was-travel-romans-005189

Lane, J., & Woodworth, M. (2016). *An Analysis of Airbnb in the United States*. Retrieved from http://cbrepkfcprod.blob.core.windows.net/downloads/store/12Samples/An_Analysis_of_Airbnb_in_the_United_States.pdf

Laundrapp. (2018). *Laundry Services for Airbnb Hosts in London | Laundrapp*. Retrieved from August 4, 2016 on https://www.laundrapp.com/business-services-airbnb

Lodging_in_Greece. (2015, Sept. 17). *Euromonitor International*.

Los Angeles Tourism. (2016). Los Angeles Sets New Tourism Record, Announces 45.5 Million Visitors in 2015. *Discover Los Angeles*. Retrieved from http://www.discoverlosangeles.com/press-releases/los-angeles-sets-new-tourism-record-announces-455-million-visitors-2015

mabpfi. (n.d.). *International Organisations and Institutions Based in Austria*. Accessed July 26, 2016. https://www.wien.gv.at/english/politics/international/organisations/overview.html

Market Pipeline Report: Los Angeles County, CA. (2016, July 12). *STR*.

MarketLine. (2015a). *Global Hotels & Motels - MarketLine Industry Profile*. Retrieved from https://store.marketline.com/report/mlohme7151--global-hotels-motels/

MarketLine. (2015b). *Hotel and Motel Industry in Europe*. Author.

MarketLine. (2015c). *Hotel Industry Analysis - Spain.pdf*. Author.

MarketLine. (2015d). *Industry Analysis - Hotels and Motels in U.S.* Author.

Marriott. (n.d.). *About Marriott Hotels | Marriott Corporate Business Information*. Accessed July 29, 2016. http://www.marriott.com/marriott/aboutmarriott.mi

Martin, H. (2016a). *Airbnb and Other Home-Sharing Businesses Have Hotels Worried - LA Times*. Retrieved from http://www.latimes.com/business/la-fi-hotels-airbnb-20160317-story.html

Martin, H. (2016b). Orange County Breaks Tourism Record with 47 Million Visitors in 2015. *Latimes.com*. Retrieved on July 3, 2016 from http://www.latimes.com/business/la-fi-orange-county-tourism-record-20160205-story.html

Mazza, S. (2016). Homeowner Sues Hermosa Beach for Short-Term Rental Ban. *The Beach Reporter*. Retrieved on July 12, 2016 from http://tbrnews.com/news/hermosa_beach/homeowner-sues-hermosa-beach-for-short-term-rental-ban/article_44528f46-4782-11e6-8476-0f90ea03013c.html

Morningstar. (2015). *Hilton Worldwide Holdings Inc. -HLT*. Retrieved from http://api40.10kwizard.com/cgi/convert/pdf/HLT-20160226-10K-20151231.pdf?ipage=10774498&xml=1&quest=1&rid=23&section=1&sequence=-1&pdf=1&dn=1

NAIHellas. (2015). *2015 Hospitality Report*. Retrieved from http://www.nai-hellas.gr/reports/CommercialRealEstate/Hospitality_Report_2015_NAIHellas.pdf

Newcomer, E. (2016). Airbnb Sues San Francisco to Prevent New Rules That Enforce a Law Airbnb Wrote. *Skift*. Retrieved from https://skift.com/2016/06/27/airbnb-sues-san-francisco-to-prevent-new-rules-that-enforce-a-law-airbnb-wrote/

Operation and Impact of the Sharing Economy on the Greek Hotel Industry. (2015). Grant Thornton.

Orcutt, A. (2012). *No. 7 Acropolis, Athens - World's Most-Visited Ancient Ruins | Travel + Leisure*. Retrieved on July 22, 2016 from http://www.travelandleisure.com/slideshows/worlds-most-visited-ancient-ruins/8

Papadimitriou, P. (2014). *In focus: Athens, Greece*. Retrieved from https://www.hvs.com/article/7121-in-focus-athens-greece

Paribas Real Estate, B. N. P. (2016). *Europe - Hotel Market Highlights*. Retrieved from http://www.hospitalitynet.org/file/152006197.pdf

Picker, L., & Isaac, M. (2016). Airbnb Is Said to Be Seeking Funding Valuing It at $30 Billion. *The New York Times*. Retrieved from http://www.nytimes.com/2016/06/29/business/dealbook/airbnb-is-said-to-be-seeking-funding-valuing-it-at-30-billion.html

PricewaterhouseCoopers (PWC). (2015). The Sharing Economy Grows up. *PwC*. Retrieved on August 1, 2016 from http://www.pwc.co.uk/issues/megatrends/collisions/sharingeconomy/the-sharing-economy-sizing-the-revenue-opportunity.html

Profile of Greece. (2016). Accessed July 24, 2016. http://www.realtor.org/intlprof.nsf/68b4eab0087e45ba86256811004ee31f/a4f509a4b38dc99b86257077004efb31?OpenDocument

PWC. (2016a). *Room for Growth: European Cities Hotel Forecast for 2015 and 2016*. Retrieved from http://www.pwc.com/gx/en/hospitality-leisure/pdf/pwc-room-to-grow-european-cities-hotel-forecast-2015-and-2016.pdf

PWC. (2016b). *Staying Power - European Cities Hotel Forecast for 2016 and 2017*. Author.

Reinhart, J. (2014). Why The Success Of 'Sharing Economy' Startups Hinges On Who Owns The Inventory. *Business Insider*. Retrieved on August 1, 2016 from http://www.businessinsider.com/the-success-of-the-sharing-economy-2014-2

San Diego Tourism Authority. (2016). *San Diego Tourism Industry Research*. Retrieved on July 3, 2016 from http://www.sandiego.org/industry-research.aspx

Sanders, S. (2015). Santa Monica Cracks Down On Airbnb, Bans 'Vacation Rentals' Under A Month. *NPR.org*. Retrieved from http://www.npr.org/sections/thetwo-way/2015/05/13/406587575/santa-monica-cracks-down-on-airbnb-bans-vacation-rentals-under-a-month

Schaal, D. (2015a). It's Beginning: Onefinestay Pilot With Hyatt Highlights Sharing Economy Collaboration To Come. *Skift*, Retrieved from https://skift.com/2015/07/21/its-beginning-onefinestay-pilot-with-hyatt-highlights-sharing-economy-collaboration-to-come/

Schaal, D. (2015b). The Inside Story Behind Expedia's Purchase of HomeAway. *Skift*. Retrieved from https://skift.com/2015/11/16/the-inside-story-behind-expedias-purchase-of-homeaway/

SEC. (n.d.). *Wyndham Worldwide Corporation-2013.12.31-10K*. Accessed July 29, 2016. https://www.sec.gov/Archives/edgar/data/1361658/000136165814000005/wyn-20131231x10k.htm

Sign Up to Drive or Tap and Ride. (n.d.). Retrieved on August 1, 2016 from https://www.uber.com/a/join?exp_hvp=1&invite_code=khzrpyyeue

Sileo, L. (2015). Independent Hotels: Can They Loosen the OTA Grip? *Tnooz*. Retrieved from https://www.tnooz.com/article/independent-lodging-break-from-OTAs-Phocuswright/

Socialdraft. (2017). How Hotels Can Beat AirBnB at Their Own Game. *Socialdraft*. Retrieved on March 4, 2014 from http://socialdraft.com/tips-for-hotels-to-deal-with-airbnb/

STATISTIK AUSTRIA - Statistiken. (n.d.). *Archive.is*. Accessed July 26, 2016. http://archive.is/jLC0r

STIFEL. (2016). *Americas Lodging Investment Summit Takeaways Part #1 2016 Outlook*. Author.

TaskRabbit. (n.d.). *TaskRabbit Connects You to Safe and Reliable Help in Your Neighborhood*. Retrieved on August 1, 2016 from https://www.taskrabbit.com/

Topic: Social Networks. (n.d.). Accessed July 21, 2016. http://www.statista.com/topics/1164/social-networks/

Tributaria, A. (2016). *New VAT Tax Rates - Spain*. Retrieved on July 12, 2016 from http://www.strongabogados.com/docs/Spain-VAT-rates-2012.pdf

Tru by Hilton. (2016). *Our Story*. Accessed July 18, 2016. http://trubyhilton.com/our-story/

Tru by Hilton. (n.d.). *Tru by Hilton*. Accessed July 8, 2016. http://trubyhilton.com/our-place/

Tugend, A. (2016). Hotel Chains and Travel Websites in a Tug of War for Customers. *The New York Times*. Retrieved from http://www.nytimes.com/2016/03/28/business/media/hotel-chains-and-travel-websites-in-a-tug-of-war-for-customers.html

Turning the Sharing Economy into a Fair Economy in Vienna - Share-Economy-in-Wien-Stadt-Wien-En.pdf. (n.d.). Retrieved on July 20, 2016 from https://www.wien.info/media/files-b2b/share-economy-in-wien-stadt-wien-en.pdf

United States Census Bureau. (2017). *NAICS (North American Industry Classification System)*. Retrieved from https://www.census.gov/eos/www/naics/

United States Securities and Exchange Commission. (2016). *MARRIOTT INTERNATIONAL INC /MD/ (Form: 10-K, Received: 02/20/2014 13:28:17)*. Retrieved from http://secfilings.nasdaq.com/edgar_conv_html%2f2014%2f02%2f20%2f0001445305-14-000495.html#FIS_BUSINESS

Vides, T. (2014). What You Never Knew About Repeat Customers and What to Do Now. *Online Marketing Institute*. Retrieved from https://www.onlinemarketinginstitute.org/blog/2014/05/what-you-never-knew-about-repeat-customers-and-what-to-do-now/

Viena Tourism Data & Facts. (n.d.). *Vienna's B2B Service for the Tourism Industry*. Accessed March 10, 2016. https://b2b.wien.info/en/statistics/data

Walton, J. K. (2018). *Tourism*. Retrieved on July 29, 2016 from https://www.britannica.com/topic/tourism

Western, B. (2002-2018). *Best Western International® | President & CEO - David Kong*. Retrieved on July 29, 2016 fromhttp://www.bestwestern.com/about-us/press-media/ceo.asp

Worgull, S. (2014). *HNN*. Retrieved on July 23, 2016 from http://www.hotelnewsnow.com/Articles/23584/Independent-vs-chain-Which-performs-better

World Travel & Tourism Council. (2016). *Travel & Tourism - Economic Impact 2016*. Retrieved from http://www.wttc.org/-/media/files/reports/economic-impact-research/2016-documents/economic-impact-summary-2016_a4-web.ashx

Wyndham Hotel Group. (2018). *About Wyndham Hotel Group: Learn about Wyndham Hotel Group Hotels at the Official Site for the Best Rate Guaranteed and Earn Wyndham Rewards Points*. Retrieved on July 29, 2016 from http://www.wyndhamhotelgroup.com/about-us/about-wyndham

Zipcar Inc. (2018). *Car Sharing: An Alternative to Car Rental with Zipcar*. Accessed August 1, 2016 from http://www.zipcar.com/

KEY TERMS AND DEFINITIONS

Airbnb: A privately held company, founded in 2008, headquartered in San Francisco, California, U.S.A. It provides an online marketplace and hospitality service, enabling people to lease or rent short-term lodging including vacation rentals, apartment rentals, homestays, hostel beds, or hotel rooms. The company does not own any lodging; it is merely a broker and receives percentage service fees (commissions) from both guests and hosts in conjunction with every booking. Airbnb employs 3,100 people, has 200,000,000 users, over 3,000,000 lodging listings in 65,000 cities and 191 countries, and the host sets the cost of lodging.

Airdna: A consultancy based in Santa Monica, California, U.S.A. This organization primarily operates in the Market Analysis, Business, and Economic Research business / industry within the Engineering, Accounting, Research, and Management Services sector. This organization has been operating for approximately 3 years. Airdna maintains a large-scale database tracking the lodging industry statistics worldwide. It is estimated to generate $300,000 in annual revenues.

Average Daily Rates (ADR): ADR is a statistical unit that is often used in the lodging industry. The number represents the average rental income per paid occupied room in a given time period.

Internet and Online Travel Agencies (OTAs): OTAs offer planning sources and booking capabilities. OTAs are travel websites that specialize in the sale of travel products to consumers. Some agencies sell a variety of travel products including flights, hotels, car rentals, cruises, activities, and packages. Other agencies specialize in the sale of tours and activities. In all cases, the travel agency has an agency agreement with tour or activity suppliers to resell their products where the agency takes payment from the consumer and pays net rates to the supplier. Major OTAs worldwide include Voyages, Expedia, Sabre, Opodo, the Priceline Group, Travelocity, Hotels.com, Travelgenio, Wotif.com, Webjet and others.

Lodging Industry: The lodging industry includes companies providing traveler accommodations, such as hotels, motels, resorts, casino hotels, and Bed-and-Breakfast Inns. In 2014, the global lodging industry grew by 4% and generated $678.6 billion in revenue. The North American Industry Classification System (NAICS) categorizes the general Accommodation and Food Sector under the industry code 721, with the following subsectors: "(1) traveler accommodation, (2) recreational accommodation, and (3) rooming and boarding houses." This study involves industry categories under NAICS codes 7211: Traveler Accommodation, 721110: Hotels (except casino hotels) and motels, 72119: Other Traveler Accommodation, and 721191: Bed-and-Breakfast Inns (NAICS, 2017).

Peer-to-Peer (P2P) Lodging: P2P property lodging rental (a.k.a. person-to-person home rental) is the process whereby an existing house owner makes their house or an empty room available for others to rent for short periods of time as an alternative form of accommodation.

Revenue Per Available Room (RevPAR): RevPAR is a performance metric in the hotel industry that is calculated by dividing a hotel's total guestroom revenue by the room count and the number of days in the period being measured. If the calculation uses total hotel revenue instead of guestroom revenue it equals TRevPAR (Total Revenue Per Available Room). TRevPAR is another closely related performance metric in the hotel industry.

Sharing Economy: Sharing economy is an umbrella term with a range of meanings, often used to describe economic activity involving online transactions. Originally growing out of the open-source community to refer to peer-to-peer based sharing of access to goods and services, the term is now sometimes used in a broader sense to describe any sales transactions that are done via online market places, even ones that are business to business (B2B), rather than peer-to-peer.

Compilation of References

2016Q1 Greek Hospitality Industry Performance. (2016). GBR Counsulting.

21.10.010 - Title. | Municipal Code | Long Beach, CA | Municode Library. (n.d.). Accessed July 27, 2016. https://www.municode.com/library/ca/long_beach/codes/municipal_code?nodeId=TIT21ZO_CH21.10GEPR_21.10.010TI

21.15.360 - Bed and Breakfast Inn. | Municipal Code | Long Beach, CA | Municode Library. (n.d.). Accessed July 27, 2016. https://www.municode.com/library/ca/long_beach/codes/municipal_code?nodeId=TIT21ZO_CH21.15DE_21.15.360BEBRIN

721: Accommodation. (n.d.). Accessed July 29, 2016. http://naics-codes.findthedata.com/l/107/Accommodation

7211: Traveler Accommodation. (n.d.). Accessed July 29, 2016. http://naics-codes.findthedata.com/l/408/Traveler-Accommodation

721110: Hotels (except Casino Hotels) and Motels. (n.d.). Accessed July 29, 2016. http://naics-codes.findthedata.com/l/1738/Hotels-except-Casino-Hotels-and-Motels

72119: Other Traveler Accommodation. (n.d.). Accessed July 29, 2016. http://naics-codes.findthedata.com/l/638/Other-Traveler-Accommodation

721191: Bed-and-Breakfast Inns. (n.d.). Accessed July 29, 2016. http://naics-codes.findthedata.com/l/1740/Bed-and-Breakfast-Inns

Aaron Smith92. (2014). 6 New Facts about Facebook. *Pew Research Center*. Retrieved from http://www.pewresearch.org/fact-tank/2014/02/03/6-new-facts-about-facebook/

Abell, D. F. (1980). *Defining the Business: The Starting Point of Strategic Planning*. Englefield Cliffs, NJ: Prentice Hall.

Abernathy, W. J., & Clark, K. B. (1985). Mapping the winds of creative destruction. *Research Policy*, *14*(1), 3–22. doi:10.1016/0048-7333(85)90021-6

Adner, R. (2002). When are technologies disruptive? A demand-based view of the emergence of competition. *Strategic Management Journal*, *23*(8), 667–688. doi:10.1002mj.246

Ahn, S.-H., Montero, M., Odell, D., Roundy, S., & Wright, P. K. (2002). Anisotropic material properties of fused deposition modeling ABS. *Rapid Prototyping Journal*, *8*(4), 248–257. doi:10.1108/13552540210441166

Airbnb Market Report - April 2016 (AirDNA). (2016). AirDNA.

Airbnb. (2016). *Airbnb Economic Impact*. Retrieved on February 28, 2016 from http://blog.airbnb.com/economic-impact-airbnb/

Airbnb. (n.d.). *About Us - Airbnb*. Retrieved on August 2, 2016 from https://www.airbnb.com/about/about-us

Airbnb. (n.d.). Who We Are, What We Stand for. *The Airbnb Blog - Belong Anywhere*. Retrieved on October 3, 2013 from http://blog.airbnb.com/who-we-are/

Akanji, O. O. (2002). Micro Finance as A Strategy for Poverty Reduction. *CBN and Financial Review*, *39*(4), 1–30.

Alexandratos, N., & Bruinsma, J. (2012). *World Agriculture Towards 2030/2050*. Retrieved from www.fao.org/docrep/016/ap106e/ap106e.pdf

Alliance for Water Efficiency. (2017). *Smart Water Application Technology (SWAT) Initiative*. Retrieved from:http://www.allianceforwaterefficiency.org/Smart_Water_Application_Technology_(SWAT)_Initiative.aspx

Alpkan, L., & Gemici, E. (2016). Disruption and Ambidexterity: How Innovation Strategies Evolve? *Procedia: Social and Behavioral Sciences*, *235*(October), 782–787. doi:10.1016/j.sbspro.2016.11.080

Altmann, V., Butzin, B., Balla, R., Golatowski, F., & Timmermann, D. (2015). A BACnet gateway for embedded Web services. *Emerging Technologies & Factory Automation (ETFA) 2015 IEEE 20th Conference on*, 1-6. 10.1109/ETFA.2015.7301563

American Hotel & Lodging Association (AHLA). (2015). *Lodging Industry Trends 2015*. Retrieved from https://www.ahla.com/sites/default/files/Lodging_Industry_Trends_2015.pdf

Anastasia. (2015). An Introduction to Sharing Economy. *Cleverism*. Retrieved from https://www.cleverism.com/introduction-to-sharing-economy/

Ansari, S., Garud, R., & Kumaraswamy, A. (2015). The Disruptor ' s Dilemma : TiVO and the U. S. Television Ecosystem. *Strategic Management Journal*. doi:10.1002mj

Ansari, S., & Krop, P. (2012). Incumbent performance in the face of a radical innovation: Towards a framework for incumbent challenger dynamics. *Research Policy*, *41*(8), 1357–1374. doi:10.1016/j.respol.2012.03.024

Ansari, S., & Munir, K. (2008). How valuable is a piece of the spectrum? Determination of value in external resource acquisition. *Industrial and Corporate Change*, *17*(2), 301–333. doi:10.1093/icc/dtn004

Arner, D. W., Barberis, J., & Buckley, R. P. (2015). The evolution of Fintech: A new post-crisis paradigm. *Geo. J. Int'l L.*, *47*, 1271.

Arnold, M., & Brunsden, J. (2017, January 31). Fintech sector fears dilution of EU 'open banking' legislation. *Financial Times*. Retrieved from https://www.ft.com/content/65342eb2-e7ab-11e6-893c-082c54a7f539

Arun, T., & Kamath, R. (2015). Financial inclusion: Policies and practices. *IIMB Management Review*, *27*(4), 267–287. doi:10.1016/j.iimb.2015.09.004

Ashforth, B. E., & Mael, F. (1989). Social identity theory and the organization. *Academy of Management Review*, *14*, 20–39.

Association of Banks. (2016). *With the launch of electronic money starts a new era in financial inclusion*. Retrieved March 10, 2017, from http://www.asbanc.com.pe/Paginas/Noticias/DetalleNoticia.aspx?ItemID=209&lang=en

Ayyar, R. (2014). The 3 Characteristics of Brands That Command Customer Loyalty. *Entrepreneur*. Retrieved from https://www.entrepreneur.com/article/237364

Baden-Fuller, C., & Morgan, M. (2010). Business models as models. *Long Range Planning*, *43*(2-3), 156–171. doi:10.1016/j.lrp.2010.02.005

Bai, X., Liu, Y., Wang, G., & Wen, C. (2017). The pattern of technological accumulation: The comparative advantage and relative impact of 3D printing technology. *Journal of Manufacturing Technology Management*, *28*(1), 39–55. doi:10.1108/JMTM-10-2016-0136

Baiyere, A. (2014). *Disrupted Disruptions: Lessons from Potential Disruptive Innovations that barely disrupted The XXV ISPIM Conference – Innovation for Sustainable Economy & Society*. Dublin, Ireland: Academic Press.

Bak, D. (2003). Rapid prototyping or rapid production? 3D printing processes move industry towards the latter. *Assembly Automation*, *23*(4), 340–345. doi:10.1108/01445150310501190

Balaji, M. S., & Roy, S. K. (2016). Value co-creation with Internet of things technology in the retail industry. *Journal of Marketing Management*, 1.

Baldwin, C. Y. (2008). Where do transactions come from? Modularity transactions and the boundaries of firms. *Industrial and Corporate Change*, *17*(1), 155–195. doi:10.1093/icc/dtm036

Banks, E. (2016). *What is a neobank and why is it our future?* Retrieved from http://eng.banks.eu/news/info/2289/

Banks.eu Staff. (2016, January 12). *What is a Neobank and Why Is It Our Future*. Retrieved from http://eng.banks.eu/news/info/2289/

Barba, R. (2017, February 3). Valuing a digital bank isn't so Simple. *American Banker*. Retrieved from https://www.americanbanker.com/news/valuing-a-digital-bank-isnt-so-simple

Barber, B. M., Lee, Y., Liu, Y., & Odean, T. (2008). Just How Much Do Individual Investors Lose by Trading. *The Society of Financial Studies, 22*(2), 610–632.

Barney, J. B. (1997). *On flipping coins* and *making technology choices: luck as an explanation of technological foresight and oversight*. In R. Garud, P. Nayyar, & Z. Shapira (Eds.), *Technological Innovation: Oversight and Foresights* (pp. 13–19). New York: Cambridge University Press. doi:10.1017/CBO9780511896613.004

Bass, F. M. (1969). A new product growth for model consumer durables. *Management Science, 15*(5), 215–227. doi:10.1287/mnsc.15.5.215

BBVA Compass Staff. (2014, February 20). BBVA acquires Simple to accelerate digital banking expansion. *BBVA Compass PR Newswire*. Retrieved from http://newsroom.bbvacompass.com/2014-02-20-BBVA-acquires-Simple-to-accelerate-digital-banking-expansion

Bechthold, L., Fischer, V., Hainzlmaier, A., Hugenroth, D., Ivanova, L., Kroth, K., . . . Sitzmann, V. (2015). *3D printing: A qualitative assessment of applications, recent trends and the technology's future potential*. Retrieved from https://www.econstor.eu/bitstream/10419/156627/1/StuDIS_2015-17.pdf

Beck, T., Demirgüç-Kunt, A., & Levine, R. (2007). Finance, inequality and the poor. *Journal of Economic Growth, 12*(1), 27–49. doi:10.100710887-007-9010-6

Beck, T., Demirgüç-Kunt, A., & Peria, M. S. M. (2007). Reaching out: Access to and use of banking services across countries. *Journal of Financial Economics, 85*(1), 234–266. doi:10.1016/j.jfineco.2006.07.002

Bengisu, M., & Nekhili, R. (2006). Forecasting emerging technologies with the aid of science and technology databases. *Technological Forecasting and Social Change, 73*(7), 835–844. doi:10.1016/j.techfore.2005.09.001

Berman, B. (2012). 3-D printing: The new industrial revolution. *Business Horizons, 55*(2), 155–162. doi:10.1016/j.bushor.2011.11.003

Bertling, J., & Rommel, S. (2016). A Critical View of 3D Printing Regarding Industrial Mass Customization Versus Individual Desktop Fabrication. In S. Dickel, J.-P. Ferdinand, & U. Petschow (Eds.), The Decentralized and Networked Future of Value Creation (pp. 75-105). Springer.

Better Finance. (2016). Long Term & Pension Savings: The Total Return, 2016 Edition. *Better Finance.* Retrieved from http://betterfinance.eu/fileadmin/user_upload/documents/Research_Reports/en/Pension_Savings_-_The_Real_Return_-_2016_Better_Finance_Presentation.pdf

Binham, C. (2015, December 4). European watchdogs to scrutinise robo-advice industry. *Financial Times.* Retrieved from https://www.ft.com/content/36d17f2c-9a7a-11e5-a5c1-ca5db4add713

Bocquet, R., Brossard, O., & Sabatier, M. (2007). Complementarities in organizational design and the diffusion of information technologies: An empirical analysis. *Research Policy, 36*(3), 367–386. doi:10.1016/j.respol.2006.12.005

Bogue, R. (2013). 3D printing: The dawn of a new era in manufacturing? *Assembly Automation*, *33*(4), 307–311. doi:10.1108/AA-06-2013-055

bornintheusa617. (2013). *Vienna - City of Dreams*. Accessed July 27, 2016. https://www.youtube.com/watch?v=AFZBOTgL_Hk

Botsman, R. (2013). The Sharing Economy Lacks A Shared Definition. *Co.Exist*. Retrieved from http://www.fastcoexist.com/3022028/the-sharing-economy-lacks-a-shared-definition

Bourell, D. L., Beaman, J. J., Leu, M. C., & Rosen, D. W. (2009). A brief history of additive manufacturing and the 2009 roadmap for additive manufacturing: looking back and looking ahead. *Proceedings of RapidTech*, 24-25.

Bourell, D. L. (2016). Perspectives on Additive Manufacturing. *Annual Review of Materials Research*, *46*(1), 1–18. doi:10.1146/annurev-matsci-070115-031606

Bourreau, M., & Valletti, T. (2015). Enabling Digital Financial Inclusion through Improvements in Competition and Interoperability: What Works and What Doesn't? *CGD Policy Paper*, *65*, 1–30.

Bower, J. (2002, May). Disruptive Change. *Harvard Business Review*, *80*(05), 95–101. PMID:12195923

Bower, J. L., & Christensen, C. M. (1996). Disruptive technologies: Catching the wave. *Journal of Product Innovation Management*, *1*(13), 75–76.

Bradley, J., Barbier, J., & Handler, D. (2013). *Embracing the Internet of Everything To Capture Your Share of $14.4 Trillion - More Relevant, Valuable Connections Will Improve Innovation, Productivity, Efficiency & Customer Experience*. Cisco. Retrieved from: http://www.cisco.com/web/about/ac79/docs/innov/IoE_Economy. pdf

Breloff, P., & Parker, S. R. (2011). Financial Inclusion in the U.S.: Spending Time In Our Own Backyard. *CGAP Blog*. Retrieved February 11, 2017 from http://www.cgap.org/blog/financial-inclusion-usspending-time-our-own-backyard

Brownlow, M. (2016). *Is There a Hotel or Tourism Tax in Vienna?* Retrieved February 4, 2016 from http://www.visitingvienna.com/accommodation/hotel-tax-vienna/

Buck, G. (2016, December 13). MasterCard: Millions of Europeans still unbanked. *GT News*. Retrieved from https://www.gtnews.com/2016/12/13/mastercard-millions-of-europeans-still-unbanked/

Buera, F. J., Kaboski, J. P., & Shin, Y. (2012). *The macroeconomics of microfinance (No. w17905)*. National Bureau of Economic Research. doi:10.3386/w17905

Bull, G. (2017). Four Drivers of Change for Financial Inclusion in 2017. *CGAP Blog*. Retrieved February 11, 2017 from http://www.cgap.org/blog/four-drivers-change-financial-inclusion-2017

Bures, T., Plasil, F., Kit, M., Tuma, P., & Hoch, N. (2016). Software Abstractions for Component Interaction in the Internet of Things. *Computer*, *49*, 50-59.

Burgers, J. H., Jansen, J. J. P., Van den Bosch, A. J. F., & Volberda, H. W. (2009). Structural differentiation and corporate venturing: The moderating role of formal and informal integration mechanisms. *Journal of Business Venturing*, *24*(3), 206–220. doi:10.1016/j.jbusvent.2009.01.006

Cable News Network (CNN). (2015). *World's Most Visited Cities*. Retrieved on July 30, 2016 from http://edition.cnn.com/2015/01/29/travel/gallery/most-visited-cities-euromonitor/

Cabrera, F. L., & de Sousa, F. R. (2016). Achieving Optimal Efficiency in Energy Transfer to a CMOS Fully Integrated Wireless Power Receiver. *Microwave Theory and Techniques IEEE Transactions on*, *64*, 3703-3713.

Callaway, S. K., & Hamilton, R. D. III. (2008). Managing disruptive technology: Internet banking ventures for traditional banks. *International Journal of Innovation and Technology Management*, *5*(1), 5–80. doi:10.1142/S0219877008001242

Calomiris, C. W. (2010). *Banking Crises Yesterday and Today. Financial History Review, 1(1)*, 3–12. Retrieved from https://jeffludwig.wordpress.com/2012/01/02/how-to-cite-white-papers/

Camerer, C., & Vepsalainen, A. (1988). The economic efficiency of corporate culture. *Strategic Management Journal*, *9*(S1), 115–126. doi:10.1002mj.4250090712

Cannon-Bowers, J., Tannenbaum, S., Salas, E., & Volpe, C. (1995). Defining competencies and establishing team training requirements. In R. Guzzo & E. Salas (Eds.), *Team effectiveness and decision making in organizations* (pp. 333–380). San Francisco, CA: Jossey-Bass.

Cao, Y., & Zhao, L. (2013). Analysis of patent management effects on technological innovation performance. *Baltic Journal of Management*, *8*(3), 286–305. doi:10.1108/BJOM-May-2012-0033

Carlo, J. L., Gaskin, J., Lyytinnen, K., & Rose, G. M. (2014). Early vs late adoption of radical information technology innovations across software development organizations: An extension of the disruptive information technology innovation model. *Information Systems Journal*, *24*(6), 537–569. doi:10.1111/isj.12039

Caulfield, B., McHugh, P., & Lohfeld, S. (2007). Dependence of mechanical properties of polyamide components on build parameters in the SLS process. *Journal of Materials Processing Technology*, *182*(1), 477–488. doi:10.1016/j.jmatprotec.2006.09.007

Cederholm, T. (2014). *Hilton's Market Share in the Domestic and International Market - Market Realist*. Retrieved from http://marketrealist.com/2014/10/hiltons-market-share-domestic-international-market/

Center for Responsible Nanotechnology (CRN). (2016). *The meaning of Nanotechnology*. Retrieved from: http://www.crnano.org/whatis.htm

Centers for Disease Control and Prevention. (2006). *Human Genetic Epidemiology Network*. Available at: www.cdc.gov/genomics/hugenet/default.htm

Central Bank of Russia. (2016). *Review of the status of financial inclusion in the Russian Federation in 2015.* Retrieved January 30, 2017 from https://www.cbr.ru/finmarkets/files/affordability/rev_fin_20161110.pdf

CGAP. (2014). *Digital Financial Inclusion: Implications for Customers, Regulators, Supervisors, and Standard-Setting Bodies.* Retrieved January 9, 2017, from https://www.cgap.org/sites/default/files/Brief-Digital-Financial-Inclusion-Feb-2015.pdf

Chandler, J. (2016). Los Angeles Is One Step Closer to Regulating Airbnb. *Curbed LA*, Retrieved from http://la.curbed.com/2016/6/23/11985210/los-angeles-airbnb-law-regulation-legal

Chang, S.-B., Lai, K.-K., & Chang, S.-M. (2009). Exploring technology diffusion and classification of business methods: Using the patent citation network. *Technological Forecasting and Social Change*, *76*(1), 107–117. doi:10.1016/j.techfore.2008.03.014

Chen, D., Heyer, S., Ibbotson, S., Salonitis, K., Steingrímsson, J. G., & Thiede, S. (2015). Direct digital manufacturing: Definition, evolution, and sustainability implications. *Journal of Cleaner Production*, *107*, 615–625. doi:10.1016/j.jclepro.2015.05.009

Cheng, Y., Huang, L., Ramlogan, R., & Li, X. (2016). Forecasting of potential impacts of disruptive technology in promising technological areas: Elaborating the SIRS epidemic model in RFID technology. *Technological Forecasting and Social Change.*

Chen, M. J., Smith, K. G., & Grimm, C. M. (1992). Action Characteristics as Predictors of Competitive responses. *Management Science*, *30*(30), 439–455. doi:10.1287/mnsc.38.3.439

Chesbrough, H. (2003). *Open Innovation: The New Imperative for Creating and Profiting from Technology*. Boston: Harvard Business School Press.

Chesbrough, H., & Rosenbloom, R. S. (2002). The role of the business model in capturing value from innovation: Evidence from Xerox Corporation's technology spin-off companies. *Industrial and Corporate Change*, *11*(3), 529–555. doi:10.1093/icc/11.3.529

Cheston, S., Conde, T., Bykere, A., & Rhyne, E. (2017). *The Business of Financial Inclusion: Insights from Banks in Emerging Markets.* Retrieved February 11, 2017 from Http://Www.Centerforfinancialinclusion.Org/Storage/Documents/Iif_Cfi_Report_Final.Pdf

Choi, C., Kim, S., & Park, Y. (2007). A patent-based cross impact analysis for quantitative estimation of technological impact: The case of information and communication technology. *Technological Forecasting and Social Change*, *74*(8), 1296–1314. doi:10.1016/j.techfore.2006.10.008

Christensen, C. M. (2006). The Ongoing Process of Building a Theory of Disruption. *Journal of Product Innovation Management*, 39–55. 10.1111/j.1540-5885.2005.00180.x

Christensen, C. M. (1993). The Rigid Disk-Drive Industry: A History of Commercial and Technological Turbulence. *Business History Review*, *67*(4), 531–588. doi:10.2307/3116804

Christensen, C. M. (1997). *The Innovator's Dilemma: When New Technologies Cause Great Firms to Fail.* Boston: Harvard Business School Press.

Christensen, C. M. (2000). *The innovator's dilemma: The revolutionary book that will change the way you do business*. New York: Haper Business.

Christensen, C. M., & Bower, J. L. (1996). Customer power, strategic investment, and the failure of leading firms. *Strategic Management Journal*, *17*(3), 197–218. doi:10.1002/(SICI)1097-0266(199603)17:3<197::AID-SMJ8043.0.CO;2-U

Christensen, C. M., & Raynor, M. (2003). *The Innovator's Solution: Creating and Sustaining Successful Growth*. Boston: Harvard Business School Press.

Christensen, C. M., & Raynor, M. E. (2003). *The Innovator's solution: Creating and sustaining successful growth*. Harvard Business School Press.

Christensen, C. M., Raynor, M. E., & McDonald, R. (2015). What Is Disruptive Innovation? *Harvard Business Review*, 44–53. PMID:17183796

Chung, G. (2016). *Justin Bieber Stays at Beyoncé's $10,000-a-Night Airbnb | Celebuzz*. Retrieved on August 11, 2016 from http://www.celebuzz.com/2016-03-18/justin-bieber-beyonce-airbnb-home-house-mansion/

CIA Factbook. (2016). *The world factbook*. Retrieved June 17, 2017 from: https://www. cia.gov/library/publications/the-world-factbook

Cirani, S., & Picone, M. (2015). Wearable Computing for the Internet of Things. *IT Professional*, *17*, 35-41.

Clampet, J. (2016). *Everything You Wanted to Know About the Hotel Industry's Gripes Against OTAs – Skift*. Retrieved on July 23, 2016 from https://skift.com/2016/04/25/everything-you-ever-wanted-to-know-about-hotel-industrys-complaints-against-otas/

Close, K. (2016). *Amazon Is Cracking Down on Fake Product Reviews - Yahoo Finance*. Retrieved from http://finance.yahoo.com/news/amazon-cracking-down-fake-product-141618091.html

COCO-MAT. (2015-2018). *About Us*. Retrieved July 18, 2016 from http://www.coco-mat.com/us_en/company/

Cohen, H., & Lefebvre, C. (Eds.). (2005). *Handbook of Categorization in Cognitive Science*. Elsevier.

Collins, D. (2008). Debt and household finance: Evidence from the Financial Diaries. *Development Southern Africa*, *25*(4), 469–479. doi:10.1080/03768350802318605

Commission, E. U. (2015a). *Digital Single Market*. Retrieved on February 16, 2016 from http://ec.europa.eu/prio-rities/digital-single-market/docs/dsmcommunication_en.pdf

Commission, E. U. (2015b). *Bitcoin Regulation*. Retrieved on February 16, 2016 from http://coinbrief.net/bitcoinr-egulation-european-commission/

Condry, M. W., & Nelson, C. B. (2016). Using smart edge IOT devices for safer rapid response with industry IOT control operations. *Proceedings of the IEEE*, *104*, 938-946.

Coon, C., Pretzel, B., Lomax, T., & Strlič, M. (2016). Preserving rapid prototypes: A review. *Heritage Science*, *4*(1), 40. doi:10.118640494-016-0097-y

Cooper, A., & Schendel, D. (1976). Strategic Responses to Technological Threats, Business Horizon. *Corporate Change*, *5*(4), 1143–1180.

Corsi, S., & Di Minin, A. (2014). Disruptive innovation in reverse: Adding a geographical dimension to disruptive innovation theory. *Creativity and Innovation Management*, *23*(1), 76–90. doi:10.1111/caim.12043

Cotropia, C. A., Lemley, M. A., & Sampat, B. (2013). Do applicant patent citations matter? *Research Policy*, *42*(4), 844–854. doi:10.1016/j.respol.2013.01.003

Cotteleer, M., & Joyce, J. (2014). 3D opportunity: Additive manufacturing paths to performance, innovation, and growth. *Deloitte Review*, *14*, 5–19.

Crockett, D. R., McGee, J. E., & Payne, G. T. (2013). Employing new business divisions to exploit disruptive innovations: The interplay between characteristics of the corporation and those of the venture management team. *Journal of Product Innovation Management*, *30*(5), 856–879. doi:10.1111/jpim.12034

Crowd funding's Potential for the Developing World. (2013). *Info Dev, Finance and Private Sector Development Department.* Washington, DC: World Bank.

Crowe, P. (2016). *It's A Deal: Marriott Is Buying Starwood Hotels for $12.4 Billion.* Retrieved from http://www.businessinsider.com/marriott-and-starwood-shareholder-vote-2016-4

Crump, S. S. (1992). *Apparatus and method for creating three-dimensional objects.* Patent No. US5121329 A.

Culumbus, L. (2015). Roundup Of Internet Of Things Forecasts And Market Estimates. *Forbes.* Retrieved from: https://www.forbes.com/sites/louiscolumbus/2016/11/27/roundup-of-internet-of-things-forecasts-and-market-estimates-2016/#29064e7e292d

Cunningham, S. W., & Kwakkel, J. (2011). Innovation forecasting: A case study of the management of engineering and technology literature. *Technological Forecasting and Social Change*, *78*(2), 346–357. doi:10.1016/j.techfore.2010.11.001

Curran, C.-S., & Leker, J. (2011). Patent indicators for monitoring convergence–examples from NFF and ICT. *Technological Forecasting and Social Change*, *78*(2), 256–273. doi:10.1016/j.techfore.2010.06.021

Dabla-Norris, M. E., Ji, Y., Townsend, R., & Unsal, D. F. (2015). Identifying constraints to financial inclusion and their impact on GDP and inequality: A structural framework for policy (No. 15-22). International Monetary Fund.

Daim, T. U., Iskin, I., & Ho, D. (2011). Technology forecasting for residential energy management devices. *Foresight*, *13*(6), 70–87. doi:10.1108/14636681111179609

Danneels, E. (2004). Disruptive technology reconsidered: A critique and research agenda. *Journal of Product Innovation Management*, *21*(4), 246–258. doi:10.1111/j.0737-6782.2004.00076.x

Deckard, C. R. (1991). *Method and apparatus for producing parts by selective sintering*. Patent No. US5017753 A.

Delbecque, B., & Carrol, A. (2017, May). *Asset Management in Europe: 9th Edition Facts and Figures*. European Fund and Asset Management Association. Retrieved from http://www.efama.org/publications/statistics/asset%20management%20report/efama%20asset%20management%20report%202017.pdf

Delis, M., Hasan, I., & Kazakis, P. (2013). Bank Regulations and Income Inequality: Empirical Evidence. *Review of Finance*.

Demirgüç-Kunt, A., Klapper, L. F., Singer, D., & Van Oudheusden, P. (2015). *The global findex database 2014: Measuring financial inclusion around the world.* Academic Press.

Demirguc-Kunt, A., Klapper, L., Singer, D., & Van Oudheusden, P. (2015). *The Global Findex Database 2014 Bringing the 2 Billion Unbanked into the Financial System.* Retrieved from http://pubdocs.worldbank.org/en/113791483565360488/N2UnbankedV5.pdf

Demirguc-Kunt, A., Klapper, L., Singer, D., & Van Oudheusden, P. (2015, April). *The Global Findex Database 2014 Bringing the 2 Billion Unbanked into the Financial System.* The World Bank Group. Retrieved from http://pubdocs.worldbank.org/en/113791483565360488/N2UnbankedV5.pdf

Desai, D. R., & Magliocca, G. N. (2013). Patents, meet Napster: 3D printing and the digitization of things. *Geological Journal*, *102*, 1691.

Desks Near Me | Find a Space to Work. (n.d.). *Desks Near Me*. Accessed August 1, 2016. https://desksnear.me/en

Dietz, M., Khanna, S., Olanrewaju, T., & Rajgopal, K. (2015). *Cutting Through the FinTech Noise: Markers of Success, Imperatives for Banks.* McKinsey & Co. Retrieved from https://www.mckinsey.com/~/media/mckinsey/industries/financial%20services/our%20insights/cutting%20through%20the%20noise%20around%20financial%20technology/cutting-through-the-fintech-noise-full-report.ashx

Discover Los Angeles. (2016). *Facts About Los Angeles*. Retrieved on March 30, 2016. from http://www.discoverlosangeles.com/content/ares-header-nav.html

Discover Los Angeles. (2016). *Research*. Accessed July 3, 2016. http://www.discoverlosangeles.com/tourism/research

Doby, D., Fessehatsion, B., & Tillipman, J. (2016). *LMU MBA Hotel & Airbnb Data Tables.* Academic Press.

Dotsika, F., & Watkins, A. (2017). Identifying potentially disruptive trends by means of keyword network analysis. *Technological Forecasting and Social Change*, *119*, 114–127. doi:10.1016/j.techfore.2017.03.020

Duflo, E., Banerjee, A., Glennerster, R., & Kinnan, C. G. (2013). *The miracle of microfinance? Evidence from a randomized evaluation (No. w18950)*. National Bureau of Economic Research.

Dunkley, E. (2016, November 16). Peer-to-peer lending pioneer seeks banking license. *Financial Times*. Retrieved from https://www.ft.com/content/53aa3dec-ac07-11e6-9cb3-bb8207902122

Dupas, P., & Robinson, J. (2013a). Savings constraints and microenterprise development: Evidence from a field experiment in Kenya. *American Economic Journal. Applied Economics*, *5*(1), 163–192. doi:10.1257/app.5.1.163

Dupas, P., & Robinson, J. (2013b). *Daily needs, income targets and labor supply: evidence from Kenya (No. w19264)*. National Bureau of Economic Research. doi:10.3386/w19264

EAI Catalyst Cleantech Sustainability. (2015). Retrieved from http://www.eai.in/blog/2015/07/big-list-of-disruptive-innovations-excellent-list.html

Edwards, L. (2016). *Privacy, security and data protection in smart cities: A critical EU law perspective*. Retrieved from http://papers.ssrn.com/sol3/papers.cfm?abstract_id=2711290

ELINET. (2015). *Literacy in Europe: Facts and Figures*. ELINET and the European Commission. Retrieved from http://www.eli-net.eu/fileadmin/ELINET/Redaktion/Factsheet-Literacy_in_Europe-A4.pdf

Ernst, H. (2003). Patent information for strategic technology management. *World Patent Information*, *25*(3), 233–242. doi:10.1016/S0172-2190(03)00077-2

EU GDPR. (2017). GDPR Key Changes. *EU GDPR*. Retrieved from http://www.eugdpr.org/key-changes.html

EU GDPR. (n.d.). *GDPR Key Changes: An overview of the main changes under GPDR and how they differ from the previous directive*. EU GPDR. Retrieved from http://www.eugdpr.org/key-changes.html

Euromonitor International. (2017). *Lodging in Greece*. Retrieved from http://www.euromonitor.com/lodging-in-greece/report

Euromonitor International. (n.d.). *Top 100 City Destinations Ranking*. Accessed June 30, 2016. http://go.euromonitor.com/rs/805-KOK-719/images/TCD-presentation_FINAL.pdf

European Banking Authority (EBA), The European Securities and Markets Authority (ESMA), & The European Insurance and Occupational Pensions Authority (EIOPA) Joint Committee. (2015, December 4). *Joint Committee Discussion Paper On Automation In Financial Advice*. JC 2015 080. European Banking Authority (EBA), the European Securities and Markets Authority (ESMA) and the European Insurance and Occupational Pensions Authority (EIOPA). Retrieved from http://www.eba.europa.eu/documents/10180/1299866/JC+2015+080+Discussion+Paper+on+automation+in+financial+advice.pdf

European Central Bank. (2014). Structural Indicators for the EU Banking Sector. *ECB*. Retrieved from https://www.ecb.europa.eu/stats/pdf/140711_ssi_table.pdf?a8644f6780aee14061e973e596141fac

European Central Bank. (2015). *Virtual currency schemes – a further analysis*. Retrieved from https://www.ecb.europa.eu/pub/pdf/other/virtualcurrencyschemesen.pdf

European Central Bank. (2017). *Single Supervisory Mechanism*. Retrieved from https://www.bankingsupervision.europa.eu/about/thessm/html/index.en.html

European Central Bank. (2017, January). *The euro area bank lending survey: Fourth quarter of 2016*. Frankfurt, Germany: ECB. Retrieved from https://www.ecb.europa.eu/stats/pdf/blssurvey_201701.pdf?6c44eff3bac4b858969b9cb71bd4a8fa

European Commission. (2015). *Literacy in Europe: Facts and Figures*. European Commission. Retrieved from http://www.eli-net.eu/fileadmin/ELINET/Redaktion/Factsheet-Literacy_in_Europe-A4.pdf

European Commission. (n.d.). *Consumer Rights and Laws*. EC. Retrieved from http://ec.europa.eu/consumers/consumer_rights/index_en.htm

European Fund and Asset Management Association. (2016). *The Pan-European Personal Pension Product*. EFAMA. Retrieved from https://www.efama.org/Publications/PEPP_flyer.pdf

Eurostat. (2016). *Tourism Statistics - Annual Results for the Accommodation Sector - Statistics Explained*. Retrieved July 17, 2016 from http://ec.europa.eu/eurostat/statistics-explained/index.php/Tourism_statistics__annual_results_for_the_accommodation_sector

Eurostat. (2017, July 20). Population pyramids, EU-28, 2001 and 2016 (% of the total population). *Eurostat*. Retrieved from http://ec.europa.eu/eurostat/statistics-explained/index.php/File:Population_pyramids,_EU-28,_2001_and_2016_(%25_of_the_total_population).png

Eurostat. (2017, March). *Asylum statistics*. Retrieved from http://ec.europa.eu/eurostat/statistics-explained/index.php/Asylum_statistics

Eurostat. (2017, March). *Migration and migrant population statistics*. Retrieved from http://ec.europa.eu/eurostat/statistics-explained/index.php/Migration_and_migrant_population_statistics#Migrant_population

EY LLP. (2017). *EY FinTech Adoption Index 2017: The Rapid Emergence of FinTech*. Retrieved July 11, 2017 from www.ey.com/Publication/vwLUAssets/ ey-fintech-adoption-index-2017/$FILE/ey-fintech-adoption-index-2017.pdf

Fallah, M. H., Fishman, E., & Reilly, R. R. (2011). Forward patent citations as predictors for patent valuation. *International Journal of Intellectual Property Management*, *4*(3), 165–178. doi:10.1504/IJIPM.2011.041081

FEVITUR. (n.d.). Salvetti & Llombart, and ESADE. *Study of Holiday Rentals*.

Feygin, M. (1988). *Apparatus and method for forming an integral object from laminations*. Patent No. US4752352 A.

Financial Conduct Authority. (2014, November 17). *FCA confirms price cap rules for payday lenders*. Financial Conduct Authority. Retrieved from https://www.fca.org.uk/news/press-releases/fca-confirms-price-cap-rules-for-payday-lenders

Financial Inclusion Commission in the U.K. website. (n.d.). *History*. Retrieved March 21, 2017, from http://www.financialinclusioncommission.org.uk/history

Financial Services User Group. (2011, March 22). *Financial Services User Group's (FSUG) opinion on interest rate restrictions in the EU*. FSUG. Retrieved from https://ec.europa.eu/info/sites/info/files/file_import/irr-2011_03_22_en_0.pdf

Finley, K. (2015). Smart meters shame you for how much water you're wasting. *Slate.com*. Retrieved from http://www.slate.com/blogs/future tense/2015/06/15/in California smart water meters help spot cases of extreme waste.html

First Research. (2015). *Lodging Industry*. Author.

Fleming, L. (2001). Recombinant uncertainty in technological search. *Management Science*, *47*(1), 117–132. doi:10.1287/mnsc.47.1.117.10671

Fleming, L., & Sorenson, O. (2001). Technology as a complex adaptive system: Evidence from patent data. *Research Policy*, *30*(7), 1019–1039. doi:10.1016/S0048-7333(00)00135-9

Flughafen Wien - Passagiere. (2018). Retrieved on July 26, 2016 from http://www.viennaairport.com/

FOM. (2017). *Dynamics of Internet Penetration in Russia. Survey Results*. Retrieved May 1, 2017 from http://fom.ru/SMI-i-internet/13585

Forbes. (2015). *20 Digital Trends For 2015*. Retrieved on February 16, 2016 from http://www.forbes.com/sites/georgedeeb/2015/01/16/the-top-20-digital-trends-of-2015/

Ford, S. (2014). Additive Manufacturing Technology: Potential Implications for U.S. Manufacturing Competitiveness. *Journal of International Commerce & Economics*.

Ford, S., & Despeisse, M. (2016). Additive manufacturing and sustainability: An exploratory study of the advantages and challenges. *Journal of Cleaner Production*, *137*, 1573–1587. doi:10.1016/j.jclepro.2016.04.150

Forer, G., & Staub, C. (2013). *The US water sector on the verge of transformation*. Ernst & Young, Global Cleantech Center White Paper. Retrieved from http://www.ey.com/Publication/vwLUAssets/Cleantech_Water_Whitepaper/$FILE/Cleantech-Water-Whitepaper.pdf

Forrest, E., & Cao, Y. (2013). Digital Additive Manufacturing: A Paradigm Shift in the Production Process and Its Socio-economic Impacts. *Engineering Management Research*, *2*(2), 66. doi:10.5539/emr.v2n2p66

Fourt, L. A., & Woodlock, J. W. (1960). Early prediction of market success for new grocery products. *Journal of Marketing*, *25*(2), 31–38. doi:10.2307/1248608

Freelancer - Hire & Find Jobs. (2016). Retrieved August 1, 2016 from https://www.freelancer.com/?t=a&utm_expid=294858-448.vN25f6toQF6x9f65mvAcgg.1&utm_referrer=https%3A%2F%2Fwww.google.com%2F

Freitag, J. (2016). Non-Branded Hotels Outperforming the Competition. *Lodging*. Retrieved from http://lodgingmagazine.com/non-branded-hotels-outperforming-the-competition/

Frost & Sullivan by Marc Einstein. (2015). *2015 Global and Japan Information and Communication Technology (ICT) Outlook*. Retrieved from http://www.frost.com/sublib/displayreport.do?id=9ABE-00-1A-00-00

Frost & Sullivan. (2014). *Mega Trends Driving Change in Business Environments*. Retrieved from http://www.investinbsr.com/ipaforum/wp-content/uploads/Iain-Jawad-IPA-Forum-2014-Presentation.pdf

Frost & Sullivan. (2016). The way of the future: Wearables poised to disrupt existing technologies and processes across industries. *PR Newswire*. Retrieved from https://search.proquest.com/docview/1758135273?accountid=14719

Ganezer, D. (2016). Rent AIRBNB, Go To Jail: Subletting Five Santa Monica Apartments, Gets Man Misdemeanor Conviction. *Santa Monica Observer*. Retrieved from http://www.smobserved.com/story/2016/07/11/news/rent-airbnb-go-to-jail-subletting-five-santa-monica-apartments-gets-man-misdemeanor-conviction/1594.html

Gao, W., Zhang, Y., Ramanujan, D., Ramani, K., Chen, Y., Williams, C. B., ... Zavattieri, P. D. (2015). The status, challenges, and future of additive manufacturing in engineering. *Computer Aided Design*, *69*, 65–89. doi:10.1016/j.cad.2015.04.001

Gartner. (2015). *Top 10 Strategic Technology Trends for 2015*. Retrieved from http://www.gartner.com/newsroom/id/2867917

Generalitat de Catalunya. Departament d'Empresa i Ocupacio. (2015). *European Tourism Forum - Luxembourg 2015*. Author.

Giannatsis, J., & Dedoussis, V. (2009). Additive fabrication technologies applied to medicine and health care: A review. *International Journal of Advanced Manufacturing Technology*, *40*(1), 116–127. doi:10.100700170-007-1308-1

Gibson, I., Rosen, D., & Stucker, B. (2015). *Additive manufacturing technologies: 3D printing, rapid prototyping, and direct digital manufacturing*. Academic Press.

Giffinger, R., & Gudrun, H. (2010). Smart cities ranking: An effective instrument for the positioning of cities? *ACE: Architecture, City and Environment, 4*(12), 7-25. Available at http://upcommons.upc.edu/revistes/bitstream/2099/8550/7/A CE_12_SA_10.pdf

Gilbert, C. (2003). The disruption opportunity. *Sloan Management Review*, *44*(4), 27–33.

Gilbert, C. (2005). Unbundling the structure of inertia: Resource versus routine rigidity. *Academy of Management Journal*, *48*(5), 741–763. doi:10.5465/AMJ.2005.18803920

Gilbert, C., & Bower, J. (2002). Disruptive change: When trying harder is part of the problem. *Harvard Business Review*, *80*(5), 94–101. PMID:12024762

Gilpin, L. (2015). *How cleantech-as-a-service will drive renewable energy adoption*. Retrieved from: http://www.techrepublic.com/article/how-cleantech-as-a-service-will-drive-renewable-energy-adoption/

Global Partnership for Financial Inclusion website. (n.d.). *About*. Retrieved March 21, 2017, from https://www.gpfi.org/about-gpfi

globalEDGE. (n.d.). *globalEDGE: Hospitality and Travel*. Retrieved from http://globaledge.msu.edu/industries/hospitality-and-travel

Goetzmann, W. N., & Jorion, P. (1999). Re-emerging markets. *Journal of Financial and Quantitative Analysis*, *34*(1), 1–32. doi:10.2307/2676244

Goldenstein, T. (2015, May 5). Smart water meters help users, agencies gauge usage. *Los Angeles Times*. Retrieved from http://www.latimes.com/local/california/la-me-smart-meter-explainer-20150505-story.html

Gordon, R. (2015). Trends in commercial 3D printing and additive manufacturing. *3D Printing and Additive Manufacturing, 2*(2), 89-90.

Govindarajan, V., & Kopalle, P. K. (2006a). Disruptiveness of Innovations: Measurement and an Assessment of Reliability and Validity. *Strategic Management Journal*, *27*(2), 189–199. doi:10.1002mj.511

Govindarajan, V., & Kopalle, P. K. (2006b). The Usefulness of Measuring Disruptiveness of Innovations. Ex-post in Making ex-ante Predictions. *Journal of Product Innovation Management*, *23*(1), 12–18. doi:10.1111/j.1540-5885.2005.00176.x

Gozzer Arias, S. (2015). Barcelona Cracks down on Illegal Subletting through Airbnb. *EL PAÍS*. Retrieved from https://elpais.com/elpais/2015/09/01/inenglish/1441115926_651764.html

Grady, B. (2016). *Driving battery storage, IoT, electric transportation mainstream*. GreenBiz Group. Retrieved from: www.techrepublic.com/article/how-cleantech-as-a-service-will-drive-renewable-energy-adoption/

Greguras, F., (2016). Water And the Internet of Things: 2016. *Wateronline*. Adopted from: https://www.wateronline.com/doc/water-and-the-internet-of-things-0001

Gross, B. C., Erkal, J. L., Lockwood, S. Y., Chen, C., & Spence, D. M. (2014). Evaluation of 3D printing and its potential impact on biotechnology and the chemical sciences. *Analytical Chemistry*, *86*(7), 3240–3253. doi:10.1021/ac403397r PMID:24432804

GSMA. (2015). *The Mobile Economy: Europe 2015*. London: GSM Association. Retrieved from https://www.gsma.com/mobileeconomy/archive/GSMA_ME_Europe_2015.pdf

GTP Editing. (2015). Athens: Hotel Occupancy Rates Decline Despite Rising Visitor Arrivals. *GTP Headlines*. Retrieved from http://news.gtp.gr/2015/12/08/athens-hotel-occupancy-rates-decline/

GTP. (2015). The 'Sharing Economy' in Tourism Deprives Greece of Millions of Euros. *GTP Headlines*. Retrieved from http://news.gtp.gr/2015/10/08/sharing-economy-tourism-deprives-greece-millions/

Gullapalli, D., Anand, S., & Maxey, D. (2008, Sept. 17). Money Fund, Hurt by Debt Tied to Lehman, Breaks the Buck. *Wall Street Journal*. Retrieved from https://www.wsj.com/articles/SB122160102128644897

Guo, N., & Leu, M. C. (2013). Additive manufacturing: Technology, applications and research needs. *Frontiers of Mechanical Engineering*, *8*(3), 215–243. doi:10.100711465-013-0248-8

Habtay, S. R. (2012). A Firm-Level Analysis on the Relative Difference between Technology-Driven and Market-Driven Disruptive Business Model Innovations. *Creativity and Innovation Management*, *21*(3), 290–303. doi:10.1111/j.1467-8691.2012.00628.x

Hagen, C., Evans, H., Ciobo, M., Miller, J., Wall, D., & Yadav, A. (2016). *Big DataArtificial intelligence and the Creative Destruction of Today's Business Models*. Adopted from ATkearney: https://www.atkearney.com/strategic-it/ideas-insights/article/-/asset_publisher/LCcgOeS4t85g/content/big-data-and-the-creative-destruction-of-today-s-business-models/10192

Hagstrom, R. (2013). *What is the difference between investing and speculation?* Retrieved from https://blogs.cfainstitute.org/investor/2013/02/27/what-is-the-difference-between-investing-and-speculation-2/

Han, E. J., & Sohn, S. Y. (2015). Patent valuation based on text mining and survival analysis. *The Journal of Technology Transfer*, *40*(5), 821–839. doi:10.100710961-014-9367-6

Hang, C. C., Garnsey, E., & Ruan, Y. (2015). Opportunities for disruption. *Technovation*, *39–40*, 83–93. doi:10.1016/j.technovation.2014.11.005

Hannagan, A., & Morduch, J. (2015). *Income gains and month-to-month income volatility: Household evidence from the US Financial Diaries*. Academic Press.

Hannig, A., & Jansen, S. (2010). *Financial inclusion and financial stability: Current policy issues*. Academic Press.

Harhoff, D., Narin, F., Scherer, F. M., & Vopel, K. (1999). Citation frequency and the value of patented inventions. *The Review of Economics and Statistics*, *81*(3), 511–515. doi:10.1162/003465399558265

Hartig, T., & Kahn, P. Jr. (2016). Living in cities, naturally. *Science*, *352*(6288), 938–940. doi:10.1126cience.aaf3759 PMID:27199417

Hartman, B., & LeBlanc, B. (2015). In pursuit of the perfect portal: Smart meters, big data, and customer engagement. *E Source*. Retrieved from https://www.esource.com/ES-WP-19/Smart-Meter-Portals https://www.tekes.fi/globalassets/julkaisut/consumer_cleantech_report.pdf

Henderson, R. M., & Clark, K. B. (1990). Architectural innovation: The reconfiguration of existing product technologies and the failures of established firms. *Administrative Science Quarterly*, *35*(1), 9–30. doi:10.2307/2393549

Hilton Worldwide. (n.d.). *Hospitality Company | About Hilton Worldwide*. Accessed July 29, 2016. http://www.hiltonworldwide.com/about/

History of Rail Transport. (2016). In *Wikipedia, the Free Encyclopedia*. Retrieved from https://en.wikipedia.org/w/index.php?title=History_of_rail_transport&oldid=729427314

History.com staff. (2010). *Automated Teller Machines*. Retrieved from http://www.history.com/topics/inventions/automated-teller-machines

HNN. (n.d.). Accessed July 21, 2016. http://www.hotelnewsnow.com/Articles/26920/Infographic-Stay-enhancing-tech-at-hotels

Horwath HTL. (2014). Retrieved from H1-2014-Hotel-and-Tourism-Markt-Report-Vienna-Salzburg-Austria (1).pdf

Hospitality Net - Hotels | A Brief History - By Jacques Levy-Bonvin. (n.d.). *Hospitality Net*. Accessed August 2, 2016. http://www.hospitalitynet.org/news/4017990.html

Hospitality Net. (n.d.). Hospitality Net - Europe Hotel Industry Passes U.S. in Growth for the First Time since 2010. *Hospitality Net*. Accessed July 17, 2016. http://www.hospitalitynet.org/news/4074750.html

Hospitality_Report_2015_NAIHellas. (2015). Naihellas.

Huang, J., Chou, T., & Lee, G. (2010). Imitative innovation strategies. *Management Decision*, *48*(6), 952–975. doi:10.1108/00251741011053488

Huang, S. H., Liu, P., Mokasdar, A., & Hou, L. (2013). Additive manufacturing and its societal impact: A literature review. *International Journal of Advanced Manufacturing Technology*, 1–13.

Huansheng, N., Hong, L., Jianhua, M., Laurence, T. Y., Yueliang, W., Xiaozhen, Y., & Runhe, H. (2015). From Internet to Smart World. *Access IEEE*, *3*, 1994-1999.

Hull, C. W. (1986). *Apparatus for production of three-dimensional objects by stereolithography*. Patent No. US4575330 A.

Hur, W. (2017). The patterns of knowledge spillovers across technology sectors evidenced in patent citation networks. *Scientometrics*, *111*(2), 595–619. doi:10.100711192-017-2329-7

Huston, G. (2015, April 30). *The internet of stupid things*. APNIC Labs. Retrieved from http://blog.apnic.net/2015/04/30/the-internet-of-stupid-things/

Hutmacher, D. W., Sittinger, M., & Risbud, M. V. (2004). Scaffold-based tissue engineering: Rationale for computer-aided design and solid free-form fabrication systems. *Trends in Biotechnology*, *22*(7), 354–362. doi:10.1016/j.tibtech.2004.05.005 PMID:15245908

Hu, X., Rousseau, R., & Chen, J. (2012). A new approach for measuring the value of patents based on structural indicators for ego patent citation networks. *Journal of the American Society for Information Science and Technology*, *63*(9), 1834–1842. doi:10.1002/asi.22632

IBISWorld. (2017a). *Global Hotels & Resorts Market Research | IBISWorld.* Retrieved on July 11, 2016 from http://www.ibisworld.com/industry/global/global-hotels-resorts.html

IBISWorld. (2017b). *Hotels & Motels in the US Market Research | IBISWorld.* Retrieved from http://www.ibisworld.com/industry/default.aspx?indid=1661

iff/ZEW. (2010). *Study on interest rate restrictions in the EU, Final Report for the EU Commission DG Internal Market and Services.* Project No. ETD/2009/IM/H3/87. Retrieved from http://ec.europa.eu/internal_market/finservices-retail/docs/credit/irr_report_en.pdf

In Focus- Athens, Greece. (2014). HVS.

ING Group (2016). *Annual Report.* Retrieved from file:///C:/Users/mcaesar/Downloads/2016%20Annual%20Report%20ING%20Groep%20N.V..pdf

Innovation: Evidence from Xerox Corporation's Technology Spin-off Companies. (2012). *Industrial and Internet of Things.*

Inside Airbnb: Los Angeles. Adding Data to the Debate. (2016). *Inside Airbnb.* Retrieved on July 6, 2016 from http://insideairbnb.com/los-angeles/

Intelligence Staff, B. I. (2017). *French Retail Giant Launches Online Bank Accounts.* Retrieved from http://www.businessinsider.com/french-retail-giant-launches-online-bank-accounts-2017-3

InterContinental Hotels Group. (2015). InterContinental Hotels Group PLC. Retrieved from http://www.ihgplc.com/files/results/prelims15/downloads/prelims15_results.pdf

InterContinental Hotels Group. (n.d.). *InterContinental Hotels Group PLC : About Us - Overview.* Accessed July 29, 2016. http://www.ihgplc.com/index.asp?pageid=16

International Monetary Fund. (2016a). *World economic outlook database.* Retrieved April 1, 2017, from https://www.imf.org/external/pubs/ft/weo/2017/01/weodata/index.aspx

International Monetary Fund. (2016b). *Financial Access Survey 2015.* Retrieved March 2, 2017 from http://data.imf.org/?sk=E5DCAB7E-A5CA-4892-A6EA-598B5463A34C

Jacobides, M., Knudsen, T., & Auger, M. (2006). Benefiting from innovation: Value creation, value appropriation and the role of industry architectures. *Research Policy*, *35*(8), 1200–1221. doi:10.1016/j.respol.2006.09.005

Jardini, A. L., Larosa, M. A., Maciel Filho, R., de Carvalho Zavaglia, C. A., Bernardes, L. F., Lambert, C. S., ... Kharmandayan, P. (2014). Cranial reconstruction: 3D biomodel and custom-built implant created using additive manufacturing. *Journal of Cranio-Maxillo-Facial Surgery*, *42*(8), 1877–1884. doi:10.1016/j.jcms.2014.07.006 PMID:25175080

Jeantette, F. P., Keicher, D. M., Romero, J. A., & Schanwald, L. P. (2000). *Method and system for producing complex-shape objects*. Patent No. US6046426 A.

Jiang, R., Kleer, R., & Piller, F. T. (2017). Predicting the future of additive manufacturing: A Delphi study on economic and societal implications of 3D printing for 2030. *Technological Forecasting and Social Change*, *117*, 84–97. doi:10.1016/j.techfore.2017.01.006

Jin, Y., Ji, S., Li, X., & Yu, J. (2017). A scientometric review of hotspots and emerging trends in additive manufacturing. *Journal of Manufacturing Technology Management*, *28*(1), 18–38. doi:10.1108/JMTM-12-2015-0114

Jun, S., Park, S., & Jang, D. (2015). A technology valuation model using quantitative patent analysis: A case study of technology transfer in big data marketing. *Emerging Markets Finance & Trade*, *51*(5), 963–974. doi:10.1080/1540496X.2015.1061387

Kabakova, O., Korovkin, V., & Plaksenkov, E. (2015). *Digital platforms and the ecosystems of financial inclusion. The Russian experience' report*. Retrieved April 3, 2017 from https://iems.skolkovo.ru/downloads/documents/SKOLKOVO_IEMS/Research_Reports/SKOLKOVO_IEMS_Research_2015-11-11_en.pdf

Kalinin, A., Shapenko, A., Korovkin, V., Zaitseva, N., Remyga, O., Novak, K., & Michael, B. (2016). *Emerging Markets Decoded: The Four Domains of Development for Growth*. Academic Press.

Kamath, R., Mukherji, A., & Ramanathan, S. (2010). Ramanagaram financial diaries: Cash patterns and repayments of microfinance borrowers. *Enterprise Development & Microfinance*, *21*(2), 101–117. doi:10.3362/1755-1986.2010.010

Kane, C. (2011, February 9). Do Credit Scores Matter Outside the US? *CNBC*. Retrieved from https://www.cnbc.com/id/41498097

Karki, M. (1997). Patent citation analysis: A policy analysis tool. *World Patent Information*, *19*(4), 269–272. doi:10.1016/S0172-2190(97)00033-1

Karvonen, M., & Kässi, T. (2013). Patent citations as a tool for analysing the early stages of convergence. *Technological Forecasting and Social Change*, *80*(6), 1094–1107. doi:10.1016/j.techfore.2012.05.006

Keller, A., & Stefan, H. (2009). Ex ante identification of disruptive innovations in the software industry applied to web applications: The case of Microsoft's vs. Google's office applications. *Technological Forecasting and Social Change*, *76*(8), 1044–1054. doi:10.1016/j.techfore.2009.03.005

Kellner, T. (2014). World's First Plant to Print Jet Engine Noozles in Mass Production. *General Electric Reports*. Retrieved from http://www.gereports.com/post/91763815095/worlds-first-plant-to-print-jet-engine-nozzles-in/

Kemp, S. (2015). *Digital, Social & Mobile Worldwide in 2015*. Retrieved from http://wearesocial.net/blog/2015/01/digital-social-mobile-worldwide-2015/

Kharpal, A. (2017, October 19). Cryptocurrencies like bitcoin are not 'mature' enough to regulate, ECB chief Mario Draghi says. *CNBC*. Retrieved from https://www.cnbc.com/2017/10/19/cryptocurrencies-are-not-mature-enough-ecb-chief-mario-draghi.html

Khouri, A. (2016). Southern California Apartment Rents Are Expected to Continue Rising through 2018. *Latimes.com*, Retrieved from http://www.latimes.com/business/realestate/la-fi-casden-forecast-20160412-story.html

Kiesling, E., Günther, M., Stummer, C., & Wakolbinger, L. M. (2012). Agent-based simulation of innovation diffusion: A review. *Central European Journal of Operations Research*, *20*(2), 183–230. doi:10.100710100-011-0210-y

Kietzmann, J., Pitt, L., & Berthon, P. (2015). Disruptions, decisions, and destinations: Enter the age of 3-D printing and additive manufacturing. *Business Horizons*, *58*(2), 209–215. doi:10.1016/j.bushor.2014.11.005

King, A., & Tucci, C. L. (2000). Responding to technology-created market niches: Do industry-specialized capabilities facilitate or hamper entry? *Academy of Management*, *TIM:* D1–D6.

King, A. A., & Baatartogtokh, B. (2015). How useful is the theory of disruptive innovation? *MIT Sloan Management Review*, *57*(1), 77–90. doi:10.1017/CBO9781107415324.004

Klenner, P., Husig, S., & Dowling, M. (2013). Ex-ante evaluation of disruptive susceptibility in established value networks - When are markets ready for disruptive innovations? *Research Policy*, *42*(4), 914–927. doi:10.1016/j.respol.2012.12.006

Korovkin, V. (2014). *Microcredit in Russia. On the eve of a boom or a crisis?* Retrieved May 24, 2017, from https:// iems.skolkovo.ru/downloads/documents/SKOLKOVO_IEMS/Research_Reports/SKOLKOVO_IEMS_Research_ 2014-12-06_en.pdf

Kostoff, R. N., Boylan, R., & Simons, G. R. (2004). Disruptive technology roadmaps. *Technological Forecasting and Social Change*, *71*(1), 141–159. doi:10.1016/S0040-1625(03)00048-9

KPMG. (2015). *KPMG Survey Reveals Top Disruptive Technologies*. Retrieved from https://home.kpmg.com/im/en/home/media/press-releases/2015/12/survey-reveals-disruptivetechnologies. html

Krueger, N., & Dickson, P. (1994). How believing in ourselves increases risk taking: Perceived self-efficacy and opportunity recognition. *Decision Sciences*, *25*(3), 385–401. doi:10.1111/j.1540-5915.1994.tb01849.x

ktv_adittrich. (n.d.). *Vermieten von Wohnungen für touristische Zwecke*. Retrieved on July 26, 2016 from https://www.wien.gv.at/arbeit-wirtschaft/privat-vermieten.html

Küchler, S. (2014). Additive technology and material cognition: A view from anthropology. *Journal of Cognition and Culture*, *14*(5), 385–399. doi:10.1163/15685373-12342133

Kumok, Z. (2016). *7 Top Non-U.S. Robo-Advisors*. Retrieved from http://www.investopedia.com/articles/financial-advisors/032216/7-top-nonus-roboadvisors.asp (Investopedia)

Labate, V. (2016). *Ancient Journeys: What Was Travel Like for the Romans? | Ancient Origins*. Retrieved on July 29, 2016 from http://www.ancient-origins.net/artifacts-ancient-technology/ancient-journeys-what-was-travel-romans-005189

Lai, K.-K., & Wu, S.-J. (2005). Using the patent co-citation approach to establish a new patent classification system. *Information Processing & Management*, *41*(2), 313–330. doi:10.1016/j.ipm.2003.11.004

Lakmaaker, M., & Goehring, M. (2017). *Taking Canada's Clean Tech Industry to the Next Level. An Opportunity for Canada's Clean Technology Sector: The Pan-Canadian Framework on Clean Growth and Climate Change*. Adopted from: http://www.national.ca/Bold-Thinking/Taking-Canadas-Clean-Tech-Industry-to-the-Next-Level.aspx

Lane, J., & Woodworth, M. (2016). *An Analysis of Airbnb in the United States*. Retrieved from http://cbrepkfcprod.blob.core.windows.net/downloads/store/12Samples/An_Analysis_of_Airbnb_in_the_United_States.pdf

Latzer, M. (2009). Information and Communication Technology Innovations: Radical and Disruptive? *New Media & Society*, *11*(4), 599–619. doi:10.1177/1461444809102964

Laundrapp. (2018). *Laundry Services for Airbnb Hosts in London | Laundrapp*. Retrieved from August 4, 2016 on https://www.laundrapp.com/business-services-airbnb

Leech, C. (2015, October 20). Direct Lending: Finding Value/Minimizing Risk. *Liberum*. Retrieved from https://www.google.com/url?sa=t&rct=j&q=&esrc=s&source=web&cd=1&cad=rja&uact=8&ved=0ahUKEwj6uKXJ_tHYAhVY8WMKHVngAT8QFggnMAA&url=https%3A%2F%2Flendingwell.com%2Fresearch-posts%2Fliberum-direct-lending-finding-value-minimising-risk%2F&usg=AOvVaw01QH7xpk-J2FNNoBGrOZB_

Leech, C. (2015, October 20). Liberum LendIt Presentation. *Liberum*. Retrieved from http://www.liberum.com/media/69233/Liberum-LendIt-Presentation.pdf

Lee, M., Kim, K., & Cho, Y. (2010). A study on the relationship between technology diffusion and new product diffusion. *Technological Forecasting and Social Change*, *77*(5), 796–802. doi:10.1016/j.techfore.2010.01.003

LeporeJ. (2014). Retrieved from http://www.newyorker.com/magazine/2014/06/23/the-disruption-machine

Levina, M. (2016). Disrupt or Die: Mobile Health and Disruptive Innovation as Body Politics. *Television & New Media*. doi:10.1177/1527476416680451

Levy, G. N. (2010). The role and future of the laser technology in the additive manufacturing environment. *Physics Procedia*, *5*, 65–80. doi:10.1016/j.phpro.2010.08.123

Levy, G. N., Schindel, R., & Kruth, J.-P. (2003). Rapid manufacturing and rapid tooling with layer manufacturing (LM) technologies, state of the art and future perspectives. *CIRP Annals-Manufacturing Technology*, *52*(2), 589–609. doi:10.1016/S0007-8506(07)60206-6

Lim, J., Nam, C., Kim, S., Rhee, H., Lee, E., & Lee, H. (2012). Forecasting 3G mobile subscription in China: A study based on stochastic frontier analysis and a Bass diffusion model. *Telecommunications Policy*, *36*(10), 858–871. doi:10.1016/j.telpol.2012.07.016

Lin, P.-C. (2015). Optimal smart gateway deployment for the Internet of Things in smart home environments. *Consumer Electronics (GCCE) 2015 IEEE 4th Global Conference on*, 273-274. 10.1109/GCCE.2015.7398715

Lodging_in_Greece. (2015, Sept. 17). *Euromonitor International.*

Lopez, T. S., Ranasinghe, D. C., Harrison, M., & McFarlane, D. (2012). Adding sense to the Internet of Things An architecture framework for smart object systems. *Personal and Ubiquitous Computing*, *16*(3), 291–308. doi:10.100700779-011-0399-8

Los Angeles Tourism. (2016). Los Angeles Sets New Tourism Record, Announces 45.5 Million Visitors in 2015. *Discover Los Angeles*. Retrieved from http://www.discoverlosangeles.com/press-releases/los-angeles-sets-new-tourism-record-announces-455-million-visitors-2015

Lyons, B. (2014). Additive manufacturing in aerospace: Examples and research outlook. *The Bridge, 44*(3).

Lyons, J. (2017). *Who Are You Calling a Challenger?* Retrieved from http://pwc.blogs.com/files/challenger-bank-2017-.pdf

Lyytinnen, K., & Rose, G. M. (2003). The disruptive nature of information technology innovations: The case of Internet computing in systems development organizations. *Management Information Systems Quarterly*, *27*(4), 557–595. doi:10.2307/30036549

mabpfi. (n.d.). *International Organisations and Institutions Based in Austria*. Accessed July 26, 2016. https://www.wien.gv.at/english/politics/international/organisations/overview.html

Malkiel, B. G. (2017, June 5). Index Funds Still Beat 'Active Portfolio Management'. *Wall Street Journal*. Retrieved from https://www.wsj.com/articles/index-funds-still-beat-active-portfolio-management-1496701157

Mansfield, E. (1961). Technical change and the rate of imitation. *Econometrica*, *29*(4), 741–766. doi:10.2307/1911817

Manyika, J., Chui, M., Bughin, J., Dobbs, R., Bisson, P., & Marrs, A. (2013). *Disruptive technologies: Advances that will transform life, business, and the global economy*. Retrieved from http://www.mckinsey.com/~/media/McKinsey/Business Functions/McKinsey Digital/Our Insights/Disruptive technologies/MGI_Disruptive_technologies_Full_report_May2013.ashx

Market Pipeline Report: Los Angeles County, CA. (2016, July 12). *STR.*

MarketLine. (2015a). *Global Hotels & Motels - MarketLine Industry Profile.* Retrieved from https://store.marketline.com/report/mlohme7151--global-hotels-motels/

MarketLine. (2015b). *Hotel and Motel Industry in Europe.* Author.

MarketLine. (2015c). *Hotel Industry Analysis - Spain.pdf.* Author.

MarketLine. (2015d). *Industry Analysis - Hotels and Motels in U.S.* Author.

Markets & Markets. (2017). *Internet of Things (IoT) in Energy Market worth 22.34 Billion USD by 2020.* Adopted from: http://www.marketsandmarkets.com/PressReleases/iot-energy.asp

Markides, C. (2006). Disruptive innovation: In need of better theory. *Journal of Product Innovation Management, 23*(1), 19–25. doi:10.1111/j.1540-5885.2005.00177.x

Markides, C. C. (2012). How Disruptive Will Innovations from Emerging Markets Be? *MIT Sloan Management Review, 54*, 23–25.

Markides, C., & Charitou, C. D. (2004). Competing with Dual Business Models – A Contingency Approach. *The Academy of Management Perspectives, 18*, 22–36.

Marriott. (n.d.). *About Marriott Hotels | Marriott Corporate Business Information.* Accessed July 29, 2016. http://www.marriott.com/marriott/aboutmarriott.mi

Martin, H. (2016a). *Airbnb and Other Home-Sharing Businesses Have Hotels Worried - LA Times.* Retrieved from http://www.latimes.com/business/la-fi-hotels-airbnb-20160317-story.html

Martin, H. (2016b). Orange County Breaks Tourism Record with 47 Million Visitors in 2015. *Latimes.com.* Retrieved on July 3, 2016 from http://www.latimes.com/business/la-fi-orange-county-tourism-record-20160205-story.html

Martin, M. (2015). *Strategic Framework: Abell's Framework for Strategic Planning.* Retrieved from https://www.cleverism.com/abells-framework-for-strategic-planning/

Martino, J. P. (2003). A review of selected recent advances in technological forecasting. *Technological Forecasting and Social Change, 70*(8), 719–733. doi:10.1016/S0040-1625(02)00375-X

Mazza, S. (2016). Homeowner Sues Hermosa Beach for Short-Term Rental Ban. *The Beach Reporter.* Retrieved on July 12, 2016 from http://tbrnews.com/news/hermosa_beach/homeowner-sues-hermosa-beach-for-short-term-rental-ban/article_44528f46-4782-11e6-8476-0f90ea03013c.html

McDermott, T., & Roxburgh, C. (2016, March 1). *Financial Advice Market Review: Final Report.* London, UK: Financial Conduct Authority. Retrieved from https://www.fca.org.uk/publication/corporate/famr-final-report.pdf

McFarlan, F. W., & Tempest, N. (1999, September). *Charles Schwab Corporation (A). Harvard Business School Case 300-024.* Harvard Business School. Retrieved from http://www.hbs.edu/faculty/Pages/item.aspx?num=26491

McKinsey Global Institute. (2017). *Jobs Lost, Jobs Gained: Workforce Transitions in A Time of Automation.* Retrieved from https://www.mckinsey.com/~/media/McKinsey/Global%20Themes/Future%20of%20Organizations/What%20the%20future%20of%20work%20will%20mean%20for%20jobs%20skills%20and%20wages/MGI-Jobs-Lost-Jobs-Gained-Report-December-6-2017.ashx

McKinsey. (2015). Disruptive Technologies: Advances that will Transform Life. *Business, and the Global Economy.* Retrieved from http://www.mckinsey.com/insi-ghts/business_technology/disruptive_technologies

Meade, N., & Islam, T. (2006). Modelling and forecasting the diffusion of innovation–A 25-year review. *International Journal of Forecasting, 22*(3), 519–545. doi:10.1016/j.ijforecast.2006.01.005

Mehrotra, A. N., & Yetman, J. (2015). *Financial inclusion-issues for central banks. BIS Quarterly Review.* Basel: Bank for International Settlements.

Mellinghoff, M. (2017). *Map of Robo Advisors in Europe and Germany.* TechFluence. Retrieved from http://www.techfluence.eu/investtech.html

Ministry of Telecommunications in Russia. (2017). *Report for the IV quarter of 2016.* Retrieved January 30, 2017 from http://minsvyaz.ru/ru/pages/statistika-otrasli/#section-403

MIT. (2016). *10 Disruptive innovations through Technologies 2016.* Adopted from: https://www.technologyreview.com/lists/technologies/2016/

Morningstar. (2015). *Hilton Worldwide Holdings Inc. -HLT.* Retrieved from http://api40.10kwizard.com/cgi/convert/pdf/HLT-20160226-10K-20151231.pdf?ipage=10774498&xml=1&quest=1&rid=23&section=1&sequence=-1&pdf=1&dn=1

Moscow School of Management SKOLKOVO. (2016). *FinTech in Russia: Analysis of Survey Results.* Author.

Mueller, B., & Kochan, D. (1999). Laminated object manufacturing for rapid tooling and patternmaking in foundry industry. *Computers in Industry, 39*(1), 47–53. doi:10.1016/S0166-3615(98)00127-4

Munz, J. O. (1956). *Photo-glyph recording.* Patent No. US2775758 A.

NAIHellas. (2015). *2015 Hospitality Report.* Retrieved from http://www.nai-hellas.gr/reports/CommercialRealEstate/Hospitality_Report_2015_NAIHellas.pdf

Nair, S. S., & Mathew, M. (2015). The dynamics between forward citations and price of singleton patents. *International Journal of Innovation and Technology Management, 12*(03), 1540003. doi:10.1142/S0219877015400039

Nam, C., Kim, S., & Lee, H. (2008). The role of WiBro: Filling the gaps in mobile broadband technologies. *Technological Forecasting and Social Change*, *75*(3), 438–448. doi:10.1016/j.techfore.2007.04.008

National Agency for Financial Studies & BDO. (2016). *The level of availability of financial services in the Russian Federation.* Report on the results of a comprehensive study among consumers and financial service providers. Retrieved May 25, 2017 from http://nafi.ru/projects/finansy/uroven-dostupnosti-finansovykh-uslug-v-rossiyskoy-federatsii/

Nayak, R., & Ketteringham, J. M. (1994). *Breakthroughs*. San Diego, CA: Pfeiffer and Co.

Nemet, G. F., & Johnson, E. (2012). Do important inventions benefit from knowledge originating in other technological domains? *Research Policy*, *41*(1), 190–200. doi:10.1016/j.respol.2011.08.009

NESTA. (2015). *Growing a Digital Social Innovation Ecosystem for Europe*. Retrieved from http:// www.nesta.org.uk/sites/default/files/dsireport.pdf

Neuworth, S. (2017, August 11). *Zopa provides update on 'disappearing money' IT glitch.* P2P Finance News. Retrieved from http://www.p2pfinancenews.co.uk/2017/08/11/zopa-update-it-glitch/

Newcomer, E. (2016). Airbnb Sues San Francisco to Prevent New Rules That Enforce a Law Airbnb Wrote. *Skift*. Retrieved from https://skift.com/2016/06/27/airbnb-sues-san-francisco-to-prevent-new-rules-that-enforce-a-law-airbnb-wrote/

Nonnecke, B., Bruch, M., & Crittenden, C. (2016). *IoT & Sustainability: Practice, Policy and Promise*. Center for Information Technology Research in the Interest of Society & the Banatao Institute University of California.

Nowodazkij, E. (2014, July 11). Data shows slight decline in number of bank branches in most EU countries. *ECB*. Retrieved from https://www.ecb.europa.eu/press/pr/date/2014/html/pr140711.en.html

OECD. (2014). *Trends in urbanizations and urban policies in OECD countries: What lessons for China?* Retrieved from http://www.oecd.org/urban/roundtable/45159707.pdf

OECD. (2015). *Digital Economy Outlook OECD*. Retrieved from http://ec.europa.eu/euro-stat/documents/42577/3222224/Digital+economy+outlook+2015/dbdec3c6-ca38-432c-82f2-1e330d9d6a24

Operation and Impact of the Sharing Economy on the Greek Hotel Industry. (2015). Grant Thornton.

Orcutt, A. (2012). *No. 7 Acropolis, Athens - World's Most-Visited Ancient Ruins | Travel + Leisure*. Retrieved on July 22, 2016 from http://www.travelandleisure.com/slideshows/worlds-most-visited-ancient-ruins/8

Oskarsson, C., & Sjöberg, N. (1994). Technology strategy and competitive analysis: The case of mobile phones. *Technology Analysis and Strategic Management*, *6*(1), 3–20. doi:10.1080/09537329408524149

P2P Finance News Staff. (2016, September 20). Ratesetter blames £4.9m loss on new fee structure. *P2P Finance News*. Retrieved from http://www.p2pfinancenews.co.uk/2016/09/20/ratesetter-loss/

Paap, J., & Katz, R. (2004). Anticipating disruptive innovation. *Research Technology Management, 47*(5), 13–22. doi:10.1080/08956308.2004.11671647

Papadimitriou, P. (2014). *In focus: Athens, Greece*. Retrieved from https://www.hvs.com/article/7121-in-focus-athens-greece

Paribas Real Estate, B. N. P. (2016). *Europe - Hotel Market Highlights*. Retrieved from http://www.hospitalitynet.org/file/152006197.pdf

Paritala, P. K., Manchikatla, S., & Yarlagadda, P. K. (2017). Digital Manufacturing-Applications Past, Current, and Future Trends. *Procedia Engineering, 174*, 982–991. doi:10.1016/j.proeng.2017.01.250

Park, J., Tari, M. J., & Hahn, H. T. (2000). Characterization of the laminated object manufacturing (LOM) process. *Rapid Prototyping Journal, 6*(1), 36–50. doi:10.1108/13552540010309868

Parussini, G. (2017). India's Massive Aadhaar Biometric Identification Program. *Wall Street Journal*. Retrieved April 29, 2017, from https://blogs.wsj.com/briefly/2017/01/13/indias-massive-aadhaar-biometric-identification-program-the-numbers/

Pfeffer, J., & Salancik, G. R. (1978). *The External Control of Organizations: A Resource Dependence Perspective*. New York: Harper & Row.

Picker, L., & Isaac, M. (2016). Airbnb Is Said to Be Seeking Funding Valuing It at $30 Billion. *The New York Times*. Retrieved from http://www.nytimes.com/2016/06/29/business/dealbook/airbnb-is-said-to-be-seeking-funding-valuing-it-at-30-billion.html

Pierrakakis, K., Gkritzali, C. D., Kandias, M., & Gritzalis, D. (2015). *3D Printing: A Paradigm Shift in Political Economy?* Paper presented at the 65th International Studies Association's Annual Convention, New Orleans, LA.

Pierrakakis, K., Kandias, M., Gritzali, C., & Gritzalis, D. (2014). *3D Printing and its regulation dynamics: The world in front of a paradigm shift.* Paper presented at the 6th International Conference on Information Law and Ethics, Thessaloniki, Greece.

Pisani, B. (2010). *Man Vs. Machine: How Stock Trading Got So Complex*. Retrieved from http://www.cnbc.com/id/38978686

PricewaterhouseCoopers (PWC). (2015). The Sharing Economy Grows up. *PwC*. Retrieved on August 1, 2016 from http://www.pwc.co.uk/issues/megatrends/collisions/sharingeconomy/the-sharing-economy-sizing-the-revenue-opportunity.html

PricewaterhouseCoopers (PwC). (2016). *Global FinTech Report 2016*. Retrieved May 4, 2017 from https://www.pwc.ru/en/banking/publications/fintech-global-report-eng.pdf

Profile of Greece. (2016). Accessed July 24, 2016. http://www.realtor.org/intlprof.nsf/68b4eab0087e45ba86256811004ee31f/a4f509a4b38dc99b86257077004efb31?OpenDocument

PwC. (2014). The new hire: How a new generation of robots is transforming manufacturing. *PwC*. Retrieved from http://www.pwc.com/us/en/industrial-products/publications/next-manufacturing-robotics.jhtml

PWC. (2016a). *Room for Growth: European Cities Hotel Forecast for 2015 and 2016*. Retrieved from http://www.pwc.com/gx/en/hospitality-leisure/pdf/pwc-room-to-grow-european-cities-hotel-forecast-2015-and-2016.pdf

PWC. (2016b). *Staying Power - European Cities Hotel Forecast for 2016 and 2017*. Author.

QIWI. (2017). *Corporate history*. Retrieved January 1, 2017 from https://corp.qiwi.com/en/company/history.action

Qu, X.L., Brame, J., Li, Q., & Alvarez, J.J.P. (2013). Nanotechnology for a safe and sustainable water supply: enabling integrated water treatment and reuse. *Accounts of Chemical Research, 46*(3), 834-843.

Rad, M. G. (2017). Disruptive innovation in media industry ecosystem and need for improving managerial cognitive capabilities in polymediation era. *Cogent Business & Management*, *23*(1), 1–24. doi:10.1080/23311975.2017.1352183

Radojičić, V. D., & Marković, G. Z. (2009). *New technology forecasting using the bass model*. Paper presented at the Telecommunication in Modern Satellite, Cable, and Broadcasting Services, 2009. TELSIKS'09. 9th International Conference on. 10.1109/TELSKS.2009.5339534

Rafi, F., & Kampas, P. (2002). How to identify your enemies before they destroy you? *Harvard Business Review*, *80*(11), 115–123. PMID:12422795

Rajan, R., & Lines, F. (2010). *How hidden fractures still threaten the world economy*. Princeton, NJ: Princeton University Press.

RateSetter. (2017, August 12). Retrieved from https://www.ratesetter.com/

Rayna, T., & Striukova, L. (2016). From rapid prototyping to home fabrication: How 3D printing is changing business model innovation. *Technological Forecasting and Social Change*, *102*, 214–224. doi:10.1016/j.techfore.2015.07.023

Reichheld, F. F. (1993). *Loyalty-Based Management. Harvard Business Review. March–April 1993 Issue*. Cambridge, MA: Harvard Business School.

Reinhardt, R., & Gurtner, S. (2015). Differences between early adopters of disruptive and sustaining innovations. *Journal of Business Research*, *68*(1), 137–145. doi:10.1016/j.jbusres.2014.04.007

Reinhart, J. (2014). Why The Success Of 'Sharing Economy' Startups Hinges On Who Owns The Inventory. *Business Insider*. Retrieved on August 1, 2016 from http://www.businessinsider.com/the-success-of-the-sharing-economy-2014-2

Resource Center of Small Business. (2016). *Statistics on Small and Medium Enterprises*. Retrieved May 10, 2017 from http://rcsme.ru/ru/statistics

Rifkin, J. (2012). The third industrial revolution: How the internet, green electricity, and 3-d printing are ushering in a sustainable era of distributed capitalism. *World Financial Review, 1*.

Ritola, M., Annala, M., Hulkkonen, S., Lahtinen, V., Latti, R., Noponen, E., ... Hietaniemi, J. (2015). Cleantech takes over consumer markets. *Demos (Mexico City, Mexico)*.

Roger, S. (2007). The disruptive potential of game technologies. *Research Technology Management*, *50*(2), 57–64. doi:10.1080/08956308.2007.11657431

Rogers, E. M. (1962). *Diffusion of innovations*. New York: Free Press.

Rogers, H., Baricz, N., & Pawar, K. S. (2016). 3D printing services: Classification, supply chain implications and research agenda. *International Journal of Physical Distribution & Logistics Management*, *46*(10), 886–907. doi:10.1108/IJPDLM-07-2016-0210

Rosenkopf, L., & Nerkar, A. (2001). Beyond local search: Boundary spanning, exploration, and impact in the optical disk industry. *Strategic Management Journal*, *22*(4), 287–306. doi:10.1002mj.160

Rosstat. (2015). *Statistics on Demographics 2015*. Retrieved May 15, 2017 from http://www.gks.ru/free_doc/new_site/population/demo/Popul2015.xls

Rosstat. (2016). *Statistics on Demographics 2016*. Retrieved May 15, 2017 from http://www.gks.ru/free_doc/new_site/population/demo/Popul2016.xls

Rosstat. (2017). *Statistics on Demographics 2017*. Retrieved May 15, 2017 from http://www.gks.ru/free_doc/new_site/population/demo/Popul2017.xls

Roth, B. (2016). Four Drivers Propelling Clean Tech's Growth. Retrieved from http://www.earth2017.com/clean-tech-ai-iot-growth/

Rothaermel, F. T., & Boeker, W. (2008). Old technology meets new technology: Complementarities, similarities, and alliance formation. *Strategic Management Journal*, *29*(1), 47–77. doi:10.1002mj.634

Roy, R. (2014). Exploring the Boundary Conditions of Disruption: Large Firms and New Product Introduction with a Potentially Disruptive Technology in the Industrial Robotics Industry. *Engineering Management*, *61*(1), 90–100.

Roy, R., & Cohen, S. K. (2015). Disruption in the US machine tool industry: The role of in house users and pre-disruption component experience in firm response. *Research Policy*, *44*(8), 1555–1565. doi:10.1016/j.respol.2015.01.004

Rubini, A. (2017). *Fintech in a Flash: Financial Technology Made Easy*. London, UK: Banking Innovations.

Runde, D. (2015) *M-Pesa And The Rise Of The Global Mobile Money Market*. Retrieved from https://www.forbes.com/sites/danielrunde/2015/08/12/m-pesa-and-the-rise-of-the-global-mobile-money-market/#62172bcd5aec

Rysdal, K. (2017). *PayPal's Schulman says the future may not be cashless, but it's definitely mobile*. Retrieved from https://www.marketplace.org/2017/05/24/business/corner-office/paypals-schulman-says-future-may-not-be-cashless-its-definitely-mobile

Ryssdal, K. (2017, March 24). PayPal's Schulman says the future may not be cashless, but it's definitely mobile. *Marketplace*. Retrieved from https://www.marketplace.org/2017/05/24/business/corner-office/paypals-schulman-says-future-may-not-be-cashless-its-definitely-mobile

Sachs, E. M., Haggerty, J. S., Cima, M. J., & Williams, P. A. (1993). *Three-dimensional printing techniques*. Patent No. US5204055 A.

Sahay, R., Čihák, M., N'Diaye, P., & Barajas, A. (2015). Rethinking financial deepening: Stability and growth in emerging markets. *Revista de Economia Institucional*, *17*(33), 73–107.

Samans, R., Blanke, J., Hanouz, M. D., & Corrigan, G. (2017). The inclusive growth and development report 2017. *In Geneva: World Economic Forum*. Retrieved February 11, 2017 from http://www3.weforum.org/docs/WEF_Forum_IncGrwth_2017.pdf

Samuelson, R. J. (2000). The internet and gutenberg. *Newsweek (Atlantic Edition)*, *135*, 2-2.

San Diego Tourism Authority. (2016). *San Diego Tourism Industry Research*. Retrieved on July 3, 2016 from http://www.sandiego.org/industry-research.aspx

Sanders, S. (2015). Santa Monica Cracks Down On Airbnb, Bans 'Vacation Rentals' Under A Month. *NPR.org*. Retrieved from http://www.npr.org/sections/thetwo-way/2015/05/13/406587575/santa-monica-cracks-down-on-airbnb-bans-vacation-rentals-under-a-month

Sandström, C. (2011). High-End Disruptive Technologies with an Inferior Performance. *International Journal of Technology Management*, *56*(2/3/4), 109–122. doi:10.1504/IJTM.2011.042977

Sandström, Ch., Berglund, H., & Magnusson, M. (2014). Symmetric Assumptions in the Theory of Disruptive Innovation: Theoretical and Managerial Implications. *Creativity and Innovation Management*, *23*(4), 472–483. doi:10.1111/caim.12092

Sangeetha, C., & Koushik, K. S. (2015). *Catalyst for financial inclusion. Bank the unbanked with digital channels. White paper*. Retrieved March 30, 2017, from https://www.infosys.com/industries/financial-services/white-papers/Documents/catalyst-financial-inclusion.pdf

Sanyal, P. (2014). *Credit to Capabilities: A Sociological Study of Microcredit Groups in India*. Cambridge, UK: Cambridge University Press. doi:10.1017/CBO9781139939461

Saviotti, P. P., & Metcalfe, J. S. (1984). A theoretical approach to the construction of technological output indicators. *Research Policy*, *13*(3), 141–151. doi:10.1016/0048-7333(84)90022-2

Schaal, D. (2015a). It's Beginning: Onefinestay Pilot With Hyatt Highlights Sharing Economy Collaboration To Come. *Skift*, Retrieved from https://skift.com/2015/07/21/its-beginning-onefinestay-pilot-with-hyatt-highlights-sharing-economy-collaboration-to-come/

Schaal, D. (2015b). The Inside Story Behind Expedia's Purchase of HomeAway. *Skift*. Retrieved from https://skift.com/2015/11/16/the-inside-story-behind-expedias-purchase-of-homeaway/

Schmidt, G. M., & Druehl, C. T. (2008). When is a disruptive innovation disruptive? *Journal of Product Innovation Management*, *25*(4), 347–369. doi:10.1111/j.1540-5885.2008.00306.x

Schulman, S., & Thomas, L. (2014). New Frontiers of the Innovation Economy and Education. *Entrepreneurship Research Journal*, *4*(1), 137–146.

Schumpeter, J. A. (1942). Capitalism, Socialism, and Democracy. Harper Perennial.

Schumpeter, J. A. (2003). *Capitalism, Socialism, and Democracy*. London: Routledge.

SEC. (n.d.). *Wyndham Worldwide Corporation-2013.12.31-10K*. Accessed July 29, 2016. https://www.sec.gov/Archives/edgar/data/1361658/000136165814000005/wyn-20131231x10k.htm

Shaw, A. (2014). Accelerating sustainability: Demonstrating the benefits of transportation technology. *The Intelligent Transportation Society of America*. Retrieved from http://digitalenergysolutions.org/dotAsset/933052fc-0c81-43cf-a061-6f76a44459d6.pdf

Shepherd, D. A., & Krueger, N. (2002). An intentions-based model of entrepreneurial teams' social cognitions. *Entrepreneurship Theory and Practice*, *27*(2), 167–185. doi:10.1111/1540-8520.00005

Shoster, S. (2016, June 24). The U.K.'s Old Decided for the Young in the Brexit Vote. *Time Magazine*. Retrieved from http://time.com/4381878/brexit-generation-gap-older-younger-voters/

SIFMA. (2017a, July 14). *US ABS Issuance and Outstanding*. Retrieved from https://www.sifma.org/resources/research/us-abs-issuance-and-outstanding/

SIFMA. (2017b, April 13). *European Structured Finance Issuance and Outstanding*. Retrieved from https://www.sifma.org/resources/research/europe-structured-finance-issuance-and-outstanding/

Siggelkow, N., & Levinthal, D. A. (2003). Temporarily divide to conquer: Centralized, decentralized, and reintegrated organizational approaches to exploitation and adaptation. *Organization Science*, *14*(6), 650–669. doi:10.1287/orsc.14.6.650.24840

Sign Up to Drive or Tap and Ride. (n.d.). Retrieved on August 1, 2016 from https://www.uber.com/a/join?exp_hvp=1&invite_code=khzrpyyeue

Sileo, L. (2015). Independent Hotels: Can They Loosen the OTA Grip? *Tnooz*. Retrieved from https://www.tnooz.com/article/independent-lodging-break-from-OTAs-Phocuswright/

Si, S., Yu, X., Wu, A., Chen, S., & Su, Y. (2014). Entrepreneurship and poverty reduction: A case study of Yiwu, China. *Asia Pacific Journal of Management*, *32*(1), 119–143. doi:10.100710490-014-9395-7

Smith, K. G., & Gannon, C. M. (1991). Organizational information processing competitive response and performance in US Domestic Airline industry. *Academy of Management Journal*, *34*(Winter), 60–85. doi:10.2307/256302

Socialdraft. (2017). How Hotels Can Beat AirBnB at Their Own Game. *Socialdraft*. Retrieved on March 4, 2014 from http://socialdraft.com/tips-for-hotels-to-deal-with-airbnb/

Soffer, T., Nachmias, R., & Ram, J. (2010). Diffusion of Web Supported Instruction in Higher Education-The Case of Tel-Aviv University. *Journal of Educational Technology & Society*, *13*(3), 212–223.

Solis, B. (2015). *25 Disruptive Technology Trends For 2015–2016*. Retrieved from http://www.briansolis.com/2015/01/25-disruptive-technology-trends-2015-2016

Sonderlund, A. L., Smith, J. R., Hutton, C., & Kapelan, Z. (2014). Using smart meters for household water consumption feedback: Knowns and unknowns. *Procedia Engineering*, *89*, 990–997. doi:10.1016/j.proeng.2014.11.216

Sood, A., James, G. M., Tellis, G. J., & Zhu, J. (2012). Predicting the path of technological innovation: SAW vs. Moore, Bass, Gompertz, and Kryder. *Marketing Science*, *31*(6), 964–979. doi:10.1287/mksc.1120.0739

Speciale, A. (2016, December 8). ECB Expands Stimulus to $2.4 Trillion as Monthly Purchases Slow. *Bloomberg*. Retrieved from https://www.bloomberg.com/news/articles/2016-12-08/ecb-extends-bond-buying-at-reduced-pace-until-end-of-next-year

Staff. (2017, August 13). Ratesetter doubles new investor accounts since base rate cut. *Peer2Peer Finance News*. Retrieved from http://www.p2pfinancenews.co.uk/2016/09/16/ratesetter-doubles-investors-rate-cut/

STATISTIK AUSTRIA - Statistiken. (n.d.). *Archive.is*. Accessed July 26, 2016. http://archive.is/jLC0r

Steenhuis, H.-J., & Pretorius, L. (2016). Consumer additive manufacturing or 3D printing adoption: An exploratory study. *Journal of Manufacturing Technology Management*, *27*(7), 990–1012. doi:10.1108/JMTM-01-2016-0002

Steenhuis, H.-J., & Pretorius, L. (2017). The additive manufacturing innovation: A range of implications. *Journal of Manufacturing Technology Management*, *28*(1), 122–143. doi:10.1108/JMTM-06-2016-0081

STIFEL. (2016). *Americas Lodging Investment Summit Takeaways Part #1 2016 Outlook*. Author.

Subramaniam, M., & Youndt, M. A. (2005). The influence of intellectual capital on the types of innovative capabilities. *Academy of Management Journal*, *48*(3), 450–463. doi:10.5465/AMJ.2005.17407911

Sullivan, B., Garvey, J., Alcocer, J., & Eldridge, A. (2014). Retail Banking 2020: Evolution or Revolution. *PwC*. Retrieved from https://www.pwc.com/gx/en/banking-capital-markets/banking-2020/assets/pwc-retail-banking-2020-evolution-or-revolution.pdf

Suominen, A., & Seppänen, M. (2014). Bibliometric data and actual development in technology life cycles: Flaws in assumptions. *Foresight*, *16*(1), 37–53. doi:10.1108/FS-03-2013-0007

Swainson, W. K. (1977). *Method, medium and apparatus for producing three-dimensional figure product*. Patent No. US4041476 A.

Tang, L., & Shapira, P. H. (2011). Regional development and interregional collaboration in the growth of nanotechnology research in China. *Scientometrics*, *86*(2), 299–315. doi:10.100711192-010-0274-9

TaskRabbit. (n.d.). *TaskRabbit Connects You to Safe and Reliable Help in Your Neighborhood*. Retrieved on August 1, 2016 from https://www.taskrabbit.com/

Tellis, G. J. (2006). Disruptive technology or visionary leadership? *Journal of Product Innovation Management*, *23*(1), 34–38. doi:10.1111/j.1540-5885.2005.00179.x

The Pew Research Center. (2015, April 2). *The Future of World Religions: Population Growth Projections, 2010-2050*. The Pew Research Center. Retrieved from http://www.pewforum.org/2015/04/02/europe/

The Telegraph Staff. (2011, March 31). Ireland's Banking Crisis: Timeline. *The Telegraph*. Retrieved from http://www.telegraph.co.uk/finance/financialcrisis/8419616/Irelands-banking-crisis-timeline.html

The World Bank Group. (2017). *Global Financial Inclusion Database*. Retrieved from http://databank.worldbank.org/data/reports.aspx?source=1228#

The World Bank Group. (2017). *Global Financial Inclusion Database*. The World Bank Group. Retrieved from http://databank.worldbank.org/data/reports.aspx?source=1228#

Thomsen, M. (2015, February 19). Microsoft's Deep Learning Project Outperforms Humans In Image Recognition. *Forbes*.

Thornhill, S., & Amit, R. (2000). A dynamic perspective of internal fit in corporate venturing. *Journal of Business Venturing*, *16*(1), 25–50. doi:10.1016/S0883-9026(99)00040-3

Thorpe, J., & Keal, A. (2017, July 10). *Britons are Europe's most prolific digital banking users according to new study*. Innofact AG and commissioned by Mastercard. Retrieved from https://newsroom.mastercard.com/eu/press-releases/digital-banking-study-show-britons-are-europes-most-prolific-digital-banking-users/

Tomtom Staff. (2017). *Tomtom Traffic Index*. Retrieved from https://www.tomtom.com/en_gb/trafficindex/list?citySize=LARGE&continent=EU&country=ALL (Tomtom)

Topic: Social Networks. (n.d.). Accessed July 21, 2016. http://www.statista.com/topics/1164/social-networks/

Trajtenberg, M. (1990). A penny for your quotes: Patent citations and the value of innovations. *The Rand Journal of Economics*, *21*(1), 172–187. doi:10.2307/2555502

Tributaria, A. (2016). *New VAT Tax Rates - Spain*. Retrieved on July 12, 2016 from http://www.strongabogados.com/docs/Spain-VAT-rates-2012.pdf

Tripsas, M. (1997). Unraveling the process of creative destruction: Complementary assets and incumbent survival in the typesetter industry. *Strategic Management Journal*, *18*(S1), 119–142. doi:10.1002/(SICI)1097-0266(199707)18:1+<119::AID-SMJ9213.3.CO;2-S

Tripsas, M. (2000). Commercializing emerging technology through complementary assets. In G. S. Day, P. J. H. Schoemakers, & R. E. Gunther (Eds.), *Wharton on managing emerging technologies* (pp. 172–186). John Wiley & Sons.

Tripsas, M. (2008). Customer preference discontinuities: A trigger for radical technological change. *Managerial and Decision Economics*, *2*(9), 79–97. doi:10.1002/mde.1389

Tru by Hilton. (2016). *Our Story*. Accessed July 18, 2016. http://trubyhilton.com/our-story/

Tru by Hilton. (n.d.). *Tru by Hilton*. Accessed July 8, 2016. http://trubyhilton.com/our-place/

Tsosie, C. (2017). *The History of the Credit Card*. Retrieved from https://www.nerdwallet.com/blog/credit-cards/history-credit-card/

Tugend, A. (2016). Hotel Chains and Travel Websites in a Tug of War for Customers. *The New York Times*. Retrieved from http://www.nytimes.com/2016/03/28/business/media/hotel-chains-and-travel-websites-in-a-tug-of-war-for-customers.html

Tumpel-Gugerell, G. (2009). *Business models in banking: Is there a best practice?* Retrieved from https://www.ecb.europa.eu/press/key/date/2009/html/sp090921.en.html

Turk, T., & Trkman, P. (2012). Bass model estimates for broadband diffusion in European countries. *Technological Forecasting and Social Change*, *79*(1), 85–96. doi:10.1016/j.techfore.2011.06.010

Turning the Sharing Economy into a Fair Economy in Vienna - Share-Economy-in-Wien-Stadt-Wien-En.pdf. (n.d.). Retrieved on July 20, 2016 from https://www.wien.info/media/files-b2b/share-economy-in-wien-stadt-wien-en.pdf

Tushman, M. L., & O'Reilly, C. A. III. (1996). Ambidextrous organizations: Managing evolutionary and revolutionary change. *California Management Review*, *38*(4), 8–30. doi:10.2307/41165852

Tushman, M., & Anderson, P. (1986). Technological discontinuities and organizational environments. *Administrative Science Quarterly*, *31*(3), 439–465. doi:10.2307/2392832

U.S. Securities and Exchange Commission. (1997, July 31). *Questions and Answers About the New "Market Risk" Disclosure Rules*. U.S. S.E.C. Retrieved from https://www.sec.gov/divisions/corpfin/guidance/derivfaq.htm

UNESCO Institute for Statistics. (2016). *ICF Macro Demographic and Health Surveys and UNICEF's Multiple Indicator Cluster Surveys*. Retrieved January 15, 2017 from http://data.uis.unesco.org

United Credit Bureau. (2016). *Report on credit structure and trends in Russia*. Retrieved May 14, 2017 from http://www.bki-okb.ru

United Nations Educational, Scientific, and Cultural Organization (UNESCO) Institute for Statistics. (2017). *Literacy rate, adult total (% of people ages 15 and above)*. World Bank Database. World Bank. Retrieved from http://data.worldbank.org/indicator/SE.ADT.LITR.FE.ZS?locations=EU

United States Census Bureau. (2017). *NAICS (North American Industry Classification System)*. Retrieved from https://www.census.gov/eos/www/naics/

United States Securities and Exchange Commission. (2016). *MARRIOTT INTERNATIONAL INC /MD/ (Form: 10-K, Received: 02/20/2014 13:28:17)*. Retrieved from http://secfilings.nasdaq.com/edgar_conv_html%2f2014%2f02%2f20%2f0001445305-14-000495.html#FIS_BUSINESS

Uriondo, A., Esperon-Miguez, M., & Perinpanayagam, S. (2015). The present and future of additive manufacturing in the aerospace sector: A review of important aspects. *Proceedings of the Institution of Mechanical Engineers. Part G, Journal of Aerospace Engineering*, *229*(11), 2132–2147. doi:10.1177/0954410014568797

Vázquez, J. P., Bienenstock, M. J., & Zuckerman, E. W. (2016). Debating Disruptive Innovation. *MIT Sloan Management Review*, *57*(3), 25–31.

Vecchiato, R. (2016). Disruptive innovation, managerial cognition, and technology competition outcomes. *Technological Forecasting and Social Change*.

Vides, T. (2014). What You Never Knew About Repeat Customers and What to Do Now. *Online Marketing Institute*. Retrieved from https://www.onlinemarketinginstitute.org/blog/2014/05/what-you-never-knew-about-repeat-customers-and-what-to-do-now/

Viena Tourism Data & Facts. (n.d.). *Vienna's B2B Service for the Tourism Industry*. Accessed March 10, 2016. https://b2b.wien.info/en/statistics/data

Waleed, R., & Munam, A. S. (2016). Performance evaluation of IoT network infrastructure. *Automation and Computing (ICAC) 2016 22nd International Conference on*, 348-353.

Wall, St. Journal Staff. (2007, October 18). Why a Run on Northern Rock but Not Countrywide. *The Wall St. Journal*. Retrieved from https://blogs.wsj.com/economics/2007/10/18/why-a-run-on-northern-rock-but-not-countrywide/

Walsh, S. T. (2004). Roadmapping a disruptive technology: A case study: The emerging microsystems and top-down nanosystems industry. *Technological Forecasting and Social Change*, *71*(1), 161–185. doi:10.1016/j.techfore.2003.10.003

Walsh, S. T., & Linton, J. D. (2000). Infrastructure for emergent industries based on discontinuous innovations. *Engineering Management Journal*, *12*(2), 23–32. doi:10.1080/10429247.2000.11415071

Walton, J. K. (2018). *Tourism*. Retrieved on July 29, 2016 from https://www.britannica.com/topic/tourism

Wan, F., Williamson, P., & Yin, E. (2015). Antecedents and implications of disruptive innovation: Evidence from China. *Technovation*, *39-40*, 94–104. doi:10.1016/j.technovation.2014.05.012

Wang, F., Hu, L., Hu, J., Zhou, J., & Zhao, K. (2016). Recent Advances in the Internet of Things: Multiple Perspectives. *IETE Technical Review*, 1.

Wang, S. (2017, February 1). SoFi Buys Mobile-Banking Startup Zenbanx for About $100 Million. *Bloomberg*. Retrieved from https://www.bloomberg.com/news/articles/2017-02-01/sofi-buys-mobile-banking-startup-zenbanx-for-about-100-million

Wang, J.-C., Chiang, C., & Lin, S.-W. (2010). Network structure of innovation: Can brokerage or closure predict patent quality? *Scientometrics*, *84*(3), 735–748. doi:10.100711192-010-0211-y

Warick-Ching, L. (2017, October 20). Young People Rely on Parents and Credit Cards to Cover Costs. *Financial Times*. Retrieved from https://www.ft.com/content/d3dddbcc-b4df-11e7-aa26-bb002965bce8

Wearable technology ecosystem industry 2016 to 2030 - analysis of 6 region and 465 companies profiles. (2016, May 03). *PR Newswire*. Retrieved from https://search.proquest.com/docview/1786275049?accountid=14719

Weber, M. (2016). Our water system: what a waste. *The New York Times*. Retrieved from http://www.nytimes.com/2016/03/22/opinion/our-water-systemwhat-a-waste.html

Weeks, M. (2015). Is disruption theory wearing new clothes or just naked? Analyzing recent critiques of disruptive innovation theory. *Innovation: Management. Policy & Practice*, *17*(4), 417–428. doi:10.1080/14479338.2015.1061896

Weller, C., Kleer, R., & Piller, F. T. (2015). Economic implications of 3D printing: Market structure models in light of additive manufacturing revisited. *International Journal of Production Economics*, *164*, 43–56. doi:10.1016/j.ijpe.2015.02.020

Western, B. (2002-2018). *Best Western International® | President & CEO - David Kong*. Retrieved on July 29, 2016 fromhttp://www.bestwestern.com/about-us/press-media/ceo.asp

White, G. R. T. (2017). *Future applications of blockchain in business and management: A Delphi study*. Wiley. 10.1002/jsc.2144

Wingfield, N. (2004, January 28). Amazon Reports Annual Net Profit For the First Time. *Wall Street Journal*. Retrieved from https://www.wsj.com/articles/SB107523879334513159?mg=prod/accounts-wsj

Wolff, J. (2017, July 10). *New European Digital Banking Study by Mastercard highlights convenience as the greatest advantage of digitized banking solutions*. London: Mastercard. Retrieved from https://newsroom.mastercard.com/eu/press-releases/new-european-digital-banking-study-by-mastercard-highlights-convenience-as-the-greatest-advantage-of-digitized-banking-solutions/?utm_source=Triggermail&utm_medium=email&utm_campaign=Post%20Blast%20%28bii-fintech%29:%20Crunch%20time%20for%20UK%20neobanks%20%E2%80%94%20IBM%20runs%20another%20blockchain%20POC%20%E2%80%94%20Santander%20partners%20Tradeshift&utm_term=BII%20List%20Fintech%20ALL

Woo, D., Piralta, K., & Matthews, J. (2015). Towards sustainable water supply: Schematic development of big data collection using Internet of Things (IoT). *Procedia Engineering*, *118*, 489–497. doi:10.1016/j.proeng.2015.08.465

Worgull, S. (2014). *HNN*. Retrieved on July 23, 2016 from http://www.hotelnewsnow.com/Articles/23584/Independent-vs-chain-Which-performs-better

World Bank Group. (2013). *Global Financial Development Report 2014: Financial Inclusion*. World Bank Publications.

World Bank. (2017). *Literacy rate, adult total (% of people ages 15 and above)*. United Nations Educational, Scientific, and Cultural Organization (UNESCO) Institute for Statistics. World Bank Database. World Bank. http://data.worldbank.org/indicator/SE.ADT.LITR.FE.ZS?locations=EU

World Travel & Tourism Council. (2016). *Travel & Tourism - Economic Impact 2016*. Retrieved from http://www.wttc.org/-/media/files/reports/economic-impact-research/2016-documents/economic-impact-summary-2016_a4-web.ashx

Wyndham Hotel Group. (2018). *About Wyndham Hotel Group: Learn about Wyndham Hotel Group Hotels at the Official Site for the Best Rate Guaranteed and Earn Wyndham Rewards Points*. Retrieved on July 29, 2016 from http://www.wyndhamhotelgroup.com/about-us/about-wyndham

Yan, B., & Luo, J. (2016). Measuring technological distance for patent mapping. *Journal of the Association for Information Science and Technology*.

Yasukawa, S., & Kano, S. (2015). Comparison of examiners' forward citations in the United States and Japan with pairs of equivalent patent applications. *Scientometrics*, *102*(2), 1189–1205. doi:10.100711192-014-1466-5

Yoon, J., Park, Y., Kim, M., Lee, J., & Lee, D. (2014). Tracing evolving trends in printed electronics using patent information. *Journal of Nanoparticle Research*, *16*(7), 2471. doi:10.100711051-014-2471-6

You, H., Li, M., Hipel, K. W., Jiang, J., Ge, B., & Duan, H. (2017). Development trend forecasting for coherent light generator technology based on patent citation network analysis. *Scientometrics*, *111*(1), 297–315. doi:10.100711192-017-2252-y

Young, P. (1993). Technological growth curves: A competition of forecasting models. *Technological Forecasting and Social Change*, *44*(4), 375–389. doi:10.1016/0040-1625(93)90042-6

Yu, D., & Hang, C. C. (2010). A Reflective Review of Disruptive Innovation Theory. *International Journal of Management Reviews*, *12*(4), 435–452. doi:10.1111/j.1468-2370.2009.00272.x

Yu, D., & Hang, C. C. (2011). Creating technology candidates for disruptive innovation: Generally applicable R&D strategies. *Technovation*, *31*(8), 401–410. doi:10.1016/j.technovation.2011.02.006

Zhou, W., & Piramuthu, S. (2013). Technology regulation policy for business et-hics: An example of RFID in supply chain management. *Journal of Business Ethics*, *116*(2), 327–340. doi:10.100710551-012-1474-4

Zipcar Inc. (2018). *Car Sharing: An Alternative to Car Rental with Zipcar*. Accessed August 1, 2016 from http://www.zipcar.com/

About the Contributors

Anatoly Zhuplev is a professor of international business and entrepreneurship at Loyola Marymount University (Los Angeles, California) and former editor-in-chief at the Journal of East West Business (2011-2013). He taught for ten years at the Moscow Management Institute, and subsequently at the Advanced Training Institute of the State Committee for Printing and Publishing in Moscow; in Bonn, Germany in 1994, 1998, 2009; in Warsaw, Poland (as a Fulbright scholar) in 2005; in Paris, France in 2004-2007, and at Northeastern University in Boston, Massachusetts in 1989-1990. His books, book chapters, and articles on International Management, International Entrepreneurship, International Business, European Energy Security and Corporate Governance (around 100 overall) have been published in the U.S., Canada, Western Europe, Russia, and the former USSR. He received his PhD from the Moscow Management Institute, Russia, in 1981, and his BS from the Moscow Engineer-Economics Institute in 1974.

* * *

Serhat Burmaoglu (PhD) is an Associate Professor on Quantitative Decision Making subject. His research interest is investigating the relationship between economic growth, productivity, competitiveness, innovation and knowledge economy in macro-economic level by using multivariate statistical analysis and data mining applications for extracting usable patterns to direct development policy. He performs bibliometric analysis and patent analysis on emerging technological subjects to prepare scenario-based foresight studies.

Jonathan Dell holds an MBA from Loyola Marymount University with an emphasis in Information Decision Systems. His passion for travel and technology converged in an exciting way for this in-depth study of the impacts of the sharing economy on traditional hotel business interests. Additionally, the evolving regulatory landscape surrounding the sharing economy is of particular interest. Jon lives in Santa Clarita, CA with his wife and three young children.

DaVion Doby is a Vice President at The TCW Group, an investment management firm in Los Angeles. He began his career in 2004 working in Operations for Capital Group. Later he moved on to Portfolio Administration and Client Services with Kayne Anderson. Currently, he works in Corporate Finance developing financial models to facilitate executive-level decision making. Mr. Doby received his MBA from Loyola Marymount University in 2016 with a dual emphasis in Finance and Marketing and is a member of the Academy of International Business. He also participated in study abroad programs in Germany and various cities in Southern Europe.

Manuel Gutierrez is currently a Systems Engineer at Northrop Grumman, an aerospace firm in Redondo Beach, California. Originally from East Los Angeles, California, he has begun his higher education journey on the East Coast of the USA at Phillips Academy – Andover, Massachusetts. Manuel Gutierrez completed his undergraduate study at the University of Southern California in Los Angeles, California with degrees in Business Administration and Mechanical Engineering. Embarking on an aerospace engineering career with an emphasis on survivability for advanced system designs, Manuel completed his MBA from Los Angeles-based Loyola Marymount University with an emphasis in Finance – a lifelong passion having performed rigorous financial and securities analysis as well as research and analysis in real estate, mergers and acquisitions, credit, structured credit, analysis of banks and regulations early in his career. Over the years since 2007, Manuel has developed economic and financial models, successfully implementing his research toward his investment partnerships.

Vladimir Korovkin is the Head of Research in Innovations and Digital in the SKOLKOVO Institute for Emerging Market Studies. Vladimir is an expert in IT-management and Internet communication platforms with a focus on organizational development and psychology as well as on data management. He holds a degree in Engineering from Moscow State Technical University and a diploma in International Business from the All-Russia Academy for Foreign Trade.

Sucet Jimena Martínez-Vergara holds a MS in Business Research from the University of Barcelona (2016) where she previously received the Business Administration and Management degree. She is a certified public accountant from University of San Simón (Bolivia) with eight years of professional experience in corporate banking. Since 2017 she is involved as a researcher within the innovation research line of the University of Barcelona PhD Business program.

Başak Özdemir received her Bachelor degree in Manufacturing Systems Engineering from Faculty of Engineering and Natural Sciences, Sabanci University, Istanbul, Turkey. She received her MSc degree in Service Management and Design from University of Warwick, United Kingdom. Her research interests are Innovation, Service Design and Co-creation.

Evgeny Plaksenkov is the Professor at Moscow School of Management SKOLKOVO and the Head of the "Finance, Payments and e-Commerce" Chair. Evgeny is an expert in Corporate Finance, Strategy and Project Management fields. Previously he worked as a Director of Corporate Programs at IMISP (International Management Institute of St. Petersburg). He developed and taught corporate education programs for Russian and international companies, including, among others, Gazprom, Vympelkom (Beeline), United Technologies Corporation. Evgeny worked as Human Resources Director at OTIS (Russia), and then as CFO and CEO of MIEL-Group. From 2009 till present he serves as an Independent Director on Board of Directors of several corporations. Evgeny graduated from St. Petersburg State University of Economics and Finance earning cum laude degree in "Economic cybernetics". He did post-graduate studies at St. Petersburg State University of Economics and Finance , Executive-MBA at Higher School of Economics, DBA at International School of Business of Financial University, and PhD degree at SKEMA Business School (France). Also studied at ESC Lyon (Ecole Superieure de Commerce de Lyon, now EMLYON Business School), SDA Bocconi University, Henley Management College. Evgeny was included in «Top-50 Independent Directors» in Russia 2013-2016 list by the Association of Independent Directors. He was nominated for «Aristos-2008» award in «Financial Director» category. In 2008 Evgeny Plaksenkov went to the first line of «Top-1000 Russian Managers» ranking by the Association of Managers of Russia and «Kommersant» journal in "Professional Services – Financial Director" nomination. Evgeny Plaksenkov currently works in corporate governance, venture capital business, consulting, academia and research.

Vincent Sabourin is Full Professor of Strategic Management & Director of the Observatory project on Strategies and Commercialization of Innovation School of Management, ESG University of Quebec in Montreal. UQAM, He holds a Ph.D. from McGill University under the supervision of Henry Mintzberg.

Oksana Smirnova, as a research associate of "Finance, Payments and E-commerce" Chair at Moscow School of Management SKOLKOVO, is engaged in the research in several fields, i.e.: Financial Inclusion, Business Ecosystems, Project Management and Valuation. In 2014, she received a BSc in Economics degree and, in 2016, MSc in Financial Economics program in International College of Economics and Finance at National Research University "Higher School of Economics" (ICEF HSE) in Moscow, Russia. Her experience in research and scientific activities includes work at the Laboratory of Interdisciplinary Empirical Studies (Perm, Russia), International Laboratory of Financial Economics (Moscow, Russia). Additionally, Oksana is teaching the courses on Financial Economics and Financial Econometrics at ICEF HSE Master's program.

Joshua Tillipman grew up in Santa Monica, California and received his MBA from Loyola Marymount University with an emphasis in Management and Organizational Behavior, as well as a B.S. in Business Administration from the University of Oregon. His life-long love for travel, global study abroad experience, and group thesis on the competitive relationship between peer-to-peer lodging services and the traditional hotel industry inspired Joshua to pursue a career in the travel-tech space. Joshua is currently a Market Executive at Agoda.com, an online travel agency, under the Priceline Group.

Jaume Valls-Pasola is professor of management and business administration at the University of Barcelona (UB). He was associate professor at the Universitat Politècnica de Catalunya until 1994 when he was appointed as professor by the University of Girona. In October 2006 he joined the UB. His main research interests are in the fields of innovation management and entrepreneurship. His recent research includes a project on success and failure factors of entrepreneurship activities and the analysis of innovative ecosystems. He has recently published (as co-editor) the book "The global management of creativity" (2017, Routledge).

Haydar Yalcin (PhD) is an Assistant Professor at department of Information Management System İzmir Katip Çelebi University. His research interest is investigating open access scientific publication, digital nativeness and knowledge dissemination in scientific publications and patents etc. He performs bibliometric analysis on emerging research areas to prepare curriculum development, tought leader selection studies.

Kemal Yayla graduated from Department of Industrial Engineering, Gazi University, Turkey. In 2010, He received his MBA degree in Production Management and Marketing from Manisa Celal Bayar University and In 2017, He received PhD degree in the Business Administration department of Ege University. His research interests focus on Scientometric Analysis, Multi-criteria Decision Making Models and Information Retrieval areas.

Index

U

W

www.ingramcontent.com/pod-product-compliance
Lightning Source LLC
LaVergne TN
LVHW081537300926
849161LV00006B/1469